OIL AND IDEOLOGY

THE
LUTHER
HARTWELL
HODGES
SERIES

ON
BUSINESS,
SOCIETY,
AND THE
STATE

WILLIAM H. BECKER, EDITOR

# OIL

## AND IDEOLOGY

The
Cultural
Creation
of the
American
Petroleum
Industry

ROGER M. OLIEN & DIANA DAVIDS OLIEN

*The*
*University*
*of North*
*Carolina*
*Press*

*Chapel Hill*
*& London*

© 2000 The University of North Carolina Press

Designed by April Leidig-Higgins
Set in Carter & Cone Galliard type
by Keystone Typesetting, Inc.
Manufactured in the United States of America

The paper in this book meets the guidelines for permanence
and durability of the Committee on Production Guidelines for
Book Longevity of the Council on Library Resources.

Library of Congress Cataloging-in-Publication Data
Olien, Roger M., 1938–
Oil and ideology : the cultural creation of the American petroleum industry / by Roger M. Olien and Diana Davids Olien.
p. cm. — (Luther Hartwell Hodges series on business, society,
and the state)
Includes bibliographical references and index.
ISBN 0-8078-2523-9 (cloth: alk. paper)
ISBN 0-8078-4835-2 (pbk.: alk. paper)
1. Petroleum industry and trade — Moral and ethical aspects —
United States — Historiography.   2. Petroleum industry and
trade — Government policy — United States — Historiography.
3. Petroleum industry and trade — Moral and ethical aspects —
United States — Public opinion.   4. Standard Oil Company —
Public opinion.   5. Rockefeller, John D. (John Davison),
1839–1937 — Public opinion.   6. Public interest — United States
— Public opinion.   7. Public opinion — United States.
I. Olien, Diana Davids, 1943– .   II. Title.   III. Series.
HD9565.O6473  1999   338.2′7282′0973 — dc21   99-29765  CIP

04 03 02 01 00   5 4 3 2 1

*In memory of J. Conrad Dunagan*

# Contents

# *Preface*

Several years ago, a friend and colleague reacted to our reexamination of the frequent attacks on the Standard Oil Company by objecting, "But everybody knows what Rockefeller did!" His comment was a highly accurate summation of opinions that prevail in academic circles and in the larger American society. Everybody knows what John D. Rockefeller did. College-level textbooks commonly include his public life in the context of conventional "robber baron" chapters. Authors of recent best-selling books about him and the American petroleum industry present a more nuanced assessment, but they still tend to take many of the charges of unethical and anticompetitive behavior for granted and move on from that assumption. In *The Prize*, for example, Daniel Yergin carries forward in his story the charge that Standard received important "drawbacks," and he takes other allegations made by Rockefeller's critics at face value. His thumbnail characterization, "a ruthless competitor that would cut to kill," was taken from Ida Tarbell. Ron Chernow, in *Titan: The Life of John D. Rockefeller, Sr.*, presents versions of events taken from some of Rockefeller's shrillest critics and uses them to justify his assumption that Rockefeller was a ruthless competitor. From that he develops a central motif, explaining how Rockefeller could "square his actions with his conscience."[1] Defenders of the regulatory state, from Theodore Roosevelt onward, commonly begin with the "Standard Oil story" to justify federal intervention in corporate affairs. In short, Standard's predations and the anticompetitive and anticonsumer identity of the domestic industry are largely taken as "given."

But why? Even if one accepts the orthodox view of the sins of Standard, it is a legitimate historical question to ask why they ultimately came to be visited on a whole industry. And if, as we have seen from evidence we have found as well as evidence brought forward by other scholars, the orthodox view often seems seriously flawed, it is even more pressing to explain how it came to prevail. Who was responsible for developing an image of an industry that, while admitting its essential role in the American economy, showed it so often at odds with public interest? Why did such persons say what they did about the industry, and how did their ideas gain credibility? An examination of the origins and evolution of ideas about the American petroleum industry is long overdue.

So is consideration of what effect prevailing ideas have had, both on the petroleum industry and on policy making directed at it. In recent years numerous scholars have studied federal petroleum policies, with differing assessments of their effectiveness. Looking at the period between 1890 and 1964, Gerald D. Nash saw development of cooperation and consensus between the federal government and industry, including agreement that the industry would stay in private hands but under public supervision, that government would subsidize the industry, and that it would regulate to solve industry problems. One might assume from these conclusions that Nash identified many instances of successful and constructive policy making, yet, once past Theodore Roosevelt, most of his presidents, by his own account, in fact did very little: Hoover was inflexible and ineffective; FDR was creative in administration but made "few substantive innovations in the realm of policy"; Harry Truman's "national oil policies did not bear fruit"; the Eisenhower years were "generally barren of accomplishment."[2] Such assessments raise the question of how, in an atmosphere of growing cooperation and consensus, with government accepted as arbiter in industry affairs, so little was done. And if cooperation and consensus were well established by the mid-sixties, why was the American petroleum industry at odds with the federal government on so many fronts by the mid-seventies?

Less optimistic but more persuasive, John G. Clark has looked at the period between 1900 and 1946 and concluded that federal policies toward mineral fuels, coal as well as oil and gas, were "unsystematic, vague, and eminently minimal." At points where federal action might have been effective, government did not act constructively; federal ventures into fuel policy making had meager results. Clark explains this situation as arising from a variety of causes. Industry did not give government enough information for intelligent policy making; big oil denied the legitimacy of the federal intervention in the industry (though little oil in the form of smaller independents was equally greedy and self-seeking); and special interests derailed efforts to get things done. Clark, however, admits ideology played a role in what happened. Ideology was sometimes a "mask in the pursuit of special benefits." But it also surfaced in times of crisis when advocates of free enterprise and "champions of trust busting" faced off and "neither of these antagonists could transcend the traditional frameworks of their arguments."[3]

In a study of energy since 1945 that is both highly persuasive and eminently useful, Richard H. K. Vietor gives a broader role to ideology, concluding that conflict and confusion over ideology have been just beneath the surface in most issues of energy policy. Vietor identifies two opposed ideological positions with respect to resource allocation, the one grounded in collectivist concern for the public interest, the other grounded in reliance on market mechanisms to allocate resources. He also notes that participants in energy politics neither understood these ideologies at work nor were consistent in using them. He concludes that the clash of these ideologies "invariably muddled" issues of energy policy and produced dysfunctional policies — of which examples are abundant in the period he studies.[4]

Certainly, if one looks at the history of the American petroleum industry and governmental policy making directed at it from its beginnings to the end of World War II, one can see ideology at work and, as Vietor observes, usually appearing to muddle issues. Beyond this, however, one can ask if ideological difference was simply a matter of the collectivist impulse facing down laissez-faire, or whether there were other equally complex ideologies at work, which went beyond muddling discussion to defining both the industry and policy toward it, some ideological currents older than the industry and some emerging with its development. For that matter, did ideology frame the issues? In short, how has the petroleum industry been affected by ideology and how did the ideology develop?

What we present is neither a history of the petroleum industry nor an exhaustive history of policy making but a history of the development of discourse affecting both. We have approached this history of petroleum industry–related discourse with seven basic questions in mind:

1. Who constructed the social and cultural identity of the domestic petroleum industry?
2. From what elements did they construct that identity?
3. What interests did they advance?
4. What contexts framed what they said?
5. What contingencies shaped what they said?
6. Why were there inconsistencies and conflicts in the public discourse on the industry?
7. What effect did the social and cultural identity of the industry as presented in public discourse have in terms of public policy?

In our attempts to answer these questions, we have taken three basic approaches, each common to historical scholarship. First, we have worked with public discourse to identify and analyze the cultural context of the American petroleum industry as it developed between 1859 and 1945. Under the umbrella of public discourse, we included a wide variety of published works, popular journalistic accounts, muckraking exposés, scientific literature, legislative reports, testimony to congressional committees, cartoons, trade journal articles, and editorials. We do not claim to have included everything; we have not retrieved radio broadcasts, to which we have found tantalizing references. But we have looked at what a wide variety of observers, outside and inside the industry, familiar and unfamiliar with its operations, said about it, and we have looked at what policy makers tried to do.

Second, we present a description of the formation and effect of public discourse directed toward a single subject, an industry that has played a vital role in the modern American economy. Thus, we have aimed to identify the major participants in discourse, explain what they drew on and put together, suggest what their interests may have been in advancing ideas, and tie what they said to the actions of policy makers. Our approach in this respect is similar to that taken by Thomas K. McCraw in his *Prophets of Regulation*, for McCraw demonstrates how regulators went about framing their regulation and assesses their success and failure. McCraw recognizes that overwhelming commitment to an ideological perspective can skew a regulator's success in framing regulation; considering one of the greatest "prophets" of regulation, Louis Brandeis, McCraw points out that commitment to antitrust, virulent anti-"bigness" that for Brandeis amounted to "moral passion," flawed his perception of the revolution in business taking place in early-twentieth-century America and led to inconsistencies in his approach to policy and regulation.[5] We believe working with participants in discourse and studying both what they said and the context in which they said it makes it easier to understand why well-intentioned policy—and not just that directed at the petroleum industry—can fall short of stated objectives. As we demonstrate in the context of the petroleum industry, because at any one time there are usually a number of available explanations and interpretations of events, drawn from different channels of discourse, policy makers may put them together in a way that turns out to be dysfunctional, irrational, or irrelevant. Then members of the public may complain that there has not really been a policy at all.

Third, social scientists will recognize our implicit, if modified, acceptance of various concepts and techniques that they have employed to study the processes of formation of public discourse and public policy. Thus, we assume the relevance of the "arena" model of competition for control of public discourse, of media "framing," of information subsidies by interested parties, and of the operation of media controllers to set public agendas. We have, however, defined our topic as a problem in intellectual history because when one works on a period of time for which there are no opinion polls, no measure of public opinion, all one can study is discourse, what people said. That gives one a lot to go by. One may never know what the early-twentieth-century American public actually thought about John D. Rockefeller, but one can analyze what was written about him. Our research problems, thus, are those of textual origin, social and cultural meanings, and political impacts.[6]

As we studied what has been said about the domestic petroleum industry, we came to recognize broad rhetorical categories, or, as we prefer to call them, channels of discourse. These are identified in terms of the subject area speakers addressed, the focus they took within that area, and the themes and rhetoric they used. Each channel has specific contextual information and characteristic perspectives along with broader elements of ideology. The main channels of discourse we work with are operational, technological, economic, political, and, most general of all, moral or normative.

Operational discourse lets industry members talk about the work they actually do. Oilmen, of necessity, have a language of field and plant operations accessible to all carrying out operations. Thus, out in the field, everyone from the new roustabout up understands what is meant by drill bits, fishing jobs, tool pushing, dry holes, blowouts, casing pulling, cut oil, separators, and gathering lines — all field phenomena that no one needs a college degree to understand. The boundary between such operational discourse and either economics or petroleum science and technology has always been a blurred one, but operational discourse is less theoretical or "scientific" than that of economics, geology, or petroleum engineering, though all are brought to bear in practical aspects of doing business. Technological discourse lets scientists and technologists discuss general problems of their specializations — enhanced oil recovery, for example, or characteristics of geological formations — in theoretical and speculative dimensions as well as practical application. Depending on time-specific needs or prevailing industry conditions, technological discourse may focus intensively on problems such as unitization of production or analyzing reservoir pressure, which can give rise to what amount to subordinate channels of discourse. Obviously, technological discourse is tied to operations, but it may have normative elements within it missing from operational discourse. It may be about what scientists and technologists think the industry might or ought to do, as well as what it can do and is doing. For that reason, technological discourse can, on occasion, conflict with operational discourse.

Obviously, if one looks at any industry, a broad channel of related discourse will be economic. Within the general subject of economics relating to petroleum, one may identify a number of subordinate channels with respect to focus. One focus is economic/operational, what oilmen use when they talk about the cost of drilling a well or the market for crude oil or gasoline. But commentators such as academic economists, popular journalists, and consumer advocates have also had quite a bit to say about the petroleum industry in the perspective of political economy, developing focuses on monopoly, conservation, and consumer welfare that have produced important secondary channels of economic discourse. Distinctive themes, such as restraint of trade and predatory pricing in monopoly-oriented discourse and oil shortage and keeping oil in the ground in conservation discourse, serve to identify these channels, though, again, they commonly borrow from one another. Thus, a conservationist may argue for keeping oil in the ground as a safeguard against future predatory pricing, thereby using themes from two channels of discourse. Not uncommonly, channels of discourse have contained contradictory ideas.[7] For example, within conservationist thinking on natural gas produced with oil is the idea that such gas should not be flared in the field. But some conservationists also argued that such gas should not be put to use as cheap fuel either — thus creating a problem of limited market for such gas.

Particularly with respect to issues of public policy and regulation, the broad channel of economic discourse relating to the petroleum industry mingles and overlaps with a broad channel of political discourse. Precise definition of the political channel is difficult, to say the least, for virtually any of the ways in which petroleum could be a topic of public concern — as something every member of the public consumes directly or indirectly, now or in the future — falls within the political discourse parameter. The economic position of the industry in American society, its power real or imagined, must, like all questions of power, be part of this channel of discourse. As we will show, oilmen themselves had a great deal to do with establishing specific focuses within this broad political context, when they talked about the excessive power of Standard Oil, for example, or when they developed antiregulatory rhetoric aimed at federal or state bodies. Once state regulatory agencies were established, they also joined in political discourse, defending their immediate policies and their regulatory turf; the latter saw them resort to states' rights rhetoric against federal regulation. Many nonindustry observers such as Henry Demarest Lloyd, Ida M. Tarbell, and Robert Marion La Follette — much more recently, Ralph Nader — also developed discourse in the political context. Historically, there has been least overlap between the political and operational channels of discourse, contributing very significantly to dysfunctional policy making and inept regulation.

Underlying or perhaps interwoven with all these channels of discourse is a channel of normative or moral discourse, of cultural definitions of right and wrong. Sometimes elements of this are easy to identify, as when an industry observer singles out a practice and says it is evil or harmful. At other times, moral discourse is camouflaged, as when economists talk about waste (but waste never means something good) or when conservationists try to prioritize uses for petroleum (some uses are better socially than others). Gendered language and imagery can be ways to convey moral messages in American culture. As intellectual historians of early America have noted, references to character traits such as cunning, deviousness, and secretiveness were often tied to effeminacy, as were behavior traits such as wastefulness, living in luxury, and wanton behavior. By contrast, strength, straightforwardness, honest ambition, and self-reliance are usually tied to masculinity. One can see gendered imagery at work within respect to oil when John D. Rockefeller becomes a secretive and cunning evildoer or when mid-twentieth-century oil finders become enterprising pioneers working on a frontier under the earth. In either instance, the moral message is clear.

Over time, as we are going to describe, the various channels of discourse about the petroleum industry emerged, and out of them emerged the cultural construction, the American perception of the industry. Since the industry itself was new in world experience and the way it came to be dominated by a giant firm unprecedented, industry observers had to struggle to define what they were seeing and put it into cultural or historical context. As they did so, they came up with analogies between petroleum and other things. Sometimes these analogies worked effectively; sometimes they did not. In one respect, however, the analogy between the growth of the petroleum industry and broader national experience seemed especially effective: oil's newness, its growth, its concentration of economic power, its producing for an increasingly consumer-oriented society made it easy to let the industry symbolize what was happening in the wider arena of late-nineteenth- and early-twentieth-century America. If one saw what was happening as harmful or counter to national interest, oil could symbolize that. In any event, as we shall demonstrate, though there were differing ways of talking about oil, a great many of them were negative and tied the industry to problems such as the abuse of economic power or waste of natural resources. And, as we shall describe, such negative ideas, a staple in discourse, had a powerful influence on the impulse to regulate the industry.

# Acknowledgments

Many friends and colleagues have helped as we pursued this study over the years, listening to our ideas, patiently bearing with our theories, and responding to our questions. Of all these individuals, however, we owe special thanks to the following persons, who have gone farther than an extra mile with us.

This study was possible, first of all, because of the generous encouragement of Clyde Barton, J. Conrad Dunagan, John C. Dunagan, Arden R. Grover, John H. Hendrix, and Robert M. Leibrock. Help from these friends was essential to the completion of a lengthy and complicated undertaking; it could not have been done without them. Duane M. Leach and H. Warren Gardner, who were, respectively, president and vice president of the University of Texas of the Permian Basin during much of the time when we pursued research for this study, were most helpful in arranging released time. Project funding was administered by the Communities Foundation of Texas.

We benefited immeasurably from the insights and criticisms of many patient colleagues and friends. Among academic colleagues, Mansell Blackford and William Childs offered us invaluable extended responses to both our study and the concepts advanced within it; similarly, we benefited from the reactions of Richard H. K. Vietor and Toby Ditz, whose own scholarly work is reflected in the pages that follow. We also owe a great debt to the late J. H. Hexter, who showed us how what we wanted to do could best be seen as a problem in intellectual history, and to the late Henrietta M. Larson, who long encouraged us to take relatively untrodden paths in business and economic history. Among our Midland friends, Leo and Betty Byerley, Robert M. Leibrock, and Nicholas Taylor devoted many hours of close reading to our manuscript in its draft stages. We cannot overemphasize how much patience that effort took and how much we owe to their attention. In particular, Robert Leibrock gave us numerous essential insights into engineering complexities of oil field operation, absolutely necessary to evaluating many of the situations we studied.

Obtaining materials for this study was a major effort, assisted greatly by Anita Voorhies, interlibrary loan librarian at the University of Texas of the Permian Basin, Ken Craven of the Harry Ransom Humanities Center of the University of Texas at Austin, Florence Bartoshesky Lathrop of the Baker Library of the Harvard Business School, and the staffs of the government documents section of the Texas Tech University Library, the New York Public Library, the Library of Congress, and the Library of the United States Supreme Court.

Finally, we thank Christina, our daughter, who grew up with this project, always outshined it, and can now enjoy its completion.

# OIL AND IDEOLOGY

# *Manhood against Money*

Before one can explain how Americans looked at the emerging petroleum industry in the late nineteenth century, it is necessary to look at the historical context, at the nation in which it was created and at some of the beliefs that were part of political and economic culture. In the century prior to the industry's birth, America's founding fathers established what they projected would be an agrarian republic of virtuous freeholders, independent landowners, and craftsmen shunning great wealth for simple self-sufficiency. After 1865, however, the United States entered a period of especially widespread, rapid, and stunning economic growth, a growth marked by the proliferation of new industries and the emergence of new forms of business organization as well as by escalating national wealth. Wherever the United States was heading, it was clearly away from the earlier agrarian model of national identity. Therein was part of the problem for late-nineteenth-century Americans: where was America heading?

To an extent hard to grasp today, later-nineteenth-century growth and the changes attending it were both dazzling and alarming to the Americans who saw it happen. Changes were unfamiliar and unsettling, dislocating to some and disquieting to many. How did one come to terms with the displacement of those long dominating local and regional markets by national or international firms, with urban growth and urban crime, with labor unions and industrial giants? Where did one locate these things in terms of traditional attitudes and values? What did industrial development mean for the values of preindustrial life? At a time of volatile economic change, how did one sort out what was socially desirable from what was not? How did one protect a society facing so many unfamiliar and stressful developments from breaking apart amid civil and moral chaos?

It did not help that, as Americans confronted these questions, the traditional preindustrial thinking about economics and politics with which they were familiar tended to stress the harmfulness of many of the things that were happening. It condemned commercial growth and great accumulations of wealth that could result from it. Colonial Americans saw greed and avarice as the undesirable results of diligence and commerce, and they condemned too keen a pursuit of gain as individually sinful and collectively harmful. Monopoly was yet another aspect of the pursuit of gain; its resultant wealth, excessive prices, and unearned profits were socially harmful. Colonial Americans' perspective on economics was essentially zero sum: if someone was making profits, it must be at the expense of someone else. From that position it was an easy step to the conclusion that great profits came from sharp practice, chicanery, or extortion; as Benjamin Franklin put it, prosperity from commerce was "generally cheating."[1] Critics of the later generation of "robber barons" would adapt such traditional distrust of successful men of commerce to criticize company builders such as Jay Gould and, of course, John D. Rockefeller. But commerce was also condemned when it led to speculation, the world of shares and stockjobbing. The speculator's world was one of illusory wealth and chicanery.[2]

Critics of commerce also developed a set of themes directed toward the consumption that commerce encouraged. When prosperity allowed either merchants or consumers to live extravagantly, to sell and enjoy luxuries, to cater to self-indulgence and wasteful behavior, then it was socially harmful. Here, as other scholars have pointed out, gender was linked with extravagance, luxury, and wanton lack of self-restraint, these being seen as feminine characteristics. In collective terms, extravagance threatened the established social order; it allowed the extravagant to challenge their betters, and it encouraged socially undesirable behavior such as time wasting, gambling, or drinking to excess. In individual terms, extravagance amounted to lack of self-control and self-restraint, wanton behavior. Extravagance and luxury weakened society, making it effeminate.[3] The healthy alternative was frugality in production and consumption, thrift, self-restraint, and self-reliance. The gendered rhetoric of extravagance and luxury came to be part of the way Americans talked about effects of wealth and consumption of virtually anything, including petroleum. We shall see conservationists, for example, complaining of "wanton waste" in the oil field and, not coincidentally, "social chaos" in booming oil towns.

In eighteenth-century America as in the later Gilded Age and Progressive

Era, prosperity generated by economic growth encouraged apprehensions of moral problems tied to political behavior. Here Americans borrowed from British political discourse; commercial growth brought political theorists to focus on property and its place in polity. In Britain, country Tory and Whig radical opponents of Walpoleian Whig hegemony saw commerce undermining the traditional political order. Commerce resulted in replacement of the autonomous, self-supporting landed subject, the man whose economic independence allowed him to follow a virtuous course on public issues, with commercial nouveaux riches — financiers, stockjobbers, and speculators. These new men's personal fortunes were tied to government monopolies, concessions, contracts, and patronage; they necessarily sought to manipulate to their private interest a government on which they were dependent.[4] And at the same time the growth of commerce resulted in subversion of the constitution, it also would subvert individual virtue and social cohesion; it would turn a nation of stalwart, frugal Britons into effete, luxury-loving consumers on the one hand and destitute paupers on the other.[5]

British Opposition theorists were, essentially, at odds with modernity, with the capitalist order; they were also the minority opinion in Britain, where their jeremiads were not taken very seriously. But, their ideas had tremendous influence in America. The undesirability of wealth generated by finance and speculation, danger from moneyed interests behind the scenes, men of commerce and politicians collaborating in corruption, and conspiracy of money power to gain exploitative monopolies and subvert the state became staples in American political discourse. Thus, for founding fathers such as Benjamin Franklin and Thomas Jefferson, commerce in America should be kept in bounds, subordinate to agrarian wealth, and the American citizen of the new Republic would resemble the Opposition ideal: a man of landed property, dependent on no one, economically self-sufficient, self-reliant, virtuous, and dedicated to the public good. This ideal of manhood became a common feature in republican political discourse, the rhetoric of those who were of a Jeffersonian persuasion and of generations of political commentators following in their footsteps.[6]

For example, the arguments waged by Jacksonian Democrats against the Bank of the United States moved within the traditional discourse on politics, power, and monopoly. The national bank, as a monopoly, was seen by opponents as the result of special rights, privileges, and competitive advantages obtained from government. In the context of the dispute, these arguments were largely opportunistic defenses of local producers and of local entrepreneurs against larger and often more efficient distant competitors. Their rhetoric, however, emphasized the peril to democracy and the powerlessness of farmers, merchants, and workers against the bank.[7]

In a century prolific of reformers and utopias, most critics of the new economic order drew on the traditional paradigmatic relationship of economics, politics, and morality, whereby a change in any element would produce dangerous mutations in all. The common tendency of abolitionists and other reformers was their habit, grounded in evangelical Protestantism, of delineating social problems in moral terms; offenders became sinners, against the social, if not the divine, order, a perspective that left little room for accommodation and compromise. They cast differences in cosmic terms and readily attached tales of personal immorality to opponents in areas of public policy.[8] Reformers also tended to share the traditional hostility toward the "commercial morality" of their times. Many proponents of the temperance movement, for example, denounced individualistic profit making at the expense of community standards. For example, in Timothy Shay Arthur's classic temperance novel, *Ten Nights in a Bar-Room*, Simon Slade, the landlord of the Sickle and Sheaf, sold his mill and opened the tavern because "every man desires to make as much money as possible and with the least labor."[9] When reformers pushed for change, they worked to rouse moral indignation, to "change the moral vision of the people" in the words of abolitionist editor William Lloyd Garrison. They used the power of the press. When a heckling mob prevented Wendell Phillips from speaking to a Boston audience in 1861, Phillips directed his attention to the reporters who were close by: "While I speak to these pencils, I speak to a million of men. What, then, are those boys? We have got the press of the country in our hands. Whether they like us or not, they know that our speeches sell their papers. With five newspapers we may defy five hundred boys."[10] The abolitionist and other movements made strikingly modern use of the media to stimulate and maintain public interest in their causes and to set a moral agenda for public policy.

Thus it was, as we shall point out in numerous instances to follow, that when Americans began to talk about the new petroleum industry from an economic or political perspective, they already had a way to look at it. They commonly used venerable rhetoric, replete with themes predating the industry's emergence by well over a century. They used that rhetoric to vent moral indignation. Once Americans put certain themes and ideas together in their understandings of the petroleum industry, they constructed a specific ideology relating to it. Thereafter they repeated those constructions about industry again and again, often with little apparent regard for particular circumstances. Repetition of what was said about the industry, especially when what was said contained elements of moral discourse, reinforced the ideology and made it common knowledge. This ideology served as a political weapon, used to secure competitive advantages during the dynamic post–Civil War period and thereafter.

With the conclusion of the Civil War, America began a near decade of rapid economic expansion. As Rondo Cameron expressed it, the United States became "the most spectacular example of rapid economic growth in the nineteenth century."[11] In some measure, economic expansion rested on demographic change. The national population increased dramatically from fewer than 31 million in 1860 to about 77 million in 1900. At the same time, by 1900 many Americans were better off than ever before. Per capita incomes doubled from 1790 to 1860 and then more than doubled during the next thirty years. The lot of unskilled workers and their families changed little, in large part because of the steady influx of unskilled foreign workers, but by 1890, 7 million of the 12.5 million families in America had incomes of fifteen hundred dollars or more per year, constituting a large middle class, the richest internal market for consumer goods in the world.[12]

This economic boom was also in part the result of federal policies: American industries still enjoyed significant advantages over foreign competitors because of high tariffs on most imported goods. During the Civil War, with anti-industrial southerners out of Congress, pro-development Republicans liberalized immigration statutes, thereby ensuring a steady influx of relatively cheap unskilled labor. Banking and currency reforms created a new national banking system, making it easier for businessmen and farmers to borrow capital, while wartime inflation eased repayment of debts. These policies, together with the funding of transcontinental railroad construction and the continuation of liberal homestead laws, gave American producers privileged access to growing and protected markets, abundant and relatively unskilled labor, and cheap credit. Expanded transportation and communication facilities also fostered economic growth. The mileage of operating railroads expanded greatly, from nine thousand miles in 1860 to seventy thousand miles in 1870 and nearly two hundred thousand miles by 1900.[13] As consumers of goods, the railroads spurred the expansion of the steel, coal, and lumber industries.[14]

The greatest expansion occurred in the manufacturing sector of the economy. By 1900 the United States produced nearly one-third of the world's manufactured goods, largely because of the proliferation and growth of owner-managed manufactories of modest size. Growth also stemmed from the application of new technologies to production. One of the most dramatic examples is the production of cigarettes, which increased fortyfold, with the same labor force, with the use of James Bonsack's new machinery.[15] In the oil industry, the manual movement of oil was eliminated during the early 1870s, and multiple-stage distillation was organized more efficiently, increasing productivity in units that were large enough to employ the new technologies. The lower unit costs of these refineries gave them increasingly significant cost advantages over plants with small capacities, making it possible for them to expand beyond their local markets up to the point at which shipping costs offset economies of scale. Even then, most refiners lacked marketing organizations and thus relied on traditional "middle men."[16]

As had long been true, middlemen, commodity brokers and wholesalers, were an important sector of the expanding postwar economy; commodity dealers bought raw material from producers to sell to processors, and wholesalers purchased finished goods from manufacturers and sold them to retailers. During the 1880s and 1890s, however, some of these intermediaries were increasingly challenged by mail order sales, chain stores, and corporations that expanded their operations into commodity purchases and wholesaling. Expanding firms moved "upstream" to acquire ownership of raw materials and the means to transport them and "downstream" into wholesale and retail operations. This mode of expansion, vertical integration, provided greater security of supply of raw materials, enabled firms to monitor the sale of finished goods, and, above all, saved transaction costs by eliminating middlemen. The most complete degree of integration commonly existed in companies that could apply the technologies of continuous-process manufacturing, in chemicals, grain milling, photographic film, sewing machines, sugar, and, by the end of the century, oil. As Richard S. Tedlow described the process: "A small number of firms realized scale economies to an unprecedented degree by expanding their distribution from coast to coast and border to border. The profit strategy during this phase was to charge low prices, which permitted only small margins per unit but made possible greatly increased total profits because of high volume."[17] Typically, the first companies to integrate vertically with success enjoyed "first mover" advantages in prices and hence markets and kept them for long periods of time.[18]

Meanwhile, down on the farm, large increases in national and world populations stimulated demand for American farm products and triggered major changes in agribusiness. The number of farms increased from about 2 million in 1860 to 5.7 million in 1900.[19] With newly invented agricultural machinery to plant and harvest their crops, settlers on the Great Plains developed "bonanza wheat farms," which realized massive gains in efficiency — according to one government estimate, with about 7 percent of the time required to plant and harvest wheat by older, hand-labor methods — though requiring large increases in invested capital. Farmers were now able to realize larger incomes, but at the hazard of higher levels of risk than before.[20] Drought and flood continued to produce localized shortfalls and economic hardship from time to time, but the most efficient farmers realized spectacular gains during good years; even the lean mid-1870s were tolerable because the prices farmers paid for manufactured goods declined more than those they received for grain. But when their crops entered larger markets, farmers faced unexpected challenges because different patterns in crop specialization and variations in weather conditions around the world occasionally produced dislocations in markets and significant variations in prices of grain.[21] The cotton, grain, and dairy farmers, moreover, were increasingly dependent on the railroads to reach distant markets, whatever prices their goods might fetch. That dependence spawned grievances with railroads.

By the beginning of the depression of 1873, various farm organizations, most notable among them the Patrons of Husbandry — the Grange — entered state politics to limit farmers' risks by securing control over freight and storage rates, succeeding in Illinois, Minnesota, Wisconsin, and Iowa during the first half of the era.[22] The editors who "spoke" for farmers waged their argument in familiar terms — in defense of the small producer who was threatened and even victimized by concentrated economic power. By appropriating traditional rhetoric on wealth, power, and morality and applying it in what was, in fact, a new set of circumstances, the Grangers translated the public policy discourse over railroad rates and services into the terms and contentions of industrial and urban society.

One such contention was the traditional apprehension that commercial growth resulted in a fissured society. To many nineteenth-century commentators wealth was increasingly concentrated in the United States. The predations of railroads and "soulless corporations" produced an aristocracy "more despotic in its nature than exists in the old world." Employing the rhetoric of class warfare as it had been invoked by Jacksonian editors and their Jeffersonian predecessors, Granger D. C. Cloud described the grand new division in American society: "The country is now divided into two parties. One party is composed of the people, strong in nothing but numbers, and the determination to battle for their rights. The other side is composed of corporations, stockjobbers, brokers, and capitalists, whose strength consists in the organization and consolidation of their interests, their control of the finances of the country, and of the different departments of the government." Moneyed interests were out to subvert democracy. The recourse to conventional efforts at reform through the Grange was revolution, an outcome that D. C. Cloud, like most Grangers, saw as extreme but was willing to consider.[23]

Railroads were among the earliest industrial targets of those suspicious of growth and wealth. In response to Granger agitation, railroads lowered long-haul rates, only to incur the enmity of producers who had previously enjoyed competitive advantages in domestic and export markets. Eastern farmers and merchants fumed at what they claimed were preferential rates given to midwestern producers. Local merchant groups from Erie, Pennsylvania, to New York City met to protest "freight discrimination." New York's near monopoly of foreign trade was directly threatened as diversions to other ports became economical. Thus, in 1873, leading merchants in that city organized the New

York Cheap Transportation Association, led by Francis B. Thurber, partner in H. B. Claflin and Company, a dry goods wholesaler, and Charles Pratt, the leading petroleum refiner in the city; one member objected that it was "singular that Chicago firms should have extra good terms granted them by New York railroads." Sectional rivalries between New York, Boston, other East Coast shippers and western shippers were always at the forefront during agitation over freight rates during the early 1870s, but that was not how contenders framed the main issue. The shippers and their spokesmen translated the contest for control of transportation into a defense of small producers against irresponsible concentrated power. They appropriated and applied the traditional preindustrial discourse on economic concentration and morality to the problem of competitive advantage.[24]

Disputes between the railroads and eastern shippers heated up further in 1876, when trunk lines engaged in rate wars. Railroads commonly gave rebates to large shippers, amounting in many cases to "half of published rates to companies guaranteeing substantial shipments."[25] Groups such as the New York Cheap Transportation Association, which became the New York Board of Trade and Transportation, were quick to react by promoting schemes for the construction of additional railroads in the state and by organizing mass meetings to sway public opinion. At one meeting held in February 1878, more than eight hundred people, "mostly well-known merchants and bankers," met at Steinway Hall to be addressed by Mayor Smith Ely and attorney Simon Sterne. The mayor described the current state of the New York economy in alarming terms: "With natural advantages unsurpassed, trade has diminished for several years, and vacant shops, stores, and manufactories, and tenements stare at us in every street. Our commercial supremacy seemed so firmly established, and our natural advantages so great, that we deemed it unnecessary to do anything for ourselves, and the trade which fairly belongs to New York has been diverted to cities less favored by nature, but whose merchants are watchful and alert to take advantage of our stupidity and supineness." As the *New York Times* reported, "His remarks were frequently interrupted by warm and prolonged applause."[26]

Sterne, a young lawyer who was developing a specialty in transportation litigation, executed an important tactical maneuver for his New York clients by redirecting the attack against rival shippers, whose claims to policy preference were as good as those of the New Yorkers, to the railroads, which were already tainted by controversy and scandal. He charged that the railroads were "spending money at elections" and that they were exercising inhibitory control over the press, thereby undermining the American political order. Clearly, more was at stake than freight rates, though by inference the way to free the press and clean up American politics was to implement transportation policies that would restore the competitive advantages of New York merchants by raising long-haul rates.[27]

The speeches in New York clearly reflected the defensive response of regional economic interests to the nationalizing of the economy, and they had their defensive counterparts in many places. During the Gilded Age, states commonly erected legal barriers to interstate commerce in order to favor local businessmen, producers, and marketers. They did so through licensing, taxation, and inspection statutes, barring drummers, levying higher fees on "foreign corporations," and mandating costly on-site compliance with regulations. Thus, butchers in a number of states used meat inspection laws and ordinances to gain competitive advantage over the large national meat packers, taking advantage of "the protectionist impulse in states."[28] As Mary Yeager has pointed out, merchants in New York organized and formed shipper pools to obtain reduced rates and offset site advantages enjoyed by Chicago and St. Louis. After refrigeration gave large packers in Chicago additional price advantages, the New Yorkers assailed them fiercely as monopolists and took their battles to the media and Congress.[29] The political stratagems of local interests to counter cost advantages enjoyed by the new national corporations is well illustrated by the example of antitrust laws and proceedings in Texas. The state's antimonopoly sentiment first appeared in the constitution of the Republic of Texas in 1836 and was retained in subsequent constitutions including that of 1876, which is still in force. As in other states, the provision was understood originally to bar grants of exclusive privilege by the state, following the common-law definition of "monopoly." After the Civil War, interpretations changed, following shifts in economic and political contexts, first bringing it to bear on railroads and later on manufacturing trusts, especially Standard Oil. Beginning in 1889, states, including Texas, Kansas, Maine, Michigan, Missouri, Nebraska, and North Carolina, passed revised and expanded antitrust legislation. By the end of 1893, two territories and twenty-one states had constitutional or statutory barriers, or both, to the operation of trusts within their borders.[30]

The revised and broadened definition of "monopoly," "a combination of capital, skill, or acts by two or more persons," to cause restriction of trade and lessen competition, affecting the prices of goods or services, was well suited to the defense of local interests. Beyond barring action that produced the identified results, the Texas law, for example, specifically barred pooling, refusal to do business, sale below costs, giveaways, rebates, territorial marketing, and exclusive dealing. It effectively barred both horizontal and vertical integration under its general provisions and enjoined specific market behavior by vertically integrated companies under its enumeration of illegal activities. Leaving little doubt as to the intended beneficiaries, the law went on to exempt municipal corporations and agricultural producers, later exempting advertising agencies, title abstract companies, mortuaries, and dance studios, along with medical practitioners, lawyers, architects, and other providers of services.[31]

Politically defensive strategies on the part of local and regional producers were also encouraged by the depression of 1873. As markets declined, manufacturers formed trade associations to try to control declining prices by agreeing to production quotas. These cartels were usually short lived, however, because the weakest members cheated to stay solvent while other members were unable to enforce compliance; forestalling—withholding goods from markets to force prices upward—was illegal, and such agreements were unenforceable in court. The failure of the cartels led larger manufacturers to eliminate competition by buying out or merging with competitors. Manufacturers who could find the necessary capital commonly bought out competitors during the 1870s, though in some instances purchasers exchanged shares in their companies for their acquisitions. During the 1880s, the device of the trust was used increasingly because of its flexibility. In this approach, competing firms placed their shares in the hands of a committee of trustees, who thereafter received the profits of the component firms and distributed their profits to the members. Outright purchase, merger, or formation of a trust has been referred to commonly as "horizontal integration," a term that covers the combination of firms in the same trade.[32] Railroads adopted similar defensive strategies. In desperate attempts to remedy the economically calamitous consequences of overbuilding, railroad owners formed rate pools in the 1870s; when such agreements failed, because they were unenforceable, financiers organized highly complex legal and corporate structures, the dozen or so great railroad systems that provided most of the passenger and freight service until after the Korean War.[33]

Despite occasional downturns, like the depression of 1873, the American economy and American businesses grew, and so did cities. During the postwar period, increasing proportions of Americans lived in cities, up from 6.2 million in 1860 to 30.2 million in 1900. More Americans lived in two cities, New York and Philadelphia, in 1900 than had lived in the whole Republic in 1790. These city dwellers, most of whom produced few goods or commodities for their own use, formed the core of the growing national market for consumer goods.[34] Along with rapid urban growth came overcrowding, rising crime, and sensitivity to the increasingly dramatic contrasts between the rich and the poor. A New York City reporter, for example, described overcrowded tenements in 1869. The buildings were poorly ventilated, with outdoor toilets near them, producing disease and "a distinct odor as far up as the third floor." Many city dwellers lived in clean, comfortable housing, but they were threatened by different sorts of problems. In more pleasant neighborhoods the wealthy lived well but had become enfeebled morally: "At last it comes to such a pass that men only eat, drink, dance, and simper, while the women dress, gossip, dance, and giggle." This was just what the critics of economic development, from Jefferson onward, had feared; growing commerce and manufacturers created not only an urban poor but also a host of wealthy idle consumers of luxury. Worse yet, the men had become womanly, for like women they danced and simpered. With society divided into the oppressed poor and the idle, effeminate wealthy, New York offered an awful example of what could be seen in many places—at least, if one saw events through the eyes of a traditional antimodernist observer.

To antimodernists, the kind of moral decay commercial and urban growth brought was, of course, not limited to the wealthy. Want of self-restraint exhibited by the wealthy had its counterpart in the viciousness of poorer city dwellers. Urban wealth encouraged the growth of a large demimonde of criminals and prostitutes. The latter, according to one observer, were especially likely to haunt the city's new luxury hotels: "It is there that men of money mostly board. The rich miner from the mines, the oil speculator who has just sold his 'well', the returned Californian from the gold regions."[35] The wealth these new men of money enjoyed was thus the fruit not of labor but of speculation and windfall. That it would foster vice was inevitable in a traditional cultural point of view.

But it was not only the gold miners and speculators who undermined morality. The ostentation and extravagance of the wealthiest city dwellers were surely signs of moral decay. According to leading clerics and other social critics, large fortunes were also evidence of immoral action. Cardinal James Gibbons denounced monopolists, whose "sole aim is to realize large dividends without regard to the paramount claims of justice and Christian charity." Washington Gladden, a clerical leader of the loose-knit Social Gospel movement, opined that "no man can honestly heap up such fortunes as have been gathered by some of the great financial bandits of our time."[36] Gladden, among many others, also warned of the consequent rise of "disorderly and dangerous classes." Charity workers in the city had long warned that poverty could lead to "such scenes as have been once and again enacted in the streets of Paris."[37] For readers of the time, he evoked the Paris Commune of 1870 and its associated scenes of anarchy and bloodshed as America's future. Want of self-restraint in pursuit of wealth would lead to more general collapse of social control.

Quite apart from the presumed danger of bread riots and revolution in the streets, middle-class writers in the city were eager to destroy "the alliance of corrupt wealth with corrupt politics," as the *Independent* called it.[38] Accusations of impropriety and corruption increasingly targeted the granting of municipal franchises to the suppliers of new services and technologies, natural gas, electricity, and fixed-rail transit, as well as to public contractors. With every new service, reformers protested that bribes were taken and the public was betrayed by these "natural monopolies."[39] In particular, the heated warfare of New York liberals with the Tweed Ring and Tammany Hall emphasized and publicized the linkage of corruption and economic power; it encouraged the belief of middle-class opinion leaders that corruption was rampant and growing in postwar America. Nor was moneyed corruption merely local. Disclosures during the Grant administrations, beginning with the Credit Mobilier scandal, involving inside dealing of railroad executives with construction companies, and heightened with the disclosures of the involvement of Grant's confidential secretary with the Whiskey Ring, made corruption seem a national problem. Indeed, when they looked, middle-class reformers would find corruption on all sides.[40] Though, as Mark W. Summers has argued, corruption was far less common than "reformers" alleged, widespread belief in it turned political contests into "crusades for vindication, not for policy-making." The thrust of political activity would be in the direction of containing and limiting the new "immoral" forces in American life. As Summers puts it, the economic reform movement, given its particular "spin" by the corruption issue, thereafter "fostered reforms that confined American freedom and public power."[41]

Urban journalists were key to the growing perception of economic growth tied to urban corruption, as well as to other social problems. The advance of communication and printing technologies, taken with urban growth, prompted highly important changes in the diffusion of information in the United States. To boost the circulation of urban daily newspapers, the new information entrepreneurs such as James Gordon Bennett and Joseph Pulitzer launched a major change in journalistic style, away from "information journalism" and toward "story" elements, which emphasized "color" and personality. The favorite themes, for example, official corruption and public and domestic violence, lent themselves admirably to sensation. Papers such as the *World* and the *Journal* thrived on "crime and underwear copy": "TORTURED BY HER MANIAC HUSBAND," "CHOPPED HER SISTER ALMOST TO PIECES," "ROBBED OF WIFE BY THE DENTIST," and the like.[42] When young William Randolph Hearst launched the *New York Morning Journal* in October 1895, he included such articles as "How It Feels to Be a Murderer," "Are Sea Serpents Real?" and "Eight Stage Beauties on Broadway."[43] The increasingly frequent editorial emphasis on violence and "crisis" reinforced the new sensationalist reportorial style, but, in any event, reporting was designed to titillate, startle, and alarm rather than offer a balanced presentation of verifiable fact. It proved extremely profitable.

The rise of the new journalism in New York City, specifically, was of great consequence because principal New York newspapers and popular magazines published in New York, such as *Harper's*, *Scribner's*, and the *Century*, increasingly influenced perspectives across the country. During the Civil War, two of New York's leading daily newspapers started home deliveries in Washington, D.C., giving them access to both lawmakers and the wire services. The latter increasingly followed the lead of the New York newspapers both in determining the newsworthiness of events and in identifying the approaches taken to stories. Moreover, as the increasing importance of wire services undermined the information hegemony of major daily newspapers outside the city,[44] greater proportions of news on national issues were generated by the New York newspapers. Thus, as Thomas Bender put it, New York City, "one of the most complicated and contentious of human environments[,] . . . assumed a commanding and never again challenged position in the nation's economic and cultural life."[45] New York's writers, editors, and publishers effectively controlled large segments of cultural and public discourse by "framing" news. When postwar New York editors and reporters identified extreme social inequality as "the defining circumstance in Gilded Age New York," they heightened sensitivity to the problem across the country and, in the process, spread fears of social unrest and instability. They also enhanced focus on reform.

During the 1890s, a new group of urban journalists, younger newspaper reporters and a group of authors whose work was often published in magazine article and book form, emerged with a reformist bent. Like the more stolid reporters at the *New York Times*, the new group wrote for the urban middle class, in publications such as *Everybody's*, *Cosmopolitan*, and *McClure's* magazines.[46] Though the themes and methods of the group were developed by numerous writers, the core of the movement consisted of about one dozen men and women, including Ray Stannard Baker, David Graham Phillips, Charles Edward Russell, Upton Sinclair, Lincoln Steffens, and Ida M. Tarbell. They were "organized" into three overlapping groups, the *McClure's* writers, Robert Hunter's Wednesday discussion group, and the Liberal Club.[47] Theodore Roosevelt, the most quotable president after Lincoln, gave this group its indelible label, "muckrakers." The new reform journalists adopted many of the stylistic devices and opinions of the sensationalist journalists. Typically, their stories were exposés, developed around leading personalities and pitched as moral crusades. If their style was new, their political themes were old; favorite topics were money power, corruption in high and low places, social problems, and the defense of democratic government. Their goal, as Stanley K. Schultz has described it, was the "regeneration of the pious, self-disciplined individual who was presumed to be the cornerstone of a democratic society."[48]

Journalists sought to control opinion and move it toward reform, but so did others. The reform focus of journalists and publishers had an academic counterpart in a new generation of scholars, the first professional social scientists in America, academics who trained increasing numbers of economists and political scientists. Drawing largely on German models of social science, leading teachers at the Johns Hopkins and other universities challenged the humanities for domination of public discourse by identifying themselves as experts.[49] Though the new school contained considerable diversity, most of the academics shared two views, which they effectively defined as modern. First, they held to the "organic" notion that all aspects of life — social, economic, and political — are interconnected and hierarchical. No longer could economic activity be considered apart from its social and political contexts and consequences. Second, they believed in focusing effort on the scientific work by specialists, trained experts; this position was, in effect, an assertion of hegemony because it set aside the dedicated amateur, whether politician, lawyer, or editor. The new experts claimed the role of "informing" public opinion, a task they undertook often in cooperation with the journalists and editors as well as on their own.[50]

Like the new generation of reform journalists, the reform-oriented academics constituted a loosely knit circle based on education, opinion, and personal connections. Academic networking, as we would recognize it today, was the creation of this generation, and within it economist Richard T. Ely was exemplary as a builder of connections. His works were published in a wide variety of circles, including religious, scientific, economic, and public policy periodicals, and his students, including Albion Small, John R. Commons, Frederic C. Howe, Thorstein Veblen, and Wesley C. Mitchell, carried variants of Ely's perspectives into wider circulation. As a relatively conscientious mentor, Ely also pushed his like-minded students into print and jobs, thereby advancing his own views in the discourse of economics and ethics. Finally, Ely broadcast his opinions in a truly voluminous correspondence with colleagues, critics, and reform journalists — including Henry Demarest Lloyd and Ida M. Tarbell in the latter group. His papers at the University of Wisconsin comprise 254 boxes of materials, a veritable monument of academic networking.[51]

Of the wide range of Gilded Age problems and social concerns they could address, for the new academics, especially the economists, the question of monopoly, raised by the growing presence of trusts, was of special interest. It was an issue in which moral concern mixed with economic analysis or, as E. Benjamin Andrews put it in 1893, "Political economy abuts on ethics." It was also an issue on which the new social science experts inclined to share the perspectives of older liberal thinkers and reform journalists: monopolies threatened national stability and well-being.[52]

When the academic experts looked at the late-nineteenth-century American economy, they generally understood the broad historical causes of the consolidation movement. Writing in 1889, Charles F. Beach pointed out that the panic of 1873 and ensuing depression provided the first great impetus to consolidation. He saw the new trust organization in an evolutionary historical perspective; the trust was "the nineteenth-century offspring of over-production, small profits, competition rampant, and labor organization." It was the "protest of solvency over insolvency." Ten years later, German economist Ernest von Halle expressed the economists' consensus: "Extraordinarily low prices, grave disturbances of the market, the crushing rivalry of competitors, or vigorous association of workingmen with repeated demands and strikes give an impulse to the starting of combinations."[53] Social scientists could agree that greater productivity achieved through concentration of production offered economic benefits to American society. Writing in 1887, Henry C. Adams argued that monopolies were less speculative and more efficient than the competition envisioned by traditional laissez-faire economics. The sticking point was a moral one. For Adams and others, one problem with the growing concentration of production was that it would let businessmen impose their morally deficient outlooks on society. As he saw it, "It is the character of the worst men and not of the best that gives color to business society. . . . The plane of business morals is lower than the moral character of the great majority of men who compose it."[54]

When Adams and his fellow academics decided the level of business morality was low, they were in fact picking up much older ideas about the destructive character of commerce, the "soulless corporation." But they were also responding to anecdotal and journalistic tales of predatory competition, wily conspiracy, and attendant corruption, all framed in traditional republican values. For all their "expertise," their insistence on gathering facts and using statistical methods to support conclusions, the new social scientists were insufficiently skeptical and analytical of tales of heartless monopolists beating down small producers. Like reforming journalists they were willing to accept and retell strongly biased, even contradictory, stories in defense of what they saw as common morality. The more secularly inclined might part ways with blatantly moralizing Christian socialists, but, as David Danbom has pointed out, they shared "indignation, moralism, and righteous rhetoric."[55] Like other reformers and the crusading journalists, as Thomas L. Haskell observed, they were out to create new ways to "institutionalize sound opinion."[56]

How could one enjoy economy without suffering from the moral consequences of monopolies and trusts? The remedy, a way to realize the advantages of mass production by large corporations without suffering the moral pollution of the world of big business, was the supervision of competition by the state. It would "regulate competition to the demands of social conscience," by limiting such things as working hours and child labor and establishing watchdog commissions to inform the public and lawmakers of competitive abuses. Adams advanced these remedies with some caution, largely because he was aware of the general inefficiency of government and because he feared that watchdogs would prove to be less than faithful to the public interest.[57] He offered no panacea, but the three major threads of Adams's argument — the efficiency of mass production, the immorality of businessmen, and the supervisory role of government — dominated academic economic discourse for the rest of the nineteenth and throughout most of the twentieth century. Richard T. Ely underscored the final point in a resolution he presented at the organizing meeting of the American Economic Association in 1885: "We regard the state as an educational and ethical agency whose positive aid is an indispensable condition of human progress."[58]

The positive role of the state might be indispensable, but knowing how and when it should act was another matter. Here morality offered inadequate guidelines. At the end of the nineteenth century, Richard T. Ely tried to place the problem of monopoly in social, political, and moral perspective for the Philadelphia Ethical Society. Though there had been three decades of agitation against monopolies and both state and federal governments had passed laws that purported to control them, Ely found the basic concept increasingly elusive: "There can be no doubt that in economic literature, as well as in the periodical press, this one word-sign 'monopoly' has been made to stand for many different and more or less antagonistic ideas; and as a consequence, the controversies in which we have been engaged concerning monopoly have produced comparatively little action and even less light."[59]

Rarely couched in narrowly economic terms, over time, the monopoly issue became a recurring part of the false alarms and aborted crusades of the Gilded Age. To the present-day observer, some of these alarms have a humorous side. For example, the *New York Times* warned its readers of an anticipated increase in the price of prunes in 1887, because agents of two mercantile firms purchased twelve thousand to fifteen thousand casks of Turkish prunes, the crop of 1886. The result of their corner in prunes was a price increase of about 50 percent in one year. That caused a costive crisis; it placed consumers at a hardship during the spring, the principal time for consumption of prunes.[60] The *New York World* warned its readers about the predations of 159 trusts, including the castor oil trust, the skewer trust, the teasel trust, the tombstone trust, and the snow shovel trust.[61] In 1901 Thomas Elmer Will despaired because "most of the great dailies and magazines have been captured" and warned that the activities of the Vanderbilts, Armours, Stanfords, and Rockefellers in higher education was an attempt by the new superrich to establish a "college-trust."[62] In short, monopolies and trusts seemed to threaten Americans on every front, at least if one believed journalists. No wonder that by the end of the century even Ely found the concept of monopoly problematic. How did one know it when one saw it?

One development making it seem necessary to do something about monopoly as a social problem was growing social unrest. Though Ely and other social scientists occasionally warned of social unrest, there were numerous observers who took it as a focus. In an increasingly general and apocalyptic scenario, a wide variety of writers claimed that America's times of trial were at hand. Nor were alarms entirely groundless. Reorganization of the economy and depressions during every decade of the Gilded Age led to an escalation in industrial actions. In 1880 alone, the United States government recorded 813 strikes, more than it identified from 1741 through 1879.[63] Much of the confrontation between labor and management is now attributed to economic dislocation, as David Montgomery has done: although manufacturing output continued to rise, growth slowed after 1870, prices fell, and profits plunged. The 1880s brought a record rate of business failures, more consolidations of large-scale enterprises, and chronic conflict between employers and workers.[64] There was also a growing dissonance between the traditional republicanism of workers and their leaders and the new power relationships of big businesses. Ideology still upheld artisans and small producers, but operational reality consisted of larger shops and mechanization.[65]

The social and political perspectives of Gilded Age labor leaders consistently supported antebellum labor ideals expressed in broad traditional economic and political discourse: belief in self-improvement, republican values, hostility to concentrations of wealth, and a preference for small producers.[66] Their publications, the *Mechanics' Free Press* among them, repeatedly warned of the moral dangers of "ill-gotten abundance."[67] These values and the policy discourses in which they were embedded were remarkably continuous and durable. As one historian found, after the great railroad strike, the Knights of Labor "pictured themselves as a counterweight to rising greed, corruption, and intemperance."[68] The connection of the monopoly power of the railroads and social unrest seemed to be underscored by the increasing militancy of their workers. Unionized workers increasingly saw their organizations as in the front line of the war against monopoly and irresponsible wealth.

To prophets of doom, labor disturbances were but another symptom of looming social crisis. By the middle of the 1870s, writers were creating an image of an "age of rapacity" that began immediately after the Civil War with the railroad construction boom. Henry Adams, reflecting on the related corruption, concluded: "Our political fabric is out of joint and running wild."[69] Similar alarms came from the pulpit. Writing in the 1890s, George Herron, a Congregationalist Social Gospeler, proclaimed that the depression that began in 1893 was "the direct result of the centralization of wealth, of the investment of the control of industry in the hands of the cunning and the strong." Underscoring the moral substance of the situation, he concluded that "commercial tyranny and social caste are at war against God" and warned that America faced "the crisis of the centuries."[70] Similarly apprehensive, Washington Gladden feared the rise of "disorderly and dangerous classes," "the influx of helpless and degraded people from other countries," and the loss of "manly independence which is the substratum of all sound character."[71] Two years later, C. T. Russell, a member of the Jehovah's Witness sect, drawing on the tradition of Christian millenarianism, sounded distinctly mainstream, perhaps because his text was peppered with material from secular media, especially the *New York World*. He argued that the rise of the trusts, the appearance of the wanton extravagance of the superrich, the increasing level of violence in the workplace and in society were all "important elements in preparation for the coming fire."[72]

Whether or not they believed that the wrath of God was imminent, it is quite clear from even a brief immersion in late-nineteenth-century social commentary that the majority of the writers believed that, in effect, the times were out of joint and the country was in grave danger. At first sight, this is paradoxical. Notwithstanding occasional downturns and the increasing publicity of disorder, the American economy was prospering as never before, and more and more Americans enjoyed the material comforts of a middle-class standard of living: why, then, did so many commentators expect calamity?

The perception that the times were out of joint had an individual as well as a social resonance. In large measure, the intensity of commitment to reform, the widespread sense of powerlessness in the emerging urban-industrial order, and the fear or anticipation of catastrophic change all rested in a major social-psychological shift that was well under way during the Gilded Age. The deterioration of labor-management relations, the defensive battles of wholesalers, the campaigns of farmers' groups against the railroads, and the increasing hostility of the urban middle class to the new large producers share an important common point of view: they are in some measure reactions to the decline of self-employment. In 1860, 88 percent of all men were self-employed; by 1910 the figure had dropped to fewer than one-third.[73] In the context of the cultural construction of American manhood, this change, making men economically dependent on others, was a calamity for masculinity.

One response to the crisis of masculinity was the rapid spread of neurasthenia, or "American nervousness," as it was labeled. Significantly, urban middle-class men seem to have been most susceptible to the malady and, in some cases, most willing to do battle with the causes, as they identified them.[74] As E. Anthony Rotundo has pointed out, social and economic changes challenged the traditional cultural construction of masculinity: "Work could serve to reassure a man about his manhood and about the freedom and power that manhood betokened."[75] The common perspective that the small producers, the clergymen, and the academics shared with labor leaders and reform journalists was their commitment to preindustrial gender roles and their deeply gendered concern, as "masculine achievers," with personal efficacy in the face of new concentrations of power and new claims to moral authority. That prompted another response to the crisis of masculinity: vehement and widespread attacks on agencies perceived as destroying masculine autonomy.

It is not surprising that concerns tied to gender should have been part of the late-nineteenth-century fear of social catastrophe, as well as of the offensive of those who sought to restrain the forces of change. Cultural anthropologists have long identified gendered defenses of traditional values forming overriding elements in both discourse and social life. The basic, elemental nature of such gendered defenses for values makes them virtually irresistible to those who would wield moral authority. Late-nineteenth-century social critics simply identified manliness with moral rectitude. Writing in 1893, Washington Gladden decried the loss of efficacy: "The old-fashioned virtue to which I referred is the manly independence which is the substratum of all sound character."[76] Similarly, William D. P. Bliss, Episcopalian and Social Gospeler, boiled down the controversy over trusts in 1900: "It is a question of manhood against money."[77] Economic dependence and manhood did not mix. Thus, in 1908, Eugene V. Debs told the American Federation of Labor, "No man can rightly claim to be a man unless he is free and self-reliant."[78] If the new economic order turned the many into wage slaves of the few, it destroyed manhood—and it was morally wrong. For that matter, it was wrong if it turned voters into dependents of bosses or consumers into dependents of those who provided products to them or some producers into dependent pawns of bigger producers.

But if one was determined to challenge the new order, to constrain its progress in the interest of traditional ideology and gender roles, where did one begin? To some extent, reformers' targets tied new developments to old ideology. They looked at cities and targeted corruption, for example. They complained about consumption and attacked the sugar and tobacco trusts, both offering consumers what had traditionally been identified as luxuries. Railroads made an excellent target, symbolic of the new industrial order both in terms of impact on the national economy and in terms of generation of new wealth. And then there was oil, among the newest of the new industries. The oil industry created a product, kerosene, that consumers soon could not do without. It also created a concentration of wealth and economic power in the forms of John D. Rockefeller and Standard Oil that seemed utterly without precedent in the industrial sector of the economy. What better symbol of the inequities of the new economic order than an industrial firm that grew to unprecedented size by offering consumers a new product that it controlled. Moreover, the company was said to be in cahoots with the railroads to execute a business strategy that turned competitors into compliant dependents. If a reformer searched for the extreme case of the challenge of the new economic order to traditional values, surely that extreme case could be found with the Standard Oil Company.

## *Hasting to Get Rich*

In August 1859, Edwin L. Drake brought in America's first oil well. Several weeks later, when two additional wells confirmed the accessibility of petroleum, speculators began a feverish competition to lease up Venango County in western Pennsylvania. Local lumbermen, salt borers, and tan-bark cutters joined the action; numerous merchants, some local but many newcomers, became overnight oilmen. Sharp-eyed New Yorkers arrived to trade in leases or in the stock of newly formed companies; unemployed laborers, teamsters, carpenters, and coopers streamed in to take jobs in the newborn industry. On their heels came hotel, tavern, and boardinghouse keepers, gamblers, and prostitutes: one could not live on oil alone. The remote rural region of steep hills and heavy stands of timber, broken by narrow creek valleys containing small farms, a region with few villages, let alone cities, was swamped by thousands of new settlers and by urban problems. Housing, streets, and sanitation were suddenly and conspicuously inadequate, making new settlements visibly untidy and unclean.[1]

Early in its history, oil was a popular novelty topic with writers and their readers. Within a year of Drake's discovery, the new industry was hailed as "The Wonder of the 19th Century!" Perceptions of the industry and of the oil-producing regions were largely based on the writings of a few journalists, including J. S. Schooley, William Wright, Edmund Morris, and B. Franklin — the last-named an apparent pseudonym for a regional editor. Their newspaper copy appeared widely in daily newspapers. Reprinted in book form, their articles enjoyed even wider circulation.

As is usually the case with a new phenomenon, the observers explained the first oil boom in the terms and history of something more familiar to their readers, the California gold rushes of recent decades. For fifty years after Drake's discovery, American and European writers clung to the apparent parallels of California gold and Pennsylvania oil to emphasize the "feverish haste for coveted treasure" by "lawless prospectors."[2] "B. Franklin," writing in 1865, began his sketch for *Harper's New Weekly Magazine* with comparisons of the "treasures" of Pennsylvania and California. During that year, oil promoters in California used the common media association of oil and gold to boost what proved to be a "salted" oil find near San Diego. From these associations, the new oil industry received the recurrent images of the gold fields: treasure-seeking prospectors after quick riches, and none too particular about how they got them, living in primitive and lawless communities.[3]

As "Franklin" described the prevailing state of western Pennsylvania following the discovery of oil, the new industry destroyed the frugal but prosperous agrarian world: "Rich farms are laid waste. The plow turns no more furrows. The scythe cuts no more bending grain. The farmers' farms are no more loaded down with the fruitful harvest. The farmer himself, with his homespun clothes, is seen no more in the fields. All is changed! The farm is sold! The old man and his grown-up sons are worth millions, and the old homestead is deserted forever."[4] That western Pennsylvania had never been the kind of agrarian paradise "Franklin" conjured up did not dull his pen as he described virtuous yeoman farmers displaced and even corrupted by commerce. According to the national mythology of rural life, the Oil Region was Edenic, and, presumably, were it not for oil, it would have remained so.

For those journalists who investigated the Oil Region of Pennsylvania during the 1860s, firsthand impressions were vivid: "Whew, what smells so?" exclaimed J. S. Schooley on arriving in Oil City, Pennsylvania, in December 1864. The answer was short and direct: "Nothing but the gaseous wealth of the oily region." Offensive to his nose, Schooley's new environment was equally unpleasant to his eye: "Every thing you see is black. The soil is black, being saturated with waste petroleum. The engine-houses, pumps, and tanks are black. . . . The shanties — for there is scarcely a house in the whole seven miles of oil territory along the creek — are black. The men that work among the barrels, machinery, tanks, and teams are white men blackened. Even the trees, which timidly clung to the sides of the bluffs, wear the universal sooty covering. Their very leaves were black."[5]

Schooley, like most observers, also noticed that society in the new Oil Region was as lacking in order as its landscape. A visitor could expect to encounter all classes of people — gentlemanly company representatives, "adventurers, speculators, and peculators, bespattered men and dowdy women." In the oil fields no one bothered to regard social position or ceremony. Schooley commented, "You sit with them at table. You visit them in their offices. You walk about their works. They make no stranger of you. You take them as you find them."[6]

Life in the oil field swore at the nineteenth-century ideals of domesticity and morality preached by reformers such as Catharine Beecher; if, as one ladies' publication put it, "the domestic fireside is the great guardian of society against the excesses of human passions," there seemed to be few such moral hearths in booming oil towns. The young men flocking to them lived as they could and as they pleased.[7] Schooley was bewildered by the lack of traditional family society in the oil field, since many married men would not take their wives and children to oil towns: "If the ladies could look in upon them they would see what unkempt savages the men become when they go beyond the limits of home and the boundaries of society. . . . It is bad at any time and any where for men to leave their wives and children, and gather together hasting to get rich."[8]

Popular musical compositions of the 1860s, such as "Pa Has Struck Ile," described the new oil-rich Pennsylvanians.

I once was unknown by the happy and gay
And the friends that I sought did all turn away;
Our dwelling was plain, and simple our fare,
And nothing inviting, of course, could be there.
But now, what a change! Our house is so grand,
No one is so fine throughout the whole land.
And we can now live in the very best style,
And it's simply because my "Pa has struck ile."[9]

The theme of oil as the source of unearned riches and social irresponsibility was reinforced by the most widely told saga of oil riches, the much distorted tale of John W. Steele. As the story went, "Coal Oil Johnny" inherited $150,000 in cash and a daily income of $2,000 from oil production. The Pennsylvania backwoods bumpkin lost it all and even went into debt during a twelve-month spree in Philadelphia and New York City. Steele's lavish spending supposedly included the purchase of a minstrel troupe, hundred-dollar tips to waiters and shoeshine boys, and huge hotel and jewelers' bills — all totaling in the millions. In the end, Steele went bankrupt, the farm was sold at a sheriff's auction, and Steele resorted to earning a meager living as a baggage handler for the Oil Creek Railroad. Commenting on his failure, the *Pittsburgh Commercial* claimed, "Perhaps no man in the United States ever squandered as much money in the same space of time."[10] During the 1860s and 1870s, versions of Steele's story appeared in most urban daily newspapers. A Titusville drama society published "The Amateur Millionaire," the local version of the tale, in 1869. During the following decade, the play was adapted by a professional troupe and performed more than five hundred times as "Struck Oil," in New York, London, and most cities in the eastern section of the United States.[11]

As usually told, Steele's story was a parable of extravagance and self-indulgence, with Steele as an oil-generated prodigal son for whom there was neither redemption nor fatted calf. It was a parable of the perils of the new industrial order, avarice and luxury. Constructed on these terms, Steele's story carried a strongly conservative moral message; it was far more than the story of a foolish man who had a good time and then had to work hard to pay his debts. Behind Steele's story, as usually retold, were also a number of social concerns. Catapulted by wealth out of their usual lifestyles, the newly rich lived lives of pointless extravagance and idleness, using riches to no constructive end. A writer for the *Atlantic Monthly* lumped all the newly wealthy of the region together as "vulgar families, the millionaires of a month, tricked out in the unaccustomed trappings of wealth, like Sandwich-Islanders in civilized hats and trousers." To the "Carpetbagger in Pennsylvania," they personified "the mammon of unrighteousness."[12] What they did indicated that oil industry–generated prosperity reduced life to the crassly materialistic world of, as Charles Eliot Norton put it, "shoddy and petroleum."[13]

One of the most perceptive and influential oil field visitors was a *New York Times* reporter, William Wright, who spent seven weeks in Pennsylvania during March and April 1865; his reports, published later in the year as *The Oil Regions of Pennsylvania*, were widely read. He stated his purpose clearly in his preface: "Petrolia needed a searching examination and a scathing exposure; it has got both."[14]

Wright was bewildered by the disorder of oil field life: "Everything betokens disorder, disarray, indifference to all except the one grand object of pursuit." His description of the reactions of a visitor reinforced the untidy image of the oil fields established in the public eye by earlier writers: "The objects he is too apt to touch, in spite of all precautions, have a greasy, clammy feel. His nostrils are assailed by gaseous odors, such as they probably never before inhaled in the open air. Into his ears is continually poured a stream of speech, in a dialect essentially different from that taught in Webster of Worcester."[15]

In the oil fields the laws of man and nature were both bent and broken. The new petroleum industry was unscientific, unpredictable. Oil, for example, had been found in such a variety of depths, situations, and amounts that there seemed no natural laws explaining its location or quantity.[16] In its early days, the industry was an intellectual anomaly. It was also an offense to common morality. At best, oil was a genuinely risky business. There were notable successes, most of them well publicized. A well drilled by Orange Noble and George Delameter in 1863, for example, cost barely $4,000 and returned more than $5 million, more than $7,000 for every dollar invested, during each of the first two years of production.[17] Ordinary experience was less exhilarating. Few ventures ever matched this performance, and most failed to return the capital invested. Titusville newspapers often counseled would-be plungers to risk no more than they could afford to lose.

As Wright saw it, in the petroleum industry swindling was "a system which has been reduced to both a science and an art."[18] He thus devoted a whole chapter entitled "How Strangers Are Taken In" to describing the plunder of investors. Promoters would doctor wells to make them gush abundantly for visitors, making even "shrewd, sharp, intelligent Eastern financiers" think they could buy into a bonanza. As for ordinary folk, they were so enthralled at meeting "a living, moving, talking *millionaire* — perhaps only a teamster three years ago" — that their faculties of judgment were completely suspended. True, people were getting rich, but most of those doing so were speculators, "preying on the gullible; unsuspecting men and women, even widows and orphans, are stripped and scalped, and flayed, and picked to the bone by a generation of sharpers."[19]

As many observers pointed out, an abundance of new oil concerns were fraudulent, a matter finding ready reflection in popular culture. "Famous Oil Firms," a song of the late 1860s, contained an illustrative list:

There's Ketchum & Cheatum, and Lure 'um & Bleed 'um,
and Swindle 'um all in a row.
Then Coax 'um and Leadum, and Leech 'um and Bleed 'um,
And Gulle 'um, Sinkum & Co.
There's Watch 'um and Nab 'um, and Knock 'um & Grab 'um
And Lather & Shave 'um well, too;
There's Force 'um & Tie 'um, and Pump 'um & Dry 'um
And Wheedle & Soap 'um in view.
There's Pare 'um and Core 'um,
and Grind 'um and Bore 'um,
And Pinchum good, and Scrape 'um, and Friend,
With Done 'um & Brown 'um and Finish & Drown 'um
And thus I might go to the end.[20]

Samuel Morris, a New Yorker who published the results of his investigation of the oil booms in a widely read and often quoted book, lamented the spread of "oil fever" through puffery, such as that offered by his fictitious firm, the Inexhaustible Moonshine Petroleum Company: "Capital $1,000,000,000 in shares of one dollar each; assets, a certain or uncertain portion of the 'Devil's Half-Acre,' a dozen highly illuminated maps, a bottle of lubricating [oil], purchased for the occasion, any amount of brass and gammon, a cartload of certificates of stock, and the complacent squirts who manipulate for an oleaginous fortune."[21]

Morris's warnings were widely repeated in America's financial center. The *National Quarterly Review*, published in New York City, claimed that of the hundreds of new and highly touted oil companies, "half of them have not as much as an office, and those that have cannot afford decent furniture for it."[22]

Investments in oil were speculative at best. Honest producers and refiners faced substantial hazards in the way of making profits, let alone fortunes. Yet William Wright thought prudent investment could be made in the oil field — if the investor made sure company agents and employees were honest, if he learned as much as he could about the company, if he bore in mind that oil on one property did not guarantee oil on the next, if he realized most wells produced no longer than eighteen months. Even so, the uninitiated might do well to put his money into "a United States seven-thirty bond" instead: and having absorbed Wright's advice, any investor without a case of oil fever would certainly prefer it.[23]

It is clear that this perspective held sway in America's business communities. For example, when members of the modestly rich Du Pont family considered placing money in Pennsylvania oil, they collected prospectuses from dozens of new oil companies, including the Anglo-Saxon Petroleum Company of Boston, the LeBoeuf Oil Company of Philadelphia, the Reno Oil Company of New York, and the Pennsylvania Imperial Oil Company of Philadelphia. The last-named venture emphasized that it had "reserved capital of $250,000 — *much the largest held by any company.*"[24] With the colorful literature in hand, the Du Ponts solicited extensive information about promoters and prospects through their various business connections. When one of their circle investigated the Anglo-Saxon Oil Company of Boston, he reported that the company's properties were worth less than it claimed, despite supporting newspaper statements that were included in the promotional literature: "Newspaper statements and general rumors about the yield of any well are very unreliable." In the end, with the expectation that the venture would still pay out in a year or so, the Du Ponts plunged for twenty dollars![25]

By the end of the Civil War, the petroleum industry had taken on its characteristic structure and organization, though the latter was so diverse and specialized that no one at the time thought of the combined activities as constituting an "industry." The divisions and operators were as follows:

1. Oil operators raised capital, sometimes with the assistance of commission salesmen; acquired leases, sometimes from professional lease brokers; organized the drilling of wells, usually by hiring an independent contractor; purchased materials; stored produced oil for shipment; sold crude oil; and arranged transportation as required by the buyer, to a point from which he would ship it.
2. Oil purchasers bought crude from producers and arranged its transportation to refiners, who purchased it.
3. Transporters moved crude oil from the wellheads to shipment points and from there to receiving facilities of refiners and moved refined goods to wholesalers/jobbers.
4. Refiners processed crude oil to produce kerosene, lubricants, and other products.
5. Wholesalers/jobbers sold refined products to domestic and foreign retailers through hired sales agents; most of the wholesalers were in the general trade, handling many products other than petroleum products.
6. Retailers were usually general merchants, often selling kerosene as a generic product, like sugar and flour.

The postwar "industry" was thus a highly diverse and specialized industry, with special perspectives and risks in each sector of it. It was carried on in a variety of locales, which commonly resulted in special competitive advantages and disadvantages peculiar to the sites. For the most part, operating conditions and problems peculiar to each sector were not understood at some distance; data relating to crude oil production circulated in the producing regions of Pennsylvania but commonly appeared only in price lists in urban refining and trading centers. Similarly, information relating to the trade in refined products was largely limited to urban newspapers and such trade publications as the *Oil, Paint, and Drug Reporter*, published in New York City.

In the absence of generalists and shared information, the various sectors of the industry developed their own practical lore as well as their own negative perspectives of the other sectors of the industry. Apart from a handful of scientific publications and Wright's book, the operational discourse through which information was routinely exchanged within the industry and the public discourse in which the industry was understood within the producing regions rarely circulated as far as the urban press. Thus journalists commonly described oil "fields" as if they were subterranean lakes, and "reserves" were thought to be readily quantified and accessed assets; oil production was achieved easily, like turning on a water tap. The operational discourse, "everyday realities" of the industry, by contrast, largely circulated in print within producing regions and in conversations of producers and workers. These sources carried a high volume of trade-specific information, but, of greater importance to oilmen, they also conveyed the general "business sense" that accumulated in this limited circle of oil producers, largely at variance with the simplistic notions about exploration and production that circulated outside their circles. They comprised both the rules of the game and the definition of risk as they were generally understood.

The man who decided to become an oil producer by drilling a well near Titusville or Oil Creek in the early 1860s, for example, knew that the cost of drilling was rising. During the first year of development, oil could be brought in at such shallow depths and with such primitive equipment that the actual cost of drilling a well might be less than $2,000. Costs increased to $4,000 or more by the mid-1860s, when operators drilled deeper and began to purchase improved equipment; costs declined to about $1,500 in the Bradford field during the 1870s and 1880s with the end of boomtime demand for goods and services. The operator also knew that once production was confirmed speculators quickly drove the price of promising properties up to five and six figures for the common "fee" purchases. Under these circumstances, operators learned that it was less expensive to lease land than to buy it; they also learned how to capitalize small ventures by trading leases, selling part of what they leased, and using the proceeds to drill the remainder. Landowners also learned the financial ropes through informal networks of friends, family, and neighbors, coming early to see oilmen less as benefactors than as dubious suitors. Oilmen, thus, frequently had to offer hard-bargaining landowners up to 50 percent of their production as a royalty — an exorbitant amount, had there been any way for the landowner to monitor production. In the absence of it, landowners generally came to believe that oil operators routinely cheated them.[26] This belief was, no doubt, based on experience; in some measure, however, it also reflected the widespread negative opinions aired by Schooley, Franklin, Wright, and others.

If the operator was lucky, through word-of-mouth communication he might find an experienced driller, always in short supply during the brief drilling booms. More likely, the workers he could find had little or no experience. As work proceeded, water encroachment presented operational problems, and operators assumed that when they began to produce large quantities of water, significant oil production was past. By the mid-1860s, operators learned that lining wells with iron pipe could keep them from flooding but that this technological improvement raised their completion costs, and, thus, most operators who struck oil continued to produce plenty of water with it. In response, they used wooden vats to separate oil from water, with limited success.[27] Among operators, conversation frequently turned to accidents and incompetence on the floors of the drilling rigs. As drilling went on, the operator hoped inexperienced workers did not accidentally lose drilling tools down the hole, requiring an expensive delay while they were "fished" out. He also hoped the crew did not hit a pocket of gas, which could blow the tools from the hole or, worse, ignite as it encountered someone's cigar or lantern. The first flowing well, brought in near the mouth of Oil Creek, ignited when its gas reached the rig boiler, and nineteen persons died. In the first step toward blow-out technology, the fire was snuffed with mud and manure.[28]

Whether, in the end, the operator struck oil was entirely a matter of guess-work, for there was no reliable empirical basis for deciding one site was better than another. Some operators resorted to dowsing rods, and others huddled near other operators in creek beds, leading some wit to say they had located through "creekology." By 1865 operators had learned that hillsides might be as productive as creek beds. This knowledge was not infallible, giving rise to additional suspicion of oilmen among investors, whose "economics" were based largely on mercantile and manufacturing experience, which contained few current examples of the sizable and total losses that were common in oil exploration.

If the operator found oil, he confronted additional problems. When he had them, he ran crude from the wellhead into wooden tanks or barrels. Barrels, however, cost two to three dollars each and were often in short supply. If his well produced much more than fifty barrels a day, the producer usually ran out of storage in short order, so he had to dig earthen pits, from which the oil evaporated or seeped out. For that matter, it did the same from leaky wooden barrels and tanks, saturating the ground around the well and making its way into the nearest creek, river, or gully. One of the most common opinions among oilmen was that if a well was shut down it would never produce again; this notion made them highly resistant to shut down for more than a few days at a time. So did the realization that if they did not produce their oil their neighbors might get it instead. Thus, wells produced even if oil was wasted or prices declined sharply. Unless all producers in a field shut down, efforts to curtail production to raise prices failed after brief trial. In any event, for mid-nineteenth-century operators, lacking large steel tanks, tank batteries, and efficient gathering lines, substantial production inevitably meant spills, leaks, and loss. There was, of course, the cold comfort that the average well of the early 1860s was expected to play out within eighteen months.[29]

The major focus of a successful producer's activity was the sale of crude oil. Here again, the newspapers in the Oil Region and casual conversation carried most of the technical information necessary for these transactions: current prices, prospective buyers, and transportation costs. When it came to selling crude oil, producers of the early 1860s had a number of options. Some pur-chased the crude of others, to sell with their own to roving agents of refiners, commission merchants, brokers, and wholesale dealers in oil. Others, "dump" men, managed to build enough storage to buy crude from many producers and speculate in it; these speculators may have been the prime movers in setting regional prices. During the first two decades of the oil industry, everyone's business calculations were complicated by highly variable supplies of crude, a matter beyond control. Variable supply was the single industry condition whose impact each sector of the industry worked to minimize. In the first year of production, when no one knew how long Pennsylvania crude production would last, prices rose from $12.60 per barrel in February 1860 to $26.60 in April. Thereafter, as more wells came in, supplies of crude rose and pressed prices down to $9.68 at the beginning of December. Still, at $9.68, most producers could make profits, and the region boomed. Boom, however, as quickly turned to bust. Supplies of crude swamped demand—which could have been met with daily production of two thousand barrels per day—and prices tumbled to five cents a barrel. With every steep decline in the price of crude oil, hordes of producers were wiped out. Such calamities became part of the legend-oriented discourse within the industry, as they emphasized the precarious nature of the business and the nerve required to conduct it.

The recollected history of the industry, commonly carried in historically retrospective stories in regional newspapers, also carried more hopeful events, as prices made gradual comebacks. During 1862, prices rose from $1 in January to $6 in September and then declined to $3 by the end of the year. Following a surge to $13.75 in mid-1864, prices slid to $7.50 in October and then rose to $11 to $12 in December. Crude fetched $8–9 in 1865 but fell to $4.50 at midyear when the Pithole field produced prolifically. Prices fell to $3.25 at the beginning of the following year and then dropped to $1.50 to $1.75 by year's end.[30] This level of commodity price volatility was without precedent either then in general business or thereafter in the petroleum industry, even during the 1930s. The recurrent pattern of new discoveries, soaring volume, and declining prices continued through the nineteenth century. During the 1870s, the price of oil varied from $1 or less per barrel to as much as $3.60. During the following decade, the volume doubled, while prices declined.[31] During these volatile price cycles, the industry drew in hordes of small producers at peaks, only to have many wash out in the valleys. Indeed, price volatility was understood within the industry as one of the chief risks facing producers. As a contemporary observer put it, "The way in which oil was bounding up and plunging down was enough to make one's head whirl."[32] Within the industry, however, devastating plunges in oil prices were expected much like oil well fires and floods, beyond both prediction and individual blame. They were generally seen as natural occurrences in a risky business.

Though they came to expect volatile prices, oilmen nonetheless tried to do something about them. Oil producers attempted to stabilize the price of crude oil repeatedly during the 1860s. In 1861 and 1862, they tried to withhold oil from the market until the price rose. In 1867 and 1868, they tried to establish minimum sales prices, enforced by voluntary compliance. In all instances, their efforts came to little because producers were always eager to obtain revenue from new discoveries and small producers could not afford to hold out long without some income from production. The final attempt of the decade, mounted in 1869 by the Petroleum Producers' Association, a producers' cartel, failed because refiners had learned the wisdom of holding oil in tankage against sharp fluctuations in supply and price. When Oil Region producers and refiners tried to raise prices in New York by forestalling, withholding crude from the markets of Pittsburgh refiners, the scheme fell through because aggressively competitive Cleveland refiners leaped to supply short markets.[33] As would so often be repeated in the future, oilmen could not unite behind a single position or strategy.

The publicity that attended the various attempts at forestalling and the larger refiners' success at moderating price fluctuations by storing crude oil fostered the belief that it might well be possible to control the price of crude oil. This belief led, inexorably, to the belief that somebody was already doing it. Hence in 1869 the rumor circulated widely in the producing regions that a mysterious European financial combine was purchasing and holding large quantities of relatively cheap crude with the intention of forcing the price up. Presumably, by crowding other buyers out of the market, it was depressing crude prices, and by holding the crude back from the market, it was raising costs to refiners. Neither inconsistency nor lack of tangible corroboration slowed the rumor's rapid progress.[34] It was also believed that conspiratorial rings of product jobbers and crude oil buyers in New York City had cornered markets, to the distress of producers and refiners in oil centers, including Cleveland and Pittsburgh. The response in Titusville was to form an oil exchange in January 1872 to concentrate crude oil buying in the region.[35] No reliable evidence of these conspiracies ever came to light, but the pervasiveness of the rumor reflected the growing recourse to conspiracy theories in a new and complex industry during economically unsettled times, when relationships between producers and refiners were strained. During the early 1870s these relationships deteriorated even more as a new refiners' association accused the producers of adulterating crude and producers tried to carry out a three-month hiatus in drilling to force oil prices up.[36] In the end, they succeeded only in driving some of the smaller refiners out of business, leaving more powerful survivors for producers to contend with. Born of mutual suspicion and failed efforts to stabilize the industry, one of the commonplaces in operational discourse came to hold that the interests of producers and refiners were incompatible; whatever worked to the refiners' advantage generally worked against oil producers.

A common problem for both producers and buyers, transportation was a difficult and controversial sector of the industry from its onset. Simply getting petroleum out of the hilly, isolated western Pennsylvania oil fields proved to be a challenging task. In 1860, Titusville was some twenty miles from the nearest rail connection. That meant hauling wagonloads of barreled crude by team out of the hills and valleys of Venango County. At some seasons of the year muddy roads became impassable, virtually shutting down traffic. Alternatively, oil buyers could ship oil by flatboat or scow down the Oil Region's winding creeks to the Allegheny River, where it could be moved to Pittsburgh. In 1862 a primitive pipeline gathering system successfully linked wells on the Tarr farm to a refinery at Plumer, thus enabling producers to sell readily to that purchaser. Teamsters responded to the new technology by dynamiting early pipelines. Their resistance, however, was short lived in impact; in 1866 the first efficient gathering system linked wells to a pipeline depot. From there, oil moved to refiners in the area or into railroad cars for shipment to refiners in urban areas. By 1870 three large railroad systems and half a dozen short-line roads served the producing region, often competing keenly for traffic.[37] These developments in gathering and transportation cut buyers' costs, but either by rail or by water, there was a good chance a shipment would fail to reach its destination. Leaky barrels of crude ignited easily, and flatboats and scows frequently capsized and sank. Crude purchasers had to look at these possibilities, as well as the variable cost of drayage — two to three dollars per barrel in the early days to one dollar per barrel by the mid-1860s — in figuring what to pay producers. They had to weigh this against what most buyers of crude, initially coal oil refiners branching out into petroleum refining, were willing to pay.[38]

The wild variations in supply and price of crude oil had a destabilizing effect on the refining sector of the industry from its beginning, as did the emergence of major regional rivalries. By the end of 1860, there were fifteen refineries in the Oil Region, all of them small, processing from five to fifteen barrels per day and wasting at least half of that. They were inexpensive to build, costing as little as two hundred dollars for the smallest and enjoying the advantages of low prices caused by flush production in new fields. These tiny "teakettle" refineries, not much removed from moonshiners' stills, had the advantage of being portable, but their limited capacity meant that no one would make a fortune from one. In contrast to their smaller Oil Region counterparts, some refiners in metropolitan centers had been in business before Drake's discovery, working with coal. These refiners, such as Charles Pratt of New York or Samuel Downer of Boston, had relatively large plants, capable of running two thousand barrels of crude per day. Though Pittsburgh coal oil refiners with ready access to coal lagged behind others less fortunate, most had begun to refine petroleum in 1861. Being in metropolitan centers gave them easier access not only to materials (other than crude) and labor but to capital necessary to expansion as well. By 1863, there were more than sixty refineries in Pittsburgh alone, with a similar number distributed between Erie, Philadelphia, New York, and Cleveland.[39]

By 1865, more sizable operations had also appeared in the Oil Region, especially at shipping points such as the railway junction at Corry. The new refineries were brick structures with stills capable of processing fifty to one hundred barrels a day in batches and employing dozens of workers. The most up-to-date refinery in the region was that run by Samuel Downer, a Boston coal oil refiner who located a plant at Corry to be near crude supply; with an enviable efficiency of extracting 60 percent yield of products from each barrel of crude, Downer's refinery ran three hundred barrels a day and employed some two hundred workers. Though less efficient than Downer's, smaller plants could still get close to a 50 percent yield, much better than teakettle refiners could do.[40]

A number of important assumptions about the refining sector of the industry were already explicit in operational discourse by the mid-1860s. Small refiners as a group were at a disadvantage relative to larger manufacturers, who had more stable sources of feed stocks and also enjoyed small advantages of scale. All refiners would achieve their most important competitive advantages by cutting costs of shipping and containerization. The former would be done eventually by driving bargains with competing railroads and by building pipelines that would compete with the railroads. The latter involved the construction of can and barrel factories. Both strategies required capital, limiting the roles of small refiners and requiring that the larger ones combine their resources and influence.[41] Beyond scale of operation, the most delimiting aspect of refining was location, which bestowed special competitive advantages and disadvantages. Refiners in the producing region had high transportation costs for refined products to eastern markets, as did those in Pittsburgh; eastern refiners paid higher costs for transportation of crude oil to their refineries, and Cleveland refiners faced the largest total distance to move crude to their refineries and products to the seaboard. During the late 1860s and early 1870s, refiners in all locations worked to minimize these handicaps and to maximize their competitive advantages.

Prolific regional production was the Oil Region refiners' chief advantage. They could usually obtain plenty of crude, sometimes at distress prices. There were also recognized rivalries between refiners whose plants were located in the fields and those whose were sited at greater distances. The supply for the former was less likely to be interrupted by transport problems, and they paid less to ship crude to their plants. When many refiners were only efficient enough to get one barrel of refined products from two barrels of crude, all refiners realized that this added up to a significant competitive advantage. On the other hand, Oil Region refiners knew that they usually paid more for other raw materials, equipment, and labor than refiners in Pittsburgh and other locations, and it was costly to ship products to distant urban markets.[42]

The major advantage of nonregional refiners was proximity to the growing national and international markets. Thus, in marketing, site compensated for transportation costs to some extent. Most of all, improvements in handling and shipping — pipelines, tank cars, and railroad connections — cut the cost of transporting a barrel of crude to a refiner in Pittsburgh, Cleveland, or New York. Pittsburgh enjoyed relatively inexpensive water transportation for crude from the producing regions; and Cleveland was able to use either rail transportation from the regions or waterborne crude from Erie. The urban refineries avoided problems of Oil Region refiners caused by the quick shifts in production sites. Urban refiners could locate on major railroad lines, which tied them to the entire region and to longer-term supplies of crude oil. The urban refiners also enjoyed more stable markets, despite the increasing volume of refined products, because they sold ever larger quantities of refined products in foreign trade, thus retaining better margins on domestic sales than most Oil Region refiners. Large-scale exports to Europe began during 1862, and they increased to the point that foreign markets were 80 percent of the size of the whole domestic market by 1865. The larger urban refiners and a few of the large Oil Region refiners that participated in the export market profited even when domestic markets were depressed, as they were in 1866. Small refiners lacked the staying power of those could enter the export market.[43]

There were two principal responses by the stronger refiners to the fragility and instability of the industry, both of them fully tested before the depression began in 1873: combination through acquisition and merger, and formation of purchasing pools to stabilize prices and shipping rates. During 1867, refiners in the Oil Region, Pittsburgh, and all the other refining centers formed regional trade groups to advance their common interests. Thus, New York refiners negotiated jointly with railroads to lower their costs in 1867, prompting Philadelphia refiners to complain that the New York buyers then had them "on a string." The response of Philadelphia refiners was to negotiate lower prices with the Pennsylvania Railroad, giving them an advantage over New York competitors of twenty cents per barrel. The quarrel then reached back into the producing regions as competing refiners' associations in Cleveland and the Oil Region negotiated improved rates with Pennsylvania's competitors. When rates to New York and Philadelphia rose, demand for crude declined, bringing the Petroleum Producers' Association to register public charges of conspiracy and collusion on the part of Cleveland and Pittsburgh refiners! In short, during the 1860s, the regional refining centers competed vigorously, focusing increasingly on negotiations with railroads to enhance competitive advantages and minimize disadvantages. By the end of the decade, it was clear that Cleveland was the relative winner. Its refiners were then processing more crude than those of Pittsburgh and Philadelphia combined and almost eight times more than the remaining refiners in the producing regions.[44]

By 1872, there was also widespread industry agreement on a number of basic points. Foremost, at the time, was the perception that the entire industry suffered from excess refinery capacity, which lowered the prices both of oil producers and of refiners. The common notion held that capacity was three times greater than demand. It was also agreed that the export sector of the industry was so important that the prices paid for refined products in European markets determined American prices of both products and crude.[45] A common expectation of opposition and antagonism between all the sectors of the oil industry, moreover, was central to operational discourse in the Oil Region. Lease buyers took advantage of landowners and operators, who cheated royalty owners and were, in turn, gouged by teamsters and, later, railroads. Oil buyers speculated in the varying prices of crude oil to reap unearned profits through speculation, presumably at the expense of the producers. Refiners increased their tankage in response to moderate swings in supply and price, thereby taking this possibility out of the hands of the buyers. Finally, in cities, consumers mistrusted retail merchants, whom they believed sold diluted or otherwise adulterated products and gave short measures and light weights to increase profits at their expense. In short, everyone expected the worst of everyone else.[46] Under these circumstances it was easy for the public to believe the most heated and inflated charges brought against businessmen categorically or against individual firms.

The battles between sectors of the oil industry and the rivalry of sectional refiners and marketers, moreover, were conducted in public view. Thus it was that when readers of mass-circulation newspapers and the most influential opinion magazines found mention of oil after 1870, it was usually in highly prejudicial accounts of its battles and feuds, as journalists depicted the "lawlessness" of the social, economic, and scientific contexts of the industry. By then, various contenders were attempting to achieve competitive advantages by influencing public policy. Producers and allied refiners in the Oil Region proved most adept at arousing public opinion, since half a dozen of the most affluent producers owned the *Titusville Herald*, a Republican journal, while a group of Democratic producers and refiners, including attorney Roger Sherman, controlled the *Titusville Courier*. As controversy raged, they used these newspapers to influence opinion in the Oil Region and, through ties with urban journals, to spread their versions of disputes liberally across more widely read pages.[47]

Perhaps, commentators of the seventeenth century would have recognized the ongoing battles within the oil industry as a Hobbesian war of all against all; in the nineteenth century, conflict metaphor born of war experience and adaptations of the writings of Charles Darwin might well have framed this perception of struggle as natural. But as far as most oilmen were concerned, a world of individualistic conflict was simply a perception from their experience in narrow sectors of the industry. There was no question, for example, that the race was to the swift among lease brokers and that the best terms went to the cleverest. Among oil producers, the assertion of the "law of capture" — that oil, as a fugitive element, belonged to the person who produced it rather than to the person whose land or lease it underlay — prompted competitive drilling in the American industry until well into the 1930s. Among oil buyers, timing and skill were everything: a half cent a gallon was the basis of new fortune or crushing loss. In their sector, the overbuilt railroads vied for lucrative trade; refiners worked to improve their competitive advantage over other refiners through cheaper crude oil purchases and modest economies of scale. The only relatively tranquil segment of the industry before 1870 was marketing; as long as oil was sold as a generic commodity, the only advantage to be secured was in a larger number of tied retailers and a higher volume of sales.

In short, most segments of the oil industry tended to be keenly competitive by the end of the Civil War, but that was not a matter for rejoicing. Each segment tended to hold that it was victimized by one or more other segments, at the same time that they all labored to offset the destabilizing effects of the sharp fluctuations in supply and price of crude oil. And when various regional interests failed to offset their perceived disadvantages, they turned to promote their interests in politics. As early as 1862, when Secretary of the Treasury Salmon P. Chase advocated a wartime tax of $6.30 on crude oil and $10.50 tax on refined products, producers held increasingly ritualized protest meetings at Titusville and Oil City and elected a committee to fight the tax in Washington. Refiners made similar representations individually but lost out; in 1864, a $8.40 per barrel tax was levied on refined products, but none was placed on crude.[48] The advantages of political clout could not have been clearer. During the following year, a congressional committee proposed a $2.50 levy on crude, prompting more meetings and the arrival of another delegation of producers in Washington. In the end, lobby efforts succeeded in reducing this to $1.00 — still too high for most producers. Six months later, a delegation and its lobbyists were back in Washington, working to have the tax lowered to twenty-five cents. While the oilmen were there, they lit a fire under Pennsylvania congress-men, whom they charged with being "indifferent to the interests of the oil producers." The producers won this political battle, too, thereby honing their political skills and confirming in the halls of Congress the value of drilling.[49] Above all, they learned that the manipulation of political discourse and pro-cesses was a useful tactic.

The importance of political activity and the skills of producers in conducting it were reconfirmed at the state level during the remainder of the decade. In March 1866, oilmen were caught unawares when railroads and steamboat owners killed a bill in the Pennsylvania legislature to incorporate a regional pipeline. Thereafter they were more vigilant. Two years later, they raised the specter of monopoly when two local short-line railroads, the Farmers Railroad and the Warren and Franklin Railroad, were purchased by a third, because the small combined line was controlled by Philadelphia investors who would con-spire with the Pennsylvania Railroad to dominate local transportation. This "conspiracy of capitalists" was actually most notable for its apparent intention to compete more effectively with short lines that were owned within the Oil Region, but the local editors raised the monopoly cry, beating midwestern farmers by several years in this regard.[50] One year later, Oil Region refiners, producers, and their legal and media allies returned to Harrisburg and quashed an attempt to tax crude oil. The following year, the Petroleum Producers' Asso-ciation sent another delegation to advance a common-carrier pipeline bill, only to see it defeated in the Pennsylvania state senate, by what the oilmen claimed was the long arm of the Pennsylvania Railroad. Monopoly, money power, and conspiracy against the public good were all enlisted as supporting elements of public discourse in the accompanying media campaign.

Political involvement brought uneven results, but by 1870, oil producers and refiners were thoroughly accustomed to political activity; indeed, a signifi-cant number of the more successful had held local, state, or federal office. By 1870, producers and refiners in the Oil Region and their allies used political processes to gain competitive advantages over producers in other areas and leverage over the railroads, pipelines, and other providers of services.[51]

In 1872, attempts of the railroads to increase freight rates roused producers to renewed complaint. The *Titusville Herald*, which began to complain about railroad monopoly conspiracies in 1868, reported intense local opposition to higher rates to New York City when the three large railroads announced them. Producers also had quarrels with small regional carriers, as when Titusville independents sued the Allegheny Valley Railroad for breach of contract, alleg-ing that the railroad had promised a seven-cent-per-barrel "drawback" on all shipments and that the road had diverted tank cars to Lockhart and Frew of Pittsburgh, big shippers outside the Oil Region.[52] The loyal, if not captive, press supported local oilmen, who in some instances owned or otherwise controlled the journals.[53]

Because conditions in the oil industry during the 1870s kept producers and refiners at loggerheads, the level of acrimony in the newspapers rose. The sharp fluctuation in supplies of crude oil continued to produce dramatic surges and declines in prices, temporarily sustaining marginal refining operations at low points. With the severe depression that began in 1873, however, most components of the industry and its major suppliers were hit hard. Markets for refined products declined, destabilizing the sector of the industry with the largest excess capacity and diminishing demand for crude and transportation. The major railroads, in turn, attempted to offset declining traffic with higher rates on both crude oil and refined products. In response, the largest shippers fell back on their strategy of playing off the railroads against each other in order to lower their costs by obtaining rebates on posted freight rates. All these reactions to the depression set old adversaries against each other again. Producers attacked distant refiners, often aligning themselves with refiners in the producing regions and with refiners and wholesalers in New York City, with whom local refiners carried on most of their trade. In the newspapers, these disputes came to focus on the railroads, already subject to criticism by shippers in New York and the Midwest.

Producers and refiners in the producing regions and refiners in New York and Philadelphia obviously had a great deal at stake in these controversies. The rebates and drawbacks they received from the railroads gave them significant competitive advantages as profit margins narrowed in the industry; higher freight costs would worsen the situation by raising their costs relative to those of refiners in Pittsburgh and Cleveland. Higher freight costs could, in fact, largely concentrate refining activity in these two cities to the obvious detriment of the refiners and economies of the producing regions and New York. Thus, the cost of transportation became highly important and controversial, a matter that would be prominent in industry discourse for decades thereafter but, in the shorter run, would shape discourse on the emergence of a leader among refiners, John D. Rockefeller's Standard Oil.

During the strenuous industry competition of the 1870s, John D. Rockefeller emerged as a consistent winner. His company, the Standard Oil Company of Ohio, became both the dominant refiner in Cleveland and the largest volume refiner in the country. The growth strategies implemented by Rockefeller and his close associates made the Standard Oil Company the industry leader in capitalization and profits. The resultant success also made Standard Oil the obvious target for competitors, who exploited negative elements of traditional ideology relating to power and dominance in order to offset Standard Oil's decisive edge. By 1877, the company's rivals had made manipulation of public discourse their principal competitive strategy.

Rockefeller began as partner with Maurice B. Clark, two of his own brothers, and Samuel Andrews in the Excelsior Works in 1863, when the Atlantic and Great Western Railroad connected the Oil Region with Cleveland, setting off a rapid growth in refining in the city. Two years later, he and Andrews bought out Clark, and Rockefeller gave up his mercantile commission business to pursue oil exclusively.[54] By 1866, Rockefeller and Andrews had borrowed heavily from Cleveland bankers to add stills to their works, the largest in Cleveland, with a capacity of 505 barrels of oil per day. The firm expanded to take advantage of rising margins for refiners in 1866, a development that also encouraged a flood of new entrants into the business — twenty more in Cleveland alone. The additional refining capacity drove margins downward in 1867 and 1868, pressing the firm and driving many of its weaker competitors out of business.[55] Rockefeller responded to the challenge by seeking to improve Standard Oil's financial performance by buying an additional refinery and by bargaining for lower transportation rates, a task he brought Henry Flagler into the firm to handle.

Thereafter, Flagler earned his place at the directors' table many times over by negotiating sizable rebates from railroads. In doing so, Flagler followed the practice of other oil shippers in driving hard bargains; attorney Roger Sherman, who would later assail Standard Oil for obtaining rebates and drawbacks, routinely drew up contracts for his clients containing both advantages.[56] Flagler's achievement, however, was magnified by the scale of Standard Oil's business. As his biographer fairly put it, he "put his firm way out in front."[57] For a brief and critical period of the company's history, lower net rates provided one of the company's principal competitive advantages. Rockefeller still carried high debt service costs, which he lowered by enlisting the aid of Stephen V. Harkness, Henry Flagler's stepbrother, who provided additional capital. By 1868, the business was worth about five hundred thousand dollars, and credit reporters judged the principals to be "honest, prudent, reliable men." Rockefeller and Standard Oil had survived the most serious challenge to survival they encountered during the nineteenth century.[58] For a time, Rockefeller, Andrews, and Flagler tightened their belts, occasionally paying for crude oil more slowly than was customary, to complete a new refinery, which outside observers valued at about three hundred thousand dollars. Expansion of this magnitude was viewed as quite speculative, however, a development that made it more costly to continue to borrow funds from bankers to fund further growth in Cleveland and New York.[59] But, by the beginning of 1870, when prices for refined goods and margins were higher, credit reporters for Roy G.

Dun and Company took a fuller measure of the partnership. They judged the oil venture to be "second to none" in importance and claimed that it often controlled prices in Philadelphia and New York, where Rockefeller and his associates were recognized as "the boldest operators in their line." During four difficult years, they built the preeminent oil-refining and -export operation in the United States.[60]

During the next two years, Rockefeller and his associates undertook even more rapid and highly risky expansion. They began by reorganizing as the Standard Oil Company of Ohio in January 1870. The new corporation had nominal capital of $1 million, with ten thousand shares of $1.00 par value. They employed the additional capitalization to integrate the operation vertically and acquired barrel works, an acid plant, and transportation facilities. They aimed, clearly, to cut operating costs during a period of declining margins in refining, caused by excess national capacity that exceeded demand by more than 100 percent.[61]

With reorganization completed at the end of 1871, as margins continued to decline, Rockefeller and his associates decided that the only way to enhance the performance of Standard Oil was to increase the size of the business by boosting the volume of refined products. During hard times, when margins were small, the most economical way to accomplish this goal was through combination with existing refiners through merger or acquisition. Thus in December of that year, they increased their capital to $2.5 million and brought Clark, Payne, and Company, another large Cleveland refiner, into the company.[62] Standard Oil moved to acquire additional firms in 1872, often going to bankers in Cleveland and New York for capital. In February it bought four more Cleveland refineries, making some of its former competitors wealthy in later years as Standard Oil's stock appreciated in value.[63] By April, the company held sixteen refineries and a Jersey City transport company. The major acquisition, however, was Jabez Bostwick Company, with its large Long Island refinery and an experienced oil-purchasing office in the Oil Region. With it, Standard Oil strengthened its position in the important export market and became a major buyer of crude oil.[64] By the end of 1872, the company had integrated horizontally, acquiring thirty-four rival refiners and bringing all the Cleveland refiners into one organization. The combined capacity of its refineries was ten thousand barrels per day, equal to half of the crude oil produced in the United States. Rockefeller's idea of combination by merger or purchase, pursued largely as Standard Oil's leaders uncovered opportunities, made the company the lion of the oil industry.[65]

During the first decade of its existence, Standard Oil became the prototype of the large multidivisional corporation, integrated back into sources of supply and transportation to factory site and forward into distribution and mass marketing, a model of achieving economy and enhancing competitive position through vertical and horizontal integration. Richard S. Tedlow has aptly summarized the evolving strategy Standard Oil pursued: "With the plant moving toward rated capacity thanks to mass marketing, unit costs continue to drop. As unit costs drop, unit price drops. As unit price drops, market share expands (but the market itself may expand even more quickly). With the expansion of market share by the firms that make the fixed investment in production (the plant) and/or marketing (e.g. advertising), the number of competitors declines. There is a market 'shake out.' The industry becomes an oligopoly."[66] During this initial period, Standard Oil's successful implementation of a profit-through-volume strategy, achieved through careful coordination of production and distribution, gave it competitive economic advantages that enabled it to dominate the petroleum industry until 1911.

Through 1870, the growth of Standard Oil in Cleveland had passed largely unnoticed by the press outside Cleveland. In town, the *Cleveland Leader* provided routine information about the company's operations and expansion, with a positive approach to these developments.[67] After 1870, local expansion still brought little attention to Standard Oil outside Cleveland, but in that town newspapers were now divided over the program. The *Leader* continued to see Standard Oil's combination of local refiners as indispensable to the prosperity of the Cleveland-based refiners, while the *Herald* increasingly questioned what it saw as the anticompetitive business environment that was created by the buyouts.[68] But outside Cleveland there was still little reason for business observers to pay much attention to Standard Oil. The company's expansion strategy was not grandly innovative. Lockhard and Frew expanded into Philadelphia from its Pittsburgh base, taking over Atlantic Refining in 1870, while Charles Pratt and Company, the largest refiner in the New York City region, increased plant capacity and secured a strong export position.[69] The difference between Standard Oil and these and other aggressive competitors was the scale and alacrity of its moves. With every expansion, Standard Oil also acquired considerable political liabilities, a development neither Rockefeller nor anyone else could reasonably have foreseen. When Standard Oil grew, it collided most forcefully with determined opposition from producers and competing refiners, opposition that embodied, defended, and exploited the strong attachment of most Americans to the ideal of the small producer. As adversaries, the "independents" — meaning not part of Standard Oil — failed to block Standard Oil's growth, but they often succeeded in preserving their own niches in the oil industry by rallying opinion makers and politicians to their cause; to do so they used familiar elements of the ongoing discourse over concentrated power, the morality of great wealth, and the socially destructive and politically corrupting effects of the expansion of industrial enterprises.

When they attacked Standard Oil, opponents thereby carried discourse beyond the more widespread concern with the power of the railroads and directed it at the related but broader phenomenon of corporate expansion. Beyond successful exploitation of public discourse as a competitive strategy, the competitors and opponents of Standard Oil created the new paradigm of industrial monopoly in the Gilded Age, and they made Standard Oil the tangible and cogent model of it. As such, it represented far more in postbellum American culture than pipelines, refineries, and kerosene. Standard Oil's competitors and other critics made it both symbol and reality of unsettling change and the focus of the major fears and anxieties of the age.

The first historical stage in this process of reifying Standard Oil as "monopoly" was orchestration of the controversy aimed against a non–Oil Region refiners' pool. While Standard Oil and others pursued acquisitions during the early 1870s, some of the larger refiners in Pittsburgh and Philadelphia conceived the idea of a shippers-refiners' ring that would insist on negotiating with the railroads for favorable terms for its members. In 1872, they took their idea to Cleveland, where refiners faced high posted freight rates and won a sympathetic hearing, especially at the Standard Oil Company. Only John D. Rockefeller had initial misgivings about the ability of the group to stick together in the face of intense efforts of the railroads to divide it and in the face of predictably shrill objections from refiners in the producing region and New York City, who were not included in the South Improvement Company (SIC), the new organization. Still, Rockefeller's concern over the persistent excess refining capacity in the country and the resultant decline in the price of kerosene led him to sign on as a director of SIC.[70]

In less than one month, his worst expectations were fulfilled: the "oil wars" began. Following a week-long run of rumors, the SIC story broke in the *Oil City Derrick* on January 20 and in the two Titusville papers on the following day. Initial reports described the association as a defensive move by Cleveland and Pittsburgh refiners, necessary to keep them in competition with refiners in the Oil Region. Within twenty-four hours, however, coverage took a decidedly negative turn, emphasizing the potentially devastating effect of the shippers' pact on the regions. On the fourth day of coverage, the antimonopoly theme appeared strongly, largely orchestrated by Roger Sherman; newspapers gave extensive coverage of mass protest meetings in Titusville, Oil City, and nearby towns, meetings organized by Sherman's associates among Oil Region refiners and producers.[71] The *Petroleum Centre Record* reported a local rally against "monopoly," meaning the railroads and the large non–Oil Region refiners. The *Oil City Register* reported local rallies to stir political opposition both in that city and Titusville, where it was claimed that three thousand oilmen protested and even discussed raising funds to build their own railroad to New York City.[72]

Through February and March, editors and orators employed ever hotter rhetoric. At the beginning of March, the editor of the *Herald* expressed what became the formulaic indictment of SIC when he denounced it as "a formidable conspiracy, representing so much capital, so artfully contrived and so plausibly projected."[73] On the following day, Oil City newspapers reported and endorsed a local protest: Love's Opera House in Oil City was jammed by oilmen who organized another producers' pool, the Petroleum Producers' Association, and agreed to sell only to Oil Region refiners and to drill no new wells for sixty days, in the interest of raising the price of crude oil.[74] But the producers' cartel failed, and its organizers blamed SIC and Standard Oil, the latter accurately, because it offered premium prices for crude to supply the Cleveland refinery. Among those involved in the fray was Frank S. Tarbell, father of Ida M. Tarbell, who was approached by Standard Oil in 1872 with an offer to pay a premium price for his crude; he declined and made certain that sympathetic newspapers in the Oil Region publicized his refusal.[75] The Petroleum Producers' Association, with the support of local and New York refiners, insisted on rates, rebates, and drawbacks that would give them competitive advantages over the others and on instituting a boycott on sales of oil to SIC members. The producers lobbied the Pennsylvania legislature in Harrisburg and brought it to revoke the charter of SIC.[76] They then proceeded to New York City and roused refiners who depended on Pennsylvania crude oil. These refiners, in turn, brought additional pressure on New York City newspapers to take the side of Oil Region producers and refiners.[77]

Separately and in alliance, these businessmen turned local disputes in a single industry into a national issue by taking their cases to the public through effective exploitation of the new postwar communication network. During the first week of March, Oil Region newspapers placed their stories far beyond western Pennsylvania. Their participation in the Associated Press service, via Western Union lines, spread their stories to New York, Philadelphia, Chicago, Boston, and elsewhere. By March, Oil Region newspapers were quoting stories from distant journals that, in turn, quoted them, evidencing a network and a self-validating information loop. Thus, the *Titusville Herald*, for example, quoted articles in which it was quoted by New York journals, including the *Sun*, *News*, *Tribune*, *Times*, and *Bulletin*. A handful of Oil Region newspapers, generally owned by local oil producers and refiners, thereby controlled public discourse relating to the heated dispute within the oil industry and SIC.[78]

Through March and April, Oil Region newspapers and associations broadened their appeal outside their area by emphasizing the ideological aspects of their cause. It was no longer merely a fight for dominance within the oil industry; it was a fight for liberty and private enterprise. As the editor of the *Titusville Herald* put it, "The peril that threatens the producer and refiner fills the rest of the business community with alarm. It unsettles all values. It gives a pause to private enterprise. It blocks the wheels of incipient public improvements. Its casts a general gloom over the whole country." He also warned that SIC "may reach and ruin every branch of productive industry in the land, or subvert to its own sole emolument the entire domestic commerce of the country." Here was heady stuff, successfully contrived to serve the producers and refiners of the Oil Region and their allies, refiners and wholesalers in New York City.[79]

The New York City newspapers responded by showing increasing interest in stories from the Oil Region. The *Tribune*, the *Times*, and the *World* routinely followed the leads of Pennsylvania newspapers. Thus, on March 14, the *Daily Tribune* quoted an Oil Region refiner as identifying "a great moneyed interest [that] had them, as it were, by the throat." In this instance, it merely reprinted an earlier story from the *Titusville Morning Herald*. In other instances, however, New York papers identified special correspondents, actually Titusville or Oil City editors who simply rewrote stories they had published earlier.[80] As they attempted to relate the arcane business of freight rates to their readers, New York editors argued that they were making common cause with "producers, dealers, and consumers" in opposing SIC, because the ring would weaken producers, destroy oil dealers, and raise consumer prices. The final point was so obvious as to require no explication at the time, and few subsequent students of the industry have questioned it.[81] The press in Pittsburgh and Cleveland generally supported their regional businessmen, as did those in Philadelphia.

Near the end of March, Oil Region newspapers published an "enemies list," headed by the railroads and brokers and refiners in New York City and Pittsburgh, on which John D. Rockefeller and Standard Oil were correctly identified as minor players in SIC. After Standard Oil attempted to break up the fragile producers' boycott, however, Oil Region journalists moved it closer to the top of the list, rewriting history to place the "Cleveland conspirators" in leading roles. Shortly thereafter even the railroads were exonerated, as the target narrowed to John D. Rockefeller, who was described as having subdued the previously invincible and sinister klatch of railroads, the Erie, the Pennsylvania, and the New York Central.[82]

When the producers brought the national press behind their cause, they took pains to see that subsequent formal history would reflect their position by issuing their own "history" of the battle over the South Improvement Company. E. C. Bishop, fiery editor of the *Oil City Derrick*, compiled it, principally from his own columns and from the resolutions that were passed by various mass meetings in the Oil Region. Thereafter, large passages from the *History* were reprinted in Oil Region newspapers and, eventually, in the record of the U.S. House committee that conducted a brief investigation of the dispute. The *History* was cited both in its original form and, more often, as a government document by most authors.[83] Region oilmen continued this practice in later disputes with Standard Oil, the railroads, and others, producing the sources that sympathetic writers would mine for material. They also discredited journalists who either ignored their cause or disapproved of it, by claiming that they had been "muzzled" by monopolists.[84] Standard Oil and the other members of SIC issued their defense piecemeal, after the fact, and over a long period of time.

The verbal onslaught against SIC continued after it was defunct, because the resulting uproar was politically useful to business interests in New York and Pennsylvania. Thus, in May, the *Oil City Derrick* reported that "the subcommittee of the [United States] House Committee on Commerce agree to report that the South Improvement Company was one of the most gigantic and dangerous conspiracies ever attempted, and that if it had not been checked in time by the people of the Oil Region and by Congressional investigation it would have resulted in the absorption and arbitrary control of trade in all the great interests of the country."[85] The controversy over the South Improvement Company established in rhetoric the image of individualistic small producers opposing a diabolical monopoly, an image that still adheres to the episode.[86] Moreover, the editors and attorneys who controlled discourse on the SIC issue took a large measure of credit for the association's defeat. When SIC's charter was revoked in Harrisburg, the *Titusville Morning Herald* crowed: "The brilliant and masterly manner in which the said company was in a few short weeks stricken from the Legislative Records by the voice of the press and the people has no parallel in history." After the whole project had been abandoned, the editor seized on the occasion to project the role of the press in American society for his readers: "The modern newspaper is the brain, the heart and the soul of modern progress, though the sources of its inspiration may be uncertain and questionable."[87] His own sources of inspiration may have been questionable, but they were far from uncertain: regional business interests gave him what he used.

How could a group of noisy oilmen be so successful in exploiting the media? The producers and refiners in the Oil Region and their allies in New York City clearly understood the roles of media and media-business networks in the formation of public discourse, perhaps because their ranks included a fair number of men with experience in local, state, and national politics. Many of them also held leadership positions within the Republican and Democratic political parties and thus were experienced organizers and managers of public meetings used to secure media coverage. All these experiences and skills had come into play in the early months of 1872, as the producers and refiners in the Oil Region succeeded in translating what had been a technical dispute over freight rates into a lively moral issue. By contrast, though they received editorial support in Cleveland and Pittsburgh, the backers of SIC were far less adept at opinion and political management; they seem to have seen the contest over SIC as merely another round of commercial rivalry, a deficient vision that would only rarely be remedied in the future.

The South Improvement Company was dead by mid-1872, but the competitive conditions within the industry that gave rise to it were still alive, as were all the rivalries, suspicions, and fears that the "oil wars" had evoked. In the producing region and in New York City, refiners and journalists were quick to charge that though SIC was gone, Standard Oil had quickly taken its place.[88] As a result of the controversy over SIC and rebates on railroad freight rates, Standard Oil and Rockefeller emerged in the middle of a veritable minefield of cultural, economic, and political hostility, directed originally against the railroads. This was dangerous ground. As Maury Klein has argued, attacks on the railroads both stimulated and muddled the discourse relating to competition and monopoly. Shippers attacked both the emerging railroad systems and local lines as monopolies on the basis of traditional belief that competition would produce the lowest rates. High fixed costs and overbuilding rendered the conventional wisdom inaccurate with the railroads, but as Klein put it, "This was simply the traditional American approach of taming the unfamiliar with the familiar harness."[89] The same harness was dusted off by independent refiners and producers, who used Standard Oil's involvement with SIC to frame the same issues of competition and monopoly. In some measure, this strategy was obvious. As the largest refiner in the country and, hence, the largest crude oil purchaser and shipper, Standard Oil was the most visible target for Oil Region producers and refiners. It was easiest for small refiners to rally public sentiment against bigger businesses by targeting the largest. Moreover, Standard Oil's opportunistic strategy of expanding even while the furor over SIC was at its height made it seem that the South Improvement Company refiners' transportation ring and the Standard Oil mergers were merely different devices to the same end — monopolistic control of crude oil prices.

Standard Oil's leaders, moreover, kept themselves in the limelight at an inopportune time by traveling to the producing regions to placate producers and to reassure regional refiners that no exceptional competitive practices would be mounted against them. It was soon apparent that this was worse than a fool's errand. Negative sentiments were too strong for reasonable discussion. The *Petroleum Centre Record* offered local refiners firm but unnecessary advice:

"It is to be hoped said refiners will not allow themselves to be soft-soaped by the honeyed words of the monopolists and conspirators."[90] Rockefeller's public attempt to cope with bad relations with producers through open negotiations ended in embarrassing rebuff. Thereafter the company would be more concerned with growth than with its noncommercial relations with the public.

During the next six years, Standard Oil completed its second major expansion, giving it virtual control over refining, export, distribution, and crude oil purchases and enhancing its vertical integration. Aware that rival refining centers might continue to use transportation facilities and contracts to secure competitive advantages, Standard Oil purchased two large pipeline systems, one in 1874 and the other two years later. By the end of 1876, it could transport crude oil through nearly four hundred miles of pipe and store about a million and a half barrels of purchased crude in its tanks. This latter capacity, equal to about two months' production from Pennsylvania, afforded secure feedstocks for the company's refineries and protection from the sharp and short-term escalations in prices that followed the decline of production in the prolific but small fields of Pennsylvania and West Virginia.[91] With pipelines, Standard Oil also protected itself from future rate pools of the railroads; with its vast storage capacity, it was less vulnerable to producer cartels. It could stabilize the price it paid for crude, in the short term, because its oil buyers could purchase when prices were low, fill the tanks, and use stored crude to hold prices down even when declining production would have forced them up. Standard Oil could control the cost of its essential raw material.

In part, the acquisition of pipelines and storage facilities was a defensive tactic as Standard Oil worked to free itself from recurrence of perils it had faced. It was also clear, at least to Henry Flagler, the company's transportation expert, that there were positive advantages to the purchasers as well: with the pipelines, the already divided ranks of railroaders could be forced to grant lower rates or face the possibility of still more pipeline construction and the loss of trade. With greater assurance of price and supply, Standard Oil could also undertake longer-term supply commitments, especially important in its vital foreign commerce in refined products and crude. Relatively small refiners, dependent on short-term supplies in the Oil Region, enjoyed none of these advantages, leaving them even more vulnerable to increases in prices and rates.

Standard Oil also improved its competitive position by expanding the production of cans and barrels and by moving into marketing. In both instances, it created new enemies. The company had made barrels since 1868; in 1872 and again in 1874, it enlarged its capacity greatly. From the beginning of its activity in container production, Standard Oil pursued increased mechanization of barrel making, previously the domain of craftsmen coopers. The company realized substantial savings by changing the balance of the workforce through mechanization, to the disadvantage of unionized skilled workers. In the process, it, like refiners in New York and Philadelphia, ran head-on into the Coopers' International Union, which represented eight thousand workers during the 1870s.[92] The result was confrontation. A major strike at Standard Oil's Cleveland barrel works began in April 1877 and lasted until mid-May. In the meantime, the "Battle of Fort Standard" reached the headlines of many newspapers and of most of the organs of skilled labor unions, as a small army of policemen and firemen held back workers and their wives, who attempted to close down the works. Though production at the plant dropped, union efforts came to little because Standard Oil was able to procure enough barrels at comparable cost with little difficulty. The workers concluded that it was hard to beat big business.[93]

Their perspective was shared by editors of major union publications, including the *Workingman's Advocate*, of Chicago; the *Labor Standard*, of New York City, the *Weekly Labor Advance*, and the *National Labor Tribune*. The last-named journal concluded that there were no remaining obstacles to Standard Oil's economic and political power, "too great a power to leave in the hands of a single company." The company's defeat of the coopers was taken as an alarming and incontrovertible demonstration of the threat to republican institutions, small producers, and workers.[94] The strike also brought the company additional notoriety as a precipitator of social conflict, as increasing numbers of writers prophesied blood in the streets of American cities. Standard Oil acquired yet another set of adversaries.

If there were good defensive reasons for vertical integration at Standard Oil, there were even better ones for horizontal integration, and the company's horizontal expansion during its major period of growth, 1875–78, was almost entirely defensive, aimed at eliminating much of the excess capacity in refining. Thus, it moved into the producing region, where it already owned one refinery, and bought additional plants, closing most of them and dismantling them for scrap. The least efficient competitors were simply left to fail in due time. These mergers, acquisitions, and failures enhanced Standard Oil's position as a purchaser of crude as the number of buyers declined — a situation that producers observed early on.[95] During this same period, the company merged with some of its largest and most efficient distant competitors, including Lockhart in Pittsburgh, Warden in Philadelphia, and Pratt in New York City. It also participated in another refiners' pool, this one promoted by leading Pittsburgh refiners and including J. J. Vandergrift and other refiners in the producing region. The aim of the group was to curtail refining to fit the market by allocating crude oil to each of five regions by quotas based on refinery capacity. Rockefeller himself agreed to serve as nominal president, and Vandergrift and Charles Pratt of New York were nominated as the other officers.[96]

Producers in the Oil Region formed a parallel association, a revitalized Petroleum Producers' Association, to control the flood of crude from the fields along the Clarion River and Turkey Run. They chose Captain William Hasson, leader of the movement against SIC, as president, ensuring both energetic leadership and the sustaining of animosity against the refiners, especially Standard Oil. The PPA took steps, including beatings and well burnings, to force recalcitrant producers to observe the the association's limitations on new production. Even these heavy-handed tactics failed to sustain prices for more than a few months, and by the end of the year, production was nearly 50 percent above the level it held when the PPA was created. Producers thus failed to limit production. Refiners enjoyed greater success, in all likelihood because there were many more producers than refiners and because small operators settled for immediate income at the expense of longer-term gain.[97]

After the failure of the second Petroleum Producers' Association, a second producers' pool, the Petroleum Producers' Agency, was formed to monopolize the purchase of petroleum and thus control its price. Like the earlier association, it failed in the face of new discoveries; it also attracted unfavorable notice, especially in New York, where the Chamber of Commerce of the State of New York, the *Herald*, and the *Commercial and Shipping List* all assailed it, in support of New York refiners and — ostensibly — consumers.[98] A subsequent agreement between the refiners' pool and the new producers' agency invited similar reproach from interests that were not included in the arrangement. Reflecting the views of local refiners, the *New York World* denounced it as "an alliance between the producers and refiners for the purpose of making arbitrary figures at which petroleum shall be sold to merchants."[99] Thus, repeated efforts of producers at pooling and cooperation failed to stabilize the industry; in the end, they did little but stir sentiment against the industry and its leaders.[100]

During 1878, the final year of Standard Oil's great growth, the company fought political and legal battles and arranged mergers and acquisitions that brought it hegemony in refining. Most notably, it humbled the Pennsylvania Railroad by defeating its scheme to compete with Standard Oil through operation of Empire refineries in Philadelphia, New York, and Pittsburgh, using low freight rates — as much as eight cents per barrel below the railroad's costs — to undercut Standard Oil. After the Pennsylvania failed to cut off Standard Oil's supply of crude oil, the short but bitter contest was over.[101] Even while the battle with Empire was winding down, Standard Oil continued to make large acquisitions, including Columbia Conduit, the only remaining unaffiliated pipeline company in Pennsylvania and the operator of a 1,500-barrel refinery in the producing region. The company acquired twenty-seven additional refineries in the region, wrapped up both Pittsburgh and Philadelphia, and made an additional major acquisition in New York City, all in 1878.

By year's end, according to its calculation, Standard Oil controlled slightly more than 90 percent of the refinery capacity of the United States.[102]

Its domination of refining established beyond dispute, Standard Oil moved to secure complete control over the wholesale operations of its business as well. This was accomplished by buying into major distributor organizations in the Midwest and seaboard areas and by diverting its trade from independent distributors in New York and other cities to Standard Oil–tied companies. Like the earlier acquisition of pipelines and tankage, the move into wholesaling had both defensive and offensive aspects, and like earlier growth, it also brought new adversaries. The results were as impressive financially as they were dangerous politically. Even as product prices declined, Standard Oil's earnings soared, owing to unprecedented efficiency and to the depression of crude oil prices caused by recurrent flush production.

As the major shipper and refiner of crude, Standard Oil assumed an even more controversial relationship with the railroads. Thus, when it defeated the Pennsylvania line, it negotiated new contracts with it, the New York Central, and the Erie that confirmed the oil company's status as the "evener" of oil traffic over the competing lines. Standard Oil thus became responsible for enforcing the rate pool for crude oil shipment that the railroads had long sought to establish. It also became the rate setter, in effect, using this power to confirm its earlier principle of equal charges, removing geographical advantages and liabilities as competitive factors for refiners. The system of rebates created in 1875 to carry through this policy became the fixed feature of Standard Oil's freight contracts thereafter as it was with Oil Region and New York refiners.[103]

While Standard Oil grew, Oil Region editors continued to fire volleys of rhetoric in its direction, and Oil Region newspapers continued to print "news" of conspiracies and plots, though the company's tactics were scarcely more secretive than those pursued by businesses in that era or in later times. Joseph D. Potts, erstwhile head of the Pennsylvania Railroad's Empire Transportation Company, threw in his lot with the Oil Region producers and publicists in 1878. Under the aegis of the General Council of the Petroleum Producers Union, he wrote and published the damning *Brief History of the Standard Oil Company*.[104] Thereafter, Potts's history was accepted by most of the company's critics. In the forefront of the opposition to Standard Oil were a handful of relatively wealthy independents and, as usual, their attorney, Roger Sherman. At the same time Sherman and his clients attacked Standard Oil, they continued their campaign against transporters. In 1878 they launched a massive political and legal assault on independent pipelines and railroads serving the region, and they induced local prosecutors to file criminal charges against them, which Sherman followed up with civil suits to recover alleged damages that were caused by discriminatory rates. He then organized media attacks on local pipelines, at the same time he supplied material to sympathetic journals in New York City. Meanwhile, Sherman's law partner, who sat in the Pennsylvania legislature, drafted and introduced supportive legislation on the independents' behalf. Leaving no avenue unexplored, Sherman also wrote damaging reports on local pipeline owners for Roy G. Dun and Company, for whom he was local correspondent. His strategy worked quickly. As Sherman's biographer reported it, "So harassed, the pipeline company executives sued for peace."[105]

This approach, combining civil and criminal suits, legislative activity, and — above all — extensive and pejorative press coverage, was a part of the tried-and-true strategy of Sherman and other adversaries of the Standard Oil Company when the battle between it and regional and New York producers and refiners resumed in 1878. Most particularly, Sherman was adept at placing the claims of his clients before the public through the new mass-circulation newspapers. During a dispute between some Bradford field producers and Standard Oil near the close of 1878, for example, Sherman's position appeared on the front page of the *New York Sun* on November 13 and again ten days later. Scant attention was given to Standard Oil's arguments.[106] The immediate occasion for renewed conflict was the introduction of a series of civil suits against the Pennsylvania and other railroads filed by the Commonwealth of Pennsylvania at the instigation of leading Oil Region independent producers and refiners, represented by Roger Sherman. The common complaint in these civil cases was that the railroads had granted unlawful rebates to the Standard Oil Company, the principal target of the suits, though not actually a party to them. Sherman's strategy followed his earlier campaigns: he instituted both civil and criminal action, drafting the former as counsel for the oilmen and drawing the latter as chief associate state counsel. At the same time, legislative allies introduced an antirebate, "non-discrimination" bill in the state legislature, while sympathetic writers and editors orchestrated a press campaign against Standard and the railroads.

The railroads, the Pennsylvania in particular, were eager to provide evidence regarding the rebates, which lowered their revenue. Alexander Cassatt, an assistant vice president of the Pennsylvania, testified that Standard Oil had received several hundreds of thousands of dollars per month during the year-and-a-half period of its agreement; but he also indicated that rebates were generally granted in at least token amounts to any shipper who had the wit to ask for them. The latter point, not to the advantage of Sherman's clients, was ignored in the proceedings and in subsequent accounts of the rebates; Sherman used his control of legal proceedings to create a historical record. As his biographer, Chester M. Destler, has pointed out, he had Benjamin Campbell, a leader of the producers' organization, read lengthy anti–Standard Oil statements into the record. From there, they appeared in newspapers and in histories of the company. In the end, the "Commonwealth" suits dragged on for more than a year before they were finally settled inconclusively by the litigants.[107] During that time a tremendous amount of anti–Standard Oil material was read into the record.

Civil proceedings were only one part of Sherman's attack on Standard Oil. He used the Commonwealth proceedings to extract testimony that his clients could use in criminal proceedings against both Standard Oil and the railroads. Thus, after the civil trials began, Pennsylvania producers filed criminal charges against Standard Oil in Clarion County. They were represented by a Sherman partner, whose brother was scheduled to preside over the proceedings in the heart of the Oil Region.[108] In the end, Sherman's legal strategy misfired. Though he obtained an eight-count indictment of Standard Oil, the governor, the attorney general, and a justice of the state supreme court all agreed that he could not compel additional testimony in the Commonwealth suits until the criminal case was disposed of. The governor further declined to request the extradition of John D. Rockefeller unless convictions of Pennsylvania defendants were secured. Then, upon motion of the defendants, the Clarion County trial was moved. Thereafter Standard Oil prolonged the proceeding until all parties accepted an inconclusive negotiated settlement.[109]

For his part, however, Sherman was not done. He produced a cleverly edited version of the Commonwealth cases for a client who filed a civil case against the Pennsylvania Railroad in New York State. Thereafter, whenever Standard Oil was under attack or investigation, Sherman readily forwarded his edited version of the trials, which he paid to have printed, to potential allies, including Henry Demarest Lloyd and Ida M. Tarbell.[110] Sherman understood fully the benefits that followed from the control of the discourse. He continued to use "history" as a competitive weapon for his clients for decades to come.

Standard emerged from these attacks by Sherman and the independents with few gaping wounds, but Rockefeller and his associates were now aware of their legal and political vulnerability. Thereafter, they paid greater attention to both areas, but there were lessons that they neglected. They would continue to underestimate the potential legal and political value of Sherman's unsuccessful legal forays to other critics of their organization. Of even greater importance, the leaders of Standard Oil did not understand how its opponents had succeeded in controlling discourse. This failing, which Rockefeller later attributed to the short-term, opportunistic focus of the company's expansion, led to a serious incongruity between the company's understanding of its actions and the views that were (and still are) held by the wider public.[111]

What the company's directors seem to have misperceived was the political power of heated rhetoric. It is clear that neither Rockefeller nor his associates saw themselves as conventional monopolists. Nor were they: they did not control prices by withholding goods from markets, nor did they raise prices on refined products more than rises in crude oil warranted. As private business-men, the directors conducted their business out of common public view, but they never seem to have thought of themselves as conspirators. Most busi-nesses do not carry on their deliberations in public. Standard Oil's directors seem to have been surprised by the monopoly and conspiracy charges, but the labels stuck. The directors also refused to be deterred by orchestrated furor in the Oil Region or by editorial attacks. Rockefeller described the company's attitude in a letter he wrote to his wife in March 1872: "We do not allow the newspaper articles to trouble us, knowing by whom written and the influences that induce them. . . . We are quite right in making no answer."[112] Rockefeller's approach was probably successful for an individual who was confronted by a private insult or by a blackmailer, but it was inept in any situation with politi-cal dimensions. More to the point, Rockefeller still understood Standard Oil as the work of his — and a few associates' — hands, not as a social and political entity. In this regard, he shared a common perspective on his own business with the producers and small refiners who regularly gave him battle. He did not, in fact, fully understand the broader consequences of his firm's sudden and striking growth. Above all, he did not understand the importance of the success that his competitors and critics enjoyed in their creation of him as one of the definitive examples of social malefactor.

In large part, this failure can be explained by the fact that Rockefeller shared the tenets of private virtue, especially commitment to probity and charity, common to his era and upbringing. He believed that he acted morally and, thus, that Standard Oil did as well. The concept of social immorality — of moral man and immoral society — made no more sense to him than it did to most of his contemporaries. During his long lifetime, he never accepted the fact that actions attributed to him and his associates at Standard Oil could be used to exemplify corporate wrongdoing, to reify the social immorality of big business. Least of all did it make any sense to him to argue that Standard Oil was immoral because it was big.

For their part the company's competitors and other adversaries also missed the wider effects of their actions. When they attacked Standard Oil, they spread a general perception of the industry that agreed with the earlier negative views of it given out by writers such Wright, and they overlooked the tendency of "mud" to spread on impact: their charges of conspiracy, price gouging, political corruption, and immoral predation were adopted by writers but would eventually characterize the whole industry. Like the less mighty sword, the pen would prove to be double edged.

# Numerous Offenses against Common Morality

As Standard Oil grew during the 1880s, so did the number of its opponents and the stridency with which they attacked the company. From the business world opposition came from regional interests threatened by Standard Oil's growing national preeminence—the clients Simon Sterne represented and firms less able to compete with the efficiency of Standard Oil's operations—refiners such as George Rice. Attacking Standard Oil was effective business strategy. Opposition also mounted among journalists, publishers, and intellectuals such as Henry Demarest Lloyd and Richard T. Ely; they took oilmen's complaints and embroidered and enhanced them to emphasize the menace of Standard Oil. In terms of discourse, the result was an emerging "history" of the company, a generally accepted construction of what it had done and why it had succeeded—both emphasizing morally unacceptable actions. In terms of response to this history, there was both more discourse about Standard Oil, identifying it as "monstrous monopoly" posing an alarming threat to public welfare, and growing pressure for state and federal action to the threat seen in the company. This pressure would lead to antitrust action and, more generally, to a perception of a need for a regulatory state. Standard Oil became the symbol for what could justify regulation, but that still left many questions about what to do.

During the 1880s, Standard Oil's leaders spent less time responding to critics than they gave to consolidating the position of the company in the domestic petroleum industry. During a four-year period, 1880–83, Standard

Oil continued to pursue horizontal and vertical integration. By 1880, with about 90 percent of the country's expanding refinery capacity, the company was uniquely exposed to economic contractions. Even a modest downturn in demand left the company with costly excess refinery capacity. Smaller refiners, with relatively secure niches in markets, were less vulnerable to economic cycles than they were to developments more specific to the oil industry, especially continuing problems stemming from recurrent and sharp shifts in the supply and price of crude oil.

Standard Oil limited its vulnerability to economic cycles and shortages of crude oil, through vertical integration. It expanded its crude oil storage capacity vastly and moved into pipeline transportation. The former move was unprecedented in scale: between 1881 and 1883, the company added thirty million barrels of tankage, largely in one-hundred-acre tank farms. These facilities made it possible for the company to store as much as a one-year supply of crude oil. Bought when prices were low, this stored oil could hold back prices when they rose. In the producing regions, which came to include Ohio along with Pennsylvania and West Virginia, this strategy gave Standard Oil near total control over prices, a situation that producers and competing refiners continued to fight.[1] Standard Oil's extensive construction of long-distance pipelines commenced after the rival Tidewater company demonstrated the technical feasibility of such projects by building a line over the Alleghenies, to connect with the Reading Railroad's route to Baltimore. Thereafter, Standard Oil laid a line to Cleveland from Pennsylvania and publicized its commitment to further construction, thereby forcing the trunk-line railroads to cut rates for shipments of crude to the company's East Coast refineries from eighty-five cents to fifteen cents per barrel. The latter figure was half of the railroads' cost, according to the New York City–based *Oil, Paint, and Drug Reporter*, which ordinarily led local media campaigns of rival refiners and distributors against Standard Oil.[2] Above all, the savings in transportation costs resulting from the pipelines enabled the company to build sizable financial reserves to carry it through depressions and to pay for acquisitions. Standard Oil's lower freight rates deprived Tidewater of its competitive advantage, and the latter company's directors accepted a market sharing agreement that permitted them to ship 18.5 percent of the crude from the producing regions. Other competitors ordinarily bought about 10 percent of the region's crude, leaving Standard Oil with the dominant share, nearly three-quarters of the market, still enough to exercise effective control over crude oil prices.[3]

As John D. Rockefeller and his associates concentrated on the implementation of their growth strategy, their adversaries stepped up public attacks on Standard Oil. Public perception of the company during the 1880s was largely framed and developed by three opponents: Simon Sterne, a New York City

lawyer; George Rice, an independent oilman from Ohio; and Henry Demarest Lloyd, a journalist and social critic from Chicago. Though they added
no elements to those already current in discourse during the disputes over
the South Improvement Company and Standard Oil's subsequent expansion,
their adversarial activities reemphasized the company's "history" and updated
it by attacking Standard Oil's subsequent development. Through two decades,
in support of the company's competitors, these men cooperated to consolidate and revalidate the dominant elements of public discourse relating to the
company.

At best only modestly successful at the bar, Simon Sterne was one of the
earliest and most effective special interest group attorneys, skilled at using
publicity and the political process to his clients' advantage. Working in the
emerging field of transportation law, Sterne regularly represented jobbers and
wholesalers battling preferential long-haul freight rates and thus joined the
hue and cry against railroads.[4]

In 1874, at the behest of his clients, he drafted a bill to create a commission
to regulate railroad rates and services in New York. Though he saw it defeated
in the legislature, Sterne continued to beat the drum for the regulation of
railroads, as counsel for the New York Board of Trade (which he also served
as chairman of its transportation committee), the New York City Chamber
of Commerce, and the New York Cheap Transportation Association. Public
speaking was part of his strategy; in 1878, for example, he referred readers of
the *Nation* to his recent speech at Steinway Hall, in which he expressed "the
feelings which the merchants exhibit against our most powerful corporation,"
the New York Central Railroad. He pointed hopefully to the recent creation of
a joint committee of the New York legislature to consider "the railway management of this State, in so far as it affects its prosperity." The prosperity of the
city and the value of its real estate, argued Sterne, depended on blocking
railway practices that deprived the city of its advantages over Baltimore, Philadelphia, and other cities. Regional rivalries were vigorous, and Sterne made it
clear that New York would have to defend itself against the rise of new trading
centers and the revitalization of older cities.[5]

Unfortunately for Sterne's clients, the New York legislature failed to act on
their behalf, so Sterne shifted the focus of his campaign. In February 1879, he
drafted a resolution, duly passed by the New York City Chamber of Commerce, calling for an immediate legislative investigation into railroad discrimination against New York merchants, with special reference to the Standard Oil
Company: "The recent developments regarding contracts with the Standard
Oil Company seem almost incredible, and show to what an extent individual
effort in any branch of business may be crushed out by a combination between
our modern highways [the railroads] and favored individuals."[6] Clearly, Sterne

was out to take advantage of the dust stirred up by Roger Sherman, who had rallied New York refiners to the anti–Standard Oil cause and exploited the connection between the railroads and Standard Oil.

Three weeks after passage of the resolution in New York City, the lower house of the New York legislature, responding to orchestrated pressure by the Grange and New York City mercantile groups, authorized a special investigating committee, the Hepburn committee, to study railroad rates and practices "to protect and extend the commercial and industrial interests of the State."[7] Before it met, the committee solicited grievances from chambers of commerce, boards of trades, and the Grange and other farmers' groups. Two of the members of the committee, Charles S. Smith and Francis B. Thurber, sat as representatives of Sterne's association clients. Sterne's official role was counsel to both the New York City Board of Trade and Chamber of Commerce, but in fact he took over the investigation, which was paid for by his clients, setting its agenda and guiding its work. Apart from providing the committee's initial position paper, in advance of investigation, he did the work of subpoena and examining witnesses, much as a committee counsel would in later times. Sterne made the most of his power by soliciting supporting statements from witnesses friendly to his clients and thus placing their arguments in the proceedings of the committee. The committee traveled over the state to take testimony from a wide variety of shippers, including dairy farmers, grain farmers, and manufacturers of staves, children's sleighs, and ale.[8] As Sterne brought agrarian and upstate manufacturing interests into the investigation, by including complaints of dairy farmers and upstate manufacturers, he built a broad-based coalition that supported the narrower interests of his clients. The hearings, which were drawn out over a seven-month period, were staged in more than half a dozen sites in the state.[9]

The railroads were on the defensive from the beginning. In a joint statement, William H. Vanderbilt (of the New York Central and Hudson River railroads) and H. J. Jewett (of the New York, Lake Erie, and Western Railroad) claimed that the railroad companies had "no power to control and save to the City of New York the jobbing trade nor to its merchants their former commissions and profits on the traffic that passes through this port to foreign countries." The committee dismissed this claim without discussion and went on to grill operating officials of various railroads, who also testified about controversial practices.[10]

Samuel Goodman, assistant general freight agent of the New York Central and Hudson River railroads, defended rate-making practices by explaining that rate varied according to the market value of the commodity, quantity (by carloads during regular time periods), and costs of providing exceptional services. Shippers who agreed to exclusive traffic agreements with the roads received additional discounts. Goodman also told the committee that the volume and diversity of freight had increased so greatly that simple rate structures were a thing of the past: nearly all rates were special rates by 1879. Published rates generally applied only to infrequent shipments of less than one carload. In short, distance was often a secondary factor in the calculation of freight rates. Business arrangements with customers were casual. Agreements, especially with large shippers, were verbal, without a written contract, leaving only recollection and a scant paper trail at the railroads to determine what the agreed-upon rates actually were at any specific time. Agreements tended to be short term, covering periods of less than one year. The immediate consequence of these circumstances was to highlight the complexity of rate making. Sterne found it difficult to effect a simple remedy for his clients' complaints about freight rates.[11]

The traffic manager of the same lines, however, testified that some shippers received special treatment because their high volume offset the high fixed costs of the railroad. According to James H. Rutter, by securing contracts with high-volume shippers, the lines could double their traffic and not increase costs above 25 percent. For this reason, Standard Oil, which generally moved forty-five cars per day over the lines, received preferential rates.[12] It was simply good sense to discount for volume. Unfortunately for Sterne, many of his clients did not qualify for this consideration, and he could not carry the day as long as the committee focused on formulas and operating policies. He needed to shift attention from the operational discourse of the railroads to republican ideology; to make his case for New York interests, he had to identify their cause with that of small producers. He also needed political allies to prompt the committee toward his objectives. He found this backing among the refiners in the Oil Region and allied New York refiners and jobbers. They provided both a villain, Standard Oil, and supporting evidence for Sterne's case.

Roger Sherman, an invaluable ally, furnished Sterne with his edited version of the proceedings in the Commonwealth cases. He also sent prejudicial testimony from the Acme Oil case, in which he represented the plaintiff, for Sterne's use. Sterne obligingly read his terse summation of it into the official record: "They crushed out a refinery in Titusville." Through the lengthy hearings, he also read into the record selective and undocumented paraphrases of testimony provided by Roger Sherman from Pennsylvania cases, most notably that of Alexander Cassatt of the Pennsylvania Railroad, who was said to have testified that his road paid Standard Oil "as much as $64,000 in one month in rebates" over an eighteen-month period.[13] The maximum total, $1,152,000, was large, but not large enough to impress the committee, so Sterne took a different course and called Pennsylvania independents and New York manufacturers to testify about their opposition to the South Improvement Company and Standard Oil. In the process, a large quantity of prejudicial and unsupported allegation was read into the official record.

Even so, Sterne's witnesses often qualified their condemnation of Standard Oil or made it clear that the company had not done as much harm as Sterne implied. Josiah Lombard, the leading New York City refiner, for example, claimed that Standard Oil told the Erie not to carry crude to him from Pennsylvania, but he added that this was "partly an inference," a deflating qualification for Sterne.[14] In a similar vein, Rufus T. Bush, another New York refiner who obtained crude from the Bradford district, claimed that Standard Oil used the railroads "to crush and grind out everybody that was not in their interest, and I believe they succeeded with all except five here in New York." A few minutes later, however, Bush described the successful shift by New York refiners to canal, pipeline, and the New York, Buffalo, and Philadelphia Railroad, moving two hundred thousand barrels on the canal in three to four months alone. More usefully for Standard Oil's smaller rivals, he insisted that there were no notable economies of scale in oil refining and that Standard Oil owed its dominant position solely to its receipt of rebates from railroads: "All the profit that the Standard Oil got, they got out of the railroads in the form of rebates." Bush also read his earlier and uncorroborated testimony to the Pennsylvania legislature into the record: Standard Oil paid 65 to 70 cents per barrel to ship crude from Bradford, while Bradford refiners were paying $1.25 to ship refined products to the seacoast.[15] On balance, his testimony did not support the claim that Standard Oil had forced competitors out of business by securing preferential arrangement with the railroads, but he provided quotable accusations that were cited by most of the company's critics then and after. In particular, dominance from rebates became one of the clichés permanently included in accounts of Standard Oil's success.

Witnesses' testimony was often far from accurate, but it was often colorful. Isaac L. Hewitt, partner in Hewitt and Schofield, a small refiner that also sold refined products from the Oil Region on commission, claimed that Standard Oil made his business unprofitable by organizing the South Improvement Company in "1866 or 1867." Through lengthy, confused, and contradictory testimony, Hewitt pyramided allegations, to the point that the usually complacent chair objected: "Give us simply your personal knowledge of matters." When he did so, Hewitt added one of the durable condemnatory stories about John D. Rockefeller, who was supposed to have threatened him by saying, "I have ways of making money that you know nothing of." Fed this line by Sterne, who obviously knew that he could elicit the quotation from Hewitt, the witness was precise on this detail. Otherwise his testimony had limited immediate value because he was fuzzy on dates and more specific matters, such as his discussions of rates with William H. Vanderbilt, among others.

Hewitt's unusually clear recollection of Rockefeller's supposed remarks went unquestioned, and it remains so in written history, exemplifying the adage that famous people attract famous quotations. It is worth noting, however, that Hewitt's recollection was that Rockefeller was not insinuating that he had *secret* ways of making money; as Sterne's introductory question makes clear, Hewitt believed that Rockefeller was actually referring to industrial installations about which Hewitt was not likely to know. Hewitt saw himself as the victim not so much of a hidden conspiracy as of Standard Oil's overt dominance of exports, which largely eliminated the function of traditional commission men: Hewitt was a casualty of vertical integration. So was Charles T. Morehouse, a Cleveland refiner, who was forced to sell out to Standard Oil after it decided to carry through further vertical integration by processing waste oil (residuum) itself, once it had sufficient refinery capacity to make it economical. At the hearings, Sterne swore in Morehouse principally to testify that John D. Rockefeller threatened him by claiming that Standard Oil "had sufficient money to lay aside a fund and wipe out [possible competitors]." He encouraged the witness to put his own expansive interpretation on his interview, including an uncharacteristically pithy statement attributed to Rockefeller: "We squeeze you out, and you die." Thus was born the "anaconda" image of Standard Oil, favored thereafter by editorial cartoonists. Sterne also used Morehouse to read conspiratorial allegations into the records, though Morehouse was generally honest in conveying it as hearsay. Morehouse did, in fact, make note of his recollection that Rockefeller never claimed to enjoy a special relationship with the railroads, a point that Sterne rushed past without comment.[16] In effect, what Sterne pumped from his witnesses were what today's observer would call "sound bites"; true or false, they stood out in discourse.

Sterne not only orchestrated and led the testimony against Standard Oil, but he also attempted to exclude testimony that might weaken his case. As he put it at one grand moment: "To all of the testimony which relates to the fact or to the inference, that the Standard Oil Company did not control or guide ultimate conduct of all the trunk lines, I object." Even so, despite Sterne's efforts, Standard Oil had its "day in court," albeit a current indictment by a grand jury in the producing regions of Pennsylvania led it to minimize its exposure through highly limited disclosures; under the circumstances, what it could say was circumscribed. John Archbold, former independent and current Standard Oil director, took the lead for the company in this public hearing, as he would do in the future. Directly to the point of the investigation of railroad rates, he insisted that he had been able to secure better rates as an independent than he could now obtain for Standard Oil, a contention supported by officers for various railroads. The manager of the company's largest competitor in the producing region also testified that his employer, A. Neyhart, never paid higher rates than Standard Oil, a contention supported by an officer of the Erie. Simon Sterne tried unsuccessfully to keep the latter statement out of the record. Other independents, not apparently allies of Standard Oil, supported the main points of the company's defense. Simon Bernheimer, once a partner in the Olefin Oil Company at Greenpoint, New York, claimed that his company received a nine-cent rebate from the Pennsylvania Railroad. Even more to the point of Standard Oil's profitability, he claimed that although current refining technology did not favor large refiners, economies of scale in the use of labor meant that "the more you could refine, the cheaper you could do it." Sterne brushed this assertion aside to move on with his agenda.[17]

As Sterne might have expected, he received little help from railroad executives, but they had to be admitted as witnesses, given the ostensible purpose of the Hepburn committee. Before that body, these witnesses tended to defend Standard Oil on the grounds that its rise had been both legal and ethical, though they believed that the company had bargained so hard with them as to take the profit out of hauling crude oil and refined products. William H. Vanderbilt, of the New York Central, for example, frequently contradicted Simon Sterne's major allegations:

Sterne: Did not the [pricing] policy of the trunk lines make Standard?
Vanderbilt: Not at all.
Sterne: Didn't the advantage of having a larger capital and the advantage in rates constitute an enormous advantage to the Standard Oil Company compared with other people?
Vanderbilt: *After* they got strong it did.[18]

Vanderbilt left no doubt as to his explanation of Standard Oil's dominance of the petroleum industry: "[I] never came in contact with any class of men as smart and able as they are in their business, and I think a great deal is to be attributed to that." In Vanderbilt's opinion, the Standard Oil executives succeeded because they were superior businessmen: "They are very shrewd men; I don't believe that by any legislative enactment or anything else through any of the States or all the States, you can keep such men as them down; you can't do it; they will be on top all the time; you see if they are not." This opinion flew in the face of Sterne's public advocacy of stamping out monopoly through legislation and regulation. It is not surprising that he did not extend consideration of Vanderbilt's assertion.[19]

Vanderbilt, like Standard Oil executives and independents, also undercut Sterne's dramatic device of hanging the Standard Oil story on a solitary figure, John D. Rockefeller. Having done business with Bostwick, Archbold, and others, Vanderbilt was aware of Standard Oil's structure and style: committee decision making through strong functional leaders. As the railroader put it, "One man would hardly have been able to do it; it is a combination of men." Though his language was intended to describe the company's governance, the key phrase, "a combination of men," was highly charged in conventional republican discourse, carrying the implication of conspiracy, monopoly, and illegality.[20] For Standard Oil, Vanderbilt's choice of words was unfortunate.

When committee deliberations were concluded, Sterne appended his final argument and then controlled the final outcome by writing the committee's report: he thus had both the penultimate and last words.[21] The focus of Sterne's (and by extension of the committee's) criticism was discounting for long-haul and high-volume business by the railroads. From the beginning, the consistent stand of New York City shippers was that freight charges should be based solely on mileage and that price distinctions should not be allowed on full-car shipments. No rebates or other discounts were acceptable. The railroads, moreover, were expected to deliver full and complete service, including making rolling stock available to all shippers, regardless of the frequency and quantity of their shipments. Such concessions for small shippers appeared for the next forty years in public discourse and were extended to include petroleum in pipelines.[22]

Sterne also advanced his own and his trade association clients' preference for relatively small manufacturers by denying that stronger capitalization and more professional management were significant advantages in competition. The shift to big business, he argued, was undesirable:

> There is no tendency of the age more marked than the tendency toward centralization and monopoly. The application of machinery to all kinds of manufacturing has dried up the various manufacturing establishments heretofore existing throughout the rural districts and the hand manufacturer, seeing his business drawn to some commercial center, has been compelled to follow and apply for a post at the machine. Nothing is left to our rural districts but the production of raw material: and now come our railroads placing our agricultural products at a marked disadvantage compared with the products of the West, annihilating the advantage of proximity to market and imposing at times actually as a rule relatively a heavier tax on their movement than is levied on the movement from other states.[23]

In this comment and others in the report, Sterne recast the Jeffersonian view of commerce in Jacksonian terms, carrying powerful elements of republican discourse into the new industrial age, bringing related moral discourse to bear on the railroads and, in the end, on Standard Oil: "This monopoly [Standard Oil], remember well, was brought about entirely by a combination between comparatively small capitalists originally and the freight rate determining powers of the railway management." Standard Oil was definitively "a monstrous monopoly, which ruthlessly crushed out [competitors] by the aid of the railways."[24]

Sterne not only denounced Standard Oil but also made it clear it was a danger to the nation: "A more extreme illustration of discrimination against the State could not be given than in the history of the Standard Oil Combination." It owed its origin to corruption: "The imbecility of the conduct of railway officials in that particular [of rebates and drawbacks] surpasses even the venality and corruption in which the Standard Oil took its start." By characterizing the railway executives as foolish and Standard Oil as wicked, he conveniently let the Erie, which he had represented, and the New York Central, a power in New York City commercial and political affairs, off the hook. The villain in the piece was conveniently distant, in Cleveland.[25] In the end, the three principal complaints were against "discriminatory rates, stock watering, and Standard Oil."[26] The last was more than a threat to competition. It was a monstrous monopoly threatening the nation.

The Hepburn committee hearings and report received widespread coverage in the New York press. As Lee Benson's thorough search of archives disclosed, "Practically every journal in New York City (and a good many upstate ones) covered the hearings day by day so that it was almost impossible for newspaper readers to remain unacquainted with the evidence. . . . No overpowering competition existed for newspaper space, and public attention was not distracted from the case built up by Sterne."[27] Certainly that was true for James Gordon Bennett's *New York Herald*. It downplayed the hearings until they convened in New York City, at the Chamber of Commerce's offices, in October. Thereafter, the committee's narrowing focus on Standard Oil was directly reflected in the *Herald*'s expanded coverage, and "John D. Rockefelloe [*sic*]" was identified as the source of New York shippers' problems. In the *Herald*'s report, when the Erie's attorney objected that "John D. Rockefelloe [*sic*] does not come under an investigation of railroads," Sterne replied, "He does, by Jove! That is just what he has been doing, and this is the way we have to get at it." Though this was a passing phrase in the day's hearings, the *Herald*'s brief synopsis made it the focus of events. The testimony of Isaac L. Hewitt and Charles L. Morehouse was reported, along with the sinister threats Rockefeller was supposed to have made to them. Selective quotation made interesting copy, and the presence of a villain made a much better story than freight rates, just as monopoly made better headlines.

Thus, in summary of testimony that was largely about milk and iron shipments, the lead was "More about the Oil Monopoly." The *Herald*'s mode of covering the hearings tended to emphasize Sterne's case against Standard Oil.

Typically, a brief introductory paragraph was followed by excepts from testimony. In the interest of the newspaper's circulation, the economics of milk and iron were passed over to get to people and events. Apart from Jay Gould, William K. Vanderbilt, and their railroads, John D. Rockefeller and Standard Oil were most available to personify the issues. The editors, moreover, made scant efforts at detachment; Sterne's arguments were characterized as the "case on behalf of the people."[28]

As this observation makes clear, Sterne and the *Herald* had a useful villain in Standard Oil, whose usefulness could be stretched well beyond the oil industry. The company could represent what was wrong about a growing and changing America. Francis B. Thurber, a New York City wholesale merchant and sometime investor in oil refineries, writing as "Average Citizen," argued in 1879 that more was at stake than regional rivalries and disputes between shippers and railroads:

> The American Republic has survived the storms and troubles of a hundred years. Whether or not it will exist for another century will depend largely upon the making and execution of our laws. It is, perhaps, not strange that legislation for the protection of the public interest should have failed to keep pace with the enormous changes which steam, electricity, and machinery have wrought — all within a half century, and it is the abuses attending the employment of these great forces which have caused the manifestations popularly known as Communism, but which to a great extent, is simply the well grounded dissatisfaction of the "Average Citizen."[29]

One could not protect average citizens from steam, electricity, and machinery, but one could protect them from Standard Oil.

The Hepburn committee hearings and report rekindled the interest of Henry Demarest Lloyd, a Chicago editor, in the earlier crusade of the Pennsylvania producers against Standard Oil, and in 1880, Lloyd began work on an antimonopoly article, which he peddled to the leading opinion magazines in the East. His initial choice, the conservative *North American Review*, was sympathetic to his cause, but its editors rejected the article as inflammatory. He next submitted it to William Dean Howells, editor of the *Atlantic Monthly*, who agreed to publish the piece but insisted on fuller documentation and verification of factual statements. When the deadline for publication of the March 1881 issue arrived, Lloyd's corrections and amplifications had not arrived, but Howells went to press, prefacing the article with his own caveat: "Through a failure to receive the author's proof in season for correction, Mr. Lloyd's article goes to our readers without the strong confirmatory facts and figures which his revision embodied. His paper was written several months ago, and as printed represents the condition of things at the time of writing. Some minor errors of statement, not affecting his positions generally, have necessarily remained uncorrected."[30]

Lloyd began "The Story of a Great Monopoly" with the controversial business career of "Commodore" Cornelius Vanderbilt: "He used the finest business brain of his day and the franchise of the State to build up a kingdom within the republic, and like a king he bequeathed his wealth and power to his eldest son." In the process of making his fortune, Vanderbilt corrupted legislatures, evaded taxes, and levied tribute on American citizens. His rapacity, and that of the other railroad kings, led to eight railroad strikes from 1876 to July 1877 and hence to "social disorders we hoped never to see in America." With the railroad strike of 1877, "the country went to the verge of Panic" as barricades "in the French style" went up in the streets of Baltimore. Lloyd went on to blame the violent episodes of the strike on "these giant forces within society, outside the law."[31]

With no transition, because in his mind and in all probability in those of his readers Vanderbilt and John D. Rockefeller were birds of a feather, Lloyd went on to an extended attack on Standard Oil. He claimed that the company had fixed the world price of kerosene "for years" and that it had done so through the device of special contracts with railroads, eliminating "all the petroleum refineries of the country except five in New York, and a few of little consequence in Western Pennsylvania." In the process, John D. Rockefeller had accumulated a fortune second only to that of the Vanderbilts. With selective use of edited testimony from Commonwealth cases — as reprinted selectively in the Oil Region — and from the report of the Hepburn committee, Lloyd presented the conventional case against Standard Oil with enhancement: the testimony of Alexander Cassatt, for example, was inflated by Lloyd to support his claim that the rebates and drawbacks amounted to thirteen million dollars over a ten-month period in 1879. He went on to assert that this amount was equal to Standard Oil's cost for crude, so "the railroads of the United States virtually gave the Standard its material free."[32]

Having little understanding of the operations of various segments of the industry, Lloyd believed and passed on the most outlandish charges of Oil Region refiners, including their allegation that Standard Oil regularly accepted superior crude at pipeline in Pennsylvania and delivered inferior crude to refiners in New York. He also criticized the company for buying oil for immediate shipment only, though he acknowledged that at the time this was done, storage tanks were full, making necessary Standard Oil's policy. He also introduced highly imaginative cost and price figures for crude oil and kerosene, obtained from an unidentified expert. In the process, he confused prices in Chicago with those posted for the export market in New York and then claimed that Standard Oil would make eleven cents per gallon of kerosene if kerosene sold for twenty-five cents in Pennsylvania, which it did not do. In short, Lloyd's data made little sense, which may account for Howells's disclaimer at the beginning of the article.[33] These errors and others that he corrected in subsequent publications did not detract from the main thrusts of his argument — reinforcement of the linkage of the much criticized railroads to Standard Oil, the refocusing of debate from rebates to pipeline ownership, and transference of the political corruption charge commonly lodged against railroads to the oil company.

For Lloyd, railroads were the original transgressors: "It is the railroads that have bred the millionaires who are now buying newspapers, and getting up corners in wheat, corn, and cotton, and are making railroad consolidations that stretch across the continent." But the example of Standard Oil posed additional danger to the public: "By the same tactics that the railroads have used to build up the Standard, they can give other combinations of capitalists the control of the wheat, lumber, cotton, or any other product of the United States."[34] Presumably the example of Standard Oil would encourage other capitalists to press for such advantage.

Like the railroads, according to Lloyd, Standard Oil regularly corrupted governments, including Congress, where it controlled an investigation of it in 1876, reducing the proceedings to "a farce" and barring a report of findings. All the private suits and public investigations had come to nothing because Standard Oil had corrupted the legal and political systems: "The plundered found that the courts, the governor, and the legislature of the state [Pennsylvania] and the Congress of the United States were the tools of the plunderer." He concluded, "The time has come to face the fact that the forces of capital and industry have outgrown the forces of our government." In a dramatic conclusion, Lloyd rallied Grangers, shippers, and small producers: "The nation is the engine of the people. They must use it for their industrial life, as they used it in 1861 for their political life. The States have failed. The United States must succeed, or the people will perish." Lloyd's remedies included those of Simon Sterne, Francis Thurber, and Roger Sherman: published railroad rates ("Publicity is the great disinfectant"); fixed and reasonable railroad rates, "based on the cost of the service not on what people will stand"; nondiscriminatory rates for all shippers. But they also included a national board to investigate railroads.[35] The dimensions of the problem demanded a federal presence.

Lloyd's article, circulating throughout the country, had an immediate impact. His biographer, with scant hyperbole, later described it as "the hit of 1881." Grangers, the Anti-monopoly League, various chambers of commerce, boards of trade, and Standard Oil's competitors distributed it. The March issue of the *Atlantic Monthly* went through six reprintings in one year, largely as a result of these bulk purchases.[36] In its day, it was widely quoted and cribbed from, as "the best single analysis of the dubious ethics of a new breed of industrial monopolists," in John L. Thomas's recent assessment of its impact.[37] As a consequence of the extensive circulation of his article, Lloyd emerged as "a fighting publicist of the first rank."[38] The timeliness and power of the article also helped fix the literary conventions of the journalistic literature of exposure, moral outrage and personification of social and economic problems. In Lloyd's article, Standard Oil emerged as "the symbol of bigness, a creature of the railroad rebate, and an enemy of free enterprise capitalism." And indeed, he succeeded in making Standard Oil such a symbol in the nation.[39]

Lloyd's ability to fix a phrase, fully as adept as that of Sterne, also fastened memorable words on Standard Oil, many of which still adhere in modern histories. Thus Lloyd characterized the oil company as the agency of public corruption by a witticism: "The Standard has done everything with the Pennsylvania legislature, except refine it." To describe the company's power, he drew on the indelible Oil Region image of the company, "the Standard octopus." His characterization of Rockefeller and Standard Oil, "the greatest, wisest, and meanest monopoly known to history," proved to be equally durable. The mixture of trenchant criticism and left-handed compliment would mark sophisticated attacks on the company for the next fifty years.[40]

While Henry Demarest Lloyd carried the campaign against Standard Oil into national periodicals, company's competitors carried on crusading in court. Some of them filed a suit in Pennsylvania, *Scofield et al. v. Lake Shore and Michigan Railroad*, alleging that Standard Oil had received rebates that were illegal in that state. The plaintiffs, certain of victory, seized the occasion to load the legal record with other allegations, the most damaging of which was that Rockefeller had victimized a widow, a theme dear to the hearts of urban editors. According to the recent historians of Standard Oil, a Mrs. Backus sold her plant in Cleveland to the Standard Oil Company of Ohio shortly after her husband's death, for a price she set and John D. Rockefeller accepted without his customary hard bargaining. Thereafter, Mrs. Backus came to believe that she should have asked for more money, which she quickly translated into the notion that Rockefeller had cheated her by not offering more than her asked price. As it got into the court record of the Scofield case, however, the fact that the widow had set the final price was obscured, coming to light later when other members of her family revealed the circumstances of the bargain. In the meantime, dating from *Scofield et al.*, which dragged on in the courts and in the newspapers from 1881 to 1886, Rockefeller was depicted as a heartless victimizer of a widow, a damaging image with the general public.[41] The core of the scenario, moreover, was a gendered condemnation of Rockefeller: he was less than a man because he took advantage of a widow.

Standard Oil continued to make enemies as it grew. Its goal of eliminating excess refinery capacity took it beyond Pennsylvania, as it attempted to "unify" refiners in the Marietta, Ohio–Parkersburg, West Virginia, area, through acquisitions by Johnson Newlon Camden's Consolidated Oil Company. Camden, a longtime Democratic Party leader, sold his properties to Standard Oil in 1875 and with the company's capital set out to buy up small refineries. He found the pickings poor and appetites large: "I am discouraged about this section," he wrote. "It is so full of *debris*, both of men and old refining traps that will be as hard to keep down as weeds in a garden. The object of the whole crew of broken oil men is to pension themselves upon us."[42] Camden advised Rockefeller against an attempt to absorb all competition in Ohio and West Virginia: "I have considered the matter carefully and am fully persuaded there is no use trying to buy it and provide for the horde here, as long as we are keeping up this margin."[43]

Standard Oil persisted with uneven results. Camden's efforts in Baltimore and Parkersburg succeeded grandly, but they also made Sylvia C. Hunt famous, as the second widow victimized by Standard Oil. After Camden approached her to strike a bargain for her ninety-barrel refinery, Mrs. Hunt first decided to sell her refinery and then thought the better of it. As difficulties began to appear, Camden explained the circumstances in a lengthy letter to John D. Rockefeller:

> The poor woman does not seem to know her own mind for a day at a time. Knowing your disposition to avoid having any trouble with her, I have conceded everything that she has asked, and have been most kind and yielding with her. She even asked me if I would surrender back her refinery to her, and I consented to do that, provided it met the approbation of those with whom I was connected. She then declined to take it back, and said no more in relation to it.
>
> I do not see anything more that can be done. In fact, she has not asked of us anything that we have not done for her. I think we had better now take the position of standing on our rights.[44]

In the final settlement, Camden/Standard Oil paid seventy-five hundred dollars per year to lease a small plant for which it had no immediate use.[45] Mrs. Hunt later claimed that she had been both pressured and cheated. Though subsequent litigation upheld Camden's version of the affair, Mrs. Hunt's allegations were repeated by Pennsylvania oilman Joseph D. Potts, in *A Brief History of the Standard Oil Company*, and transmitted thereby to Ida M. Tarbell.[46]

Good stories die hard, even in the face of contrary evidence, when they involve the rich and powerful and the alleged victimization of a widow.[47]

Undaunted, Camden crossed the Ohio River to round up refiners in Marietta. With his quick acquisition of the most successful local company, Marietta Oil and Refining, his campaign began with a conspicuous victory. "A vigorously blunt man," according to John D. Rockefeller's principal biographer, Camden then put additional pressure on the remaining Marietta refiners by offering higher short-term prices to their suppliers of crude, effectively choking off stubborn negotiators. As he wrote to Rockefeller, "We will either get them or starve them."[48] Next, he secured preferential freight rates from railroads, principally the Baltimore and Ohio, the main link for Marietta producers and refiners.[49]

This tactic might have ended the campaign had it not been for a determined holdout, George Rice. Rice ventured into oil production in West Virginia after the Civil War and relocated to Marietta, Ohio, in 1872. "A quick active man," according to a local credit reporter, Rice made his mark quickly by reopening the Macksburg field with a series of four successful wells. At the end of one year, he had acquired properties and leases worth at least five thousand dollars, in addition to a three-quarters interest in the Lowell Oil Company.[50] When Camden's acquisition campaign began in Marietta, George Rice was doing well in the oil business. His investment in tankage, oil, and refinery was deemed worth about twenty-five thousand dollars, with his leases worth two to four times that amount. He produced about half of the feedstock for his sixty-barrel refinery and sold his products to western distributors, as Marietta refiners had done since the rise of the city as an oil town. According to the local credit reporter, Rice was "making money all the time."[51]

No doubt Rice gained a considerable motive to attack Standard Oil publicly in 1881, when Camden forced the local railroad, the Marietta and Cincinnati, to grant Camden Consolidated a rebate on high-volume shipments. In any event, by the end of 1882, Rice was one of the most visible opponents of Standard Oil, a situation "which seriously imperils him if a fight were to commence" in the view of the Roy G. Dun agency reporter.[52] In response to Camden and Standard Oil, however, Rice developed a three-part strategy for holding off the industrial giant. He cooperated with independents in other regions, joining the Schofield suit as a litigant. He himself filed dozens of suits against Standard Oil during the next forty years. Most of all, Rice became active politically, thereafter his major competitive tactic. In 1879, he prompted an investigation of railroad rates by the Ohio legislature and tried to interest Ohio congressman James A. Garfield in launching a broader inquiry. This initiative died when Garfield, a dark horse candidate, received the Republican presidential nomination and won the election of 1880. Rice's presumed pleasure at having an acquaintance in the White House was short lived, as Garfield was assassinated after only four months in office. But Rice could manage Standard Oil without presidential favors because he made himself such a vocal, aggressive opponent. His celebrity made him hard to handle.[53]

Despite Camden's campaign and his own diversion from business in response to it, Rice continued to prosper on a modest scale. By 1881, he had increased the capacity of his refinery, added oil properties, and pursued a successful competitive strategy by developing direct sales to retailers of kerosene in the most remote parts of the country served by the railroads. With this approach, he met Standard Oil head-on, entering markets by cutting prices, affordable because he served as his own wholesaler, and he carved out small but profitable markets in Mississippi, Louisiana, and Iowa in competition with two Standard Oil companies, Chess and Carley and Standard of Iowa.[54] He also succeeded in negotiating lower railroad rates than Standard Oil received to some points, by negotiating harder and taking advantage of his larger competitor's slower-moving management. Thus, he invaded the territory of Chess and Carley, in the South, through vigorous negotiation with the Louisville and Nashville Railroad. After nearly six months, Chess and Carley discovered Rice's achievement and urged the Louisiana and Nashville to tighten up their operations, to "turn another screw."[55]

After he learned of this eminently quotable instruction, George Rice added his considerable voice and energy to the long-standing campaign of Oil Region producers and refiners for control of railroad rates — actually for the preferential treatment of short-haul shippers — through the creation of a federal commission. When the regulatory body — the Interstate Commerce Commission — was finally created in 1887, Senator John Sherman, a frequent recipient of items from Rice, observed that the measure had "general support." Other legislators displayed more enthusiasm: Senator George F. Edmunds of Vermont saw the agency as finally limiting "the tyranny of this corporate management and corporate combination," referring in part to the widespread belief in Standard Oil's preferential treatment. For Senator James F. Wilson of Iowa, passage represented a victory for "businessmen . . . farmers . . . and the people." Thus, the creation of the first federal regulatory agency pulled Standard Oil squarely into the discourse surrounding a major public policy as the target of reform, a process that began with the "oil wars" of the 1870s and that George Rice sustained for several decades.[56] Rice himself became a frequent plaintiff against Standard Oil before the commission, as we shall see.

Rice's second response, to the turning of "another screw" by the Louisville and Nashville, was publication of *Black Death*, a bombastic attack on Chess and Carley, Standard Oil, and the railroads, reiterating the charges of skullduggery that Pennsylvania independents had leveled at Standard Oil nearly a decade earlier. He followed that salvo with *The Standard Oil Company: Its Dishonest Tricks Exposed*, which Rice published in New York City to firm up his ties with Standard Oil's New York opponents and to reach the sympathetic Manhattan press. Early in his crusade, Rice received warm support from the *Oil, Paint, and Drug Reporter*, which still spoke for independent oil jobbers and product wholesalers.[57] Thereafter, Rice's publications appeared in a steady stream until his death in 1905. *Railway Discrimination as Given to the Standard Oil Trust*, which he wrote and published in 1888, was directed specifically at the preferential treatment of large producers and shippers — read "Standard Oil" — who used and supplied tank cars over those who moved refined products in barrels, commonly in less than carload lots. Rice sidestepped the economic aspects of the rate issue by substituting ample doses of traditional antimonopoly rhetoric to justify the preferential treatment of small producers, whose enemies were "not men of intellect, or genius, or learning, or even of honest thrift, or patient industry . . . [but] cold, calculating men, who, by open bribery and naked rascality, secure the favor of railroad officials, until they wring one hundred millions of dollars annually from the mass of the people and the overburdened industry of the country."

Like Rice's other publications, *Railway Discrimination* drew on the Hepburn committee report and incorporated the writings of New York allies, such as Francis B. Thurber. It also argued for relatively lower rates for small producers to offset the advantages in capital enjoyed by Standard Oil; Rice made no bones about exploiting the moral preference accorded small producers in traditional republican discourse. For that matter, the traditional preference given to small producers in common law was sustained in transportation legislation as legislatures and Congress required common carriers to construct rate tables that minimized advantages that would otherwise have been granted to shippers because of volume or size. Courts agreed; in the landmark *Trans-Missouri Freight Association* case, the United States Supreme Court defended protection of small traders.[58] So, in arguing preferential treatment of small producers, Rice rode political currents of his time as well as contributed to them.

Rice's efforts in print were far from great art. Many of Rice's later publications, such as *The Standard Oil Company, 1872–1892*, published in Marietta in 1892, were clumsily cobbled together from pleadings Rice made before courts and regulatory agencies. What Rice may have lacked as writer or thinker, however, he more than made up for in exposure; he ensured widespread notice by opinion and policy makers by supplying copies of his works to numerous writers and public officials. Over the years, U.S. congressmen from Ohio, Pennsylvania, and New York received multiple mailings from Rice, as did writers and intellectuals, including Henry Demarest Lloyd, in both February and March 1882.[59] A man with a mission, Rice continued to assail Standard Oil in print until his death in 1905.

Once the new Interstate Commerce Commission began to hear complaints, George Rice took yet another avenue to attack Standard Oil, and he became the principal complainant before it with respect to oil-related issues. In all, Rice either filed or participated in more than two dozen cases, with favorable decisions in about half that number. Many of the complaints related to either long-haul discrimination or lowered rates for volume or tank car shipments. Rice's most significant action before the ICC came when he participated in a complaint that originated in the producing region, filed by William C. Scofield, Daniel Shurmer, John Teagle, and others, a case that went a long distance to deprive Standard Oil of the advantages of capital and volume. The commission ruled that the railroads were obliged to furnish tank cars for all shippers if they could not afford to buy them, that the roads must rent Standard Oil's tank cars for a reasonable fee and not discount charges, that these cars must be available to all shippers, and that in the absence of a sufficient number of cars, shippers who sent barrels in box cars could not be charged more than Standard Oil when it shipped in tank cars.[60] In the Scofield case and Rice's other filings, the intent of Rice and of the commission was clearly to aid small producers who could not afford the "considerable expense" of competing with Standard Oil for economies in transportation, as the regulatory body put it in *Rice v. Louisville and Nashville*.[61] Rice also advocated expansion of ICC investigative power to compel testimony from both railroads and Standard Oil.[62]

Rice was quite effective before the ICC, but he had even greater success getting coverage from New York newspapers. Thus, in November 1887, the *New York World* reported his claim that Standard Oil monopolized the industry and sought to destroy all opposition; the paper condemned the company on its editorial pages, identifying dislike for Standard Oil as the single most important force behind the creation of the ICC. The *World* also followed the line of Rice and other independents on the relation of Standard Oil's success and the railroads: "That company, through special rates of transportation from railroads, has been able to drive almost every competitor from the field. If they are placed on the same basis as other shippers, the backbone of the monopoly will be broken."[63] The next year the *World* reported favorably on Rice's suit against the Louisville and Nashville Railroad and, on its editorial page, denounced the attempts of "the Octopus and its tank cars" to sustain a ruthless monopoly. According to the *World*, Rice's cases offered "a chance for individual operators to live."[64] The *World* and other newspapers advertised Rice as the prototypical small businessman battling sinister power, thereby advancing Rice's objective, diminishing the competitive advantages of Standard Oil through the regulatory powers of government.

Above all, in newspapers and journals, by the 1880s Standard Oil was depicted as the definitive American monopoly. As had been true during the "oil wars" of the early 1870s, major battles were being fought in print, in court, and in the hearings of regulatory bodies; manipulation of public discourse to influence public policy had emerged as the independents' most effective competitive tactic in coping with the operational and financial strength of the Standard Oil Company. Moreover, the industry had become a moral battleground, in which economics was irrelevant and right action was paramount. This was turf eminently suited to sensational journalism, and at the same time Rice was attacking Standard Oil before the ICC, charges surfaced in New York that could satisfy the most determined journalist out for sensation—allegations that Standard Oil attempted to explode the refinery of a competitor in Buffalo, New York.

The essentials of the case, as the *New York Times* reported them in considerable detail, were not especially damning. Two Vacuum Lubricating Company employees left this member of the Standard Oil Trust in 1882 and organized a company in Buffalo to compete in the profitable lubricating oil market. Their plan was to challenge Standard Oil in Boston and other parts of New England. Though they lacked the technical skill to design and build a refinery, they hired away a third Vacuum employee who, presumably, possessed the requisite knowledge. Thereafter they built their plant, though the exceptional drinking habits of the designer-supervisor made the project more stressful and time-consuming than they had anticipated. Finally, on the first test, the supervisor ordered refinery laborers to stoke up hot fires under the stills and then retired to a nearby tavern—or, alternatively, in the tale, to Standard Oil's Atlas works in Buffalo. While he was gone, the safety valve blew off, releasing gasses into the air and causing the owners and refinery workers to fear an explosion, which did not occur. When a second try at refining failed, the supervisor walked off the job without notice. After two more weeks of failure, he disappeared altogether, leaving the owners of the new Buffalo refinery in the lurch, after Vacuum hired him away, presumably in an attempt to block successful operation of the competing facility.

Thereafter, the managers of the Vacuum refinery supposedly spirited the supervisor away to Boston and then to California. All the while the Buffalo refinery owners fought off creditors while they tried to bring their plant into operation. After one year, they succeeded, only to find that Vacuum underpriced them in Boston—their target market—because it received volume discounts from railroads. At about the time the failed entrepreneurs were considering giving up their business, the absent supervisor reappeared, now off Vacuum's payroll and primed with a useful tale: Standard Oil had paid him to design a refinery that would explode!

No doubt believing this convenient tale, the two refiners went to court, pressing both civil and criminal charges against the Vacuum managers and the trustees of the Standard Oil Trust. The cases dragged on for nearly five years. In the end, the civil cases were a wash, as two were won by each side: notably, Vacuum won a judgment for patent infringement, and the Buffalo refinery owners won in their allegation that Vacuum had conspired to bring an employee to break his contract. The civil cases, especially Vacuum's victories, received scant attention in the press.

The major focus of public discourse was the criminal case, which came to trial in early 1887. Without exception, the daily newspapers in New York City took up the cause of the Buffalo Refining Company, the *New York World* most shrilly. Its theme was a David-versus-Goliath battle between "the fighting President of the Buffalo Lubricating Oil company" and "all the wealth and power and influence . . . of the gigantic Standard Oil Monopoly." The *World* reprinted all the charges against Vacuum and Standard Oil, with scant mention of their defense, observing that this case was merely the most recent to demonstrate that Standard Oil's policy was "corruption, destruction of competition, and robbery." The allegations were, in the view of the *World*, incontrovertible evidence of "the iniquity which this corporation represents" and of "numerous offenses against common morality."[65]

In the end, the managers of Vacuum were convicted of one count of conspiracy and fined $250. Before that, John D. Rockefeller and the other trustees of Standard Oil were dismissed as defendants, a result that was pushed to the second page of the *World* on May 9, 1888. The front-page material was the prosecutor's reflection on the case: "It was the people's fight and The *World* helped us win it."[66] Though "the people" won only $250, Standard Oil was a major casualty in the battle. The *World*'s version and interpretation of the case became a set piece in discourse thereafter, without the strongly qualifying aspects presented during the trial. This was the last major piece of history that would frame discourse relating to the Standard Oil Trust, and it confirmed the strongest allegations of Roger Sherman, Henry Demarest Lloyd, and George Rice. Standard Oil acted as though it was above the law; its prime offense was against morality.

Rice successfully manipulated discourse against Standard Oil and came close to making a career out of attacking the company, something other scholars have observed when they characterized him as a professional litigant or a gadfly,[67] but there was a positive side to Rice's actions well worth emphasis: he stayed in business. In fact, obstreperous resistance to Standard Oil became as much a part of his strategy in competition with the industrial giant as it had earlier been among Pennsylvania Oil Region refiners. George Rice raised too big a fuss for Standard Oil to shut him down; moreover, by going before the

ICC, he got that body effectively to set aside the major competitive transportation advantages Standard Oil had over a small refiner like himself. In 1888 Rice's Globe Oil of West Virginia was, as the Dun reporter put it, "a good concern," worth about a quarter million dollars. What closed Rice down was not any direct action by Standard Oil but the decline of crude production in his area, crude he could buy cheaply. No matter: Rice continued to sue Standard Oil.[68]

By this point the reader may well be wondering if Standard Oil made any attempt to combat the attacks upon it. In 1883, Rice's adversary, the energetic Johnson Newlon Camden, did try rejoinder. Then a member of the United States Senate, Camden argued Standard Oil's case in an exchange with John C. Welch, a Pennsylvania journalist, in the *North American Review*, the oldest and one of the most widely read opinion magazines. Camden developed a position that adopted common opinions about the oil industry and turned them to Standard Oil's advantage. Thus, he argued that the early industry was marred by "prodigal waste," and he emphasized the "apparent instability of the whole business, which was hourly expected to vanish." Into this scene stepped Standard Oil as order giver to bring "firm and intelligent control" through its "genius for organization," though the company's management produced only "natural results."[69] Responding to its numerous critics, Camden claimed that Standard Oil was "the target for unlimited abuse and misrepresentation," largely through the activity of "unsuccessful oil-men, sensational writers, and persons with grievances." The principal problem of Standard Oil's competitors was that the company was selling kerosene so cheaply that inefficient operators could not compete. Masking their real difficulty, inefficiency, competitors raised what Camden called "the anti-Monopoly racket." In a rare flight of humor, he decried the tendency of adversaries to depict the company as the corrupter of everyone and everything—even of the railroads![70]

In rebuttal, John C. Welch simply went to Simon Sterne's account of the Hepburn committee investigations. Standard Oil conducted its business in secret, dominated the industry through control of transportation, manipulated the prices of raw and finished materials, and bought journalists and politicians to accomplish its dirty public work. Translated into traditional republican rhetoric, Standard Oil was secretive, overpowerful, and corrupt. The moral to be drawn from the story was the peril of money power, which would seize control of the entire country, through its "power in the press, in the national and state legislatures, in the courts, in official life, in political parties, in social ramifications, in literature, in the pulpit."[71] Once again, the sins of Standard Oil reached well beyond the petroleum industry; Standard Oil showed the peril of what lay ahead.

There was little new information in either article. Standard Oil's defense of its domination as the result of efficiency had been made before, and Welch's criticisms were familiar. But because they could exploit traditional ideas, critics such as Welch seem to have enjoyed more credibility than defenders such as Camden. Even such moderate newspapers as the *New York Times* routinely endorsed the most extreme claims of the company's local competitors and their allies. For example, after Standard Oil won a tax dispute with the State of Pennsylvania, the *Times* opined that the company "bent the legislature, the executives, and even the courts of the State of Pennsylvania to its will," describing it as "the most odious and grasping of all monopolies." It routinely referred to "the Standard Oil gang" along with conventional lists of the company's sins: according to the *Times*, "Bullying and bribery have been the favorite methods of the Standard Oil Company from the days when it first began to levy forced contributions upon the oil consuming world."[72]

Apart from republican rhetoric, what contributed additional credibility to the case against Standard Oil was the mounting volume of literature repeating charges against it. In *The Railways and the Republic*, a widely quoted work appearing in 1886, J. F. Hudson got all the anti–Standard Oil discourse between two covers. Drawing on his own interpretation of Alexander Cassatt's testimony in the Commonwealth suit, he argued that Standard Oil had received one hundred million dollars in rebates and drawbacks.[73] Since Hudson's amount was the estimated value of Standard Oil, it stood to reason, he argued, that none of the company's gains could be attributed to managerial skills. Hudson deftly pulled Standard Oil's most outspoken critics into his history, borrowing on Sterne, George Rice, Colonel Potts's *Brief History*, the prosecution in the Buffalo refinery case, the litigants in the *Scofield et al.* case, and the newspapers of the producing region.[74] He offered no new evidence. His rhetoric was also conventional, drawn largely from moral discourse: "These conspiracies are but illustrations of the corruption and unscrupulousness which marked the rise of the Standard Oil Company," which had thereby received "extortionate immense profits." The consequences of Standard Oil's corrupt alliance with the railroads went far beyond the oil industry, in Hudson's view: it left the country caught squarely between "the domination of a privileged class in its great corporations" and "an irresponsible and reckless proletariat." Unless the railroads and other large corporations were subjected to the restraints of law, particularly the proposed Interstate Commerce Act, America would be faced with "an outbreak of the destructive and ruinous spirit of revolution." There was much more at stake than George Rice and the price of kerosene.[75]

Hudson's book, published in New York City by Harper Brothers, one of the most successful book and magazine publishers, enjoyed widespread circulation and prompted economist Richard T. Ely to seize on it for material for a series of articles in *Harper's Magazine*. Ely began by identifying strongly with small producers, equating their interests with "economic liberty." He illustrated his major arguments with references to the South Improvement Company, drawn from Colonel Potts's *Brief History* and the report of the Hepburn committee, acknowledging his debt to Simon Sterne in strong terms: "The American people owe a debt of gratitude to Mr. Sterne for the ability, fearlessness, and self-sacrificing fidelity with which he conducted the difficult inquiry." Ely also leaned heavily on J. F. Hudson, by both attribution and style. Thus, he concluded his first article: "In short, the abuses which have given rise to the problem of the railway are germinal in character. They drag their slimy length over our country, and every turn in their progress is marked by a progeny of evils. Thus is our land cursed!"[76]

As this sample of Ely's prose shows, the language employed in the attacks on Standard Oil and in debate over trusts steadily conveyed heavily moralistic perspectives, advanced by businessmen, journalists, and scholars in defense of preindustrial market systems and of higher moral values, as one scholar put it, "not unlike the ones they had known in the small towns of their youth."[77] That discourse also reflects an interactive process in which businessmen, politicians, writers, and intellectuals took adversarial responses to growth on the part of Standard Oil and reshaped them into a cultural construction of the company. That so many disparate actors could emerge with a remarkably coherent cultural construction is explained by the homespun morality underlying their various responses as well as constant borrowing from one another. State lawmakers and judges, writers, some social scientists, and Standard Oil's competitors defended major assumptions of classical economics, especially the belief that monopoly would create its own competition in the medium term, if monopolists did not resort to unfair competitive practices. Thus, the concepts of "ruinous competition" and "predatory pricing," both heavily laden with moral judgment and strongly supportive of the business interests of Standard Oil's competitors, emerged as definitive. Standard Oil's practices sustained monopoly and were, thus, unfair. As reformers applied antitrust doctrine, dominance of a market was prima facie evidence of wrongdoing, and large concentrations of capital were taken as adequate evidence of the restraint of trade.[78] Moreover, in court and legislative investigative proceedings, the testimony of small producers and regional interests regarding alleged anticompetitive action enjoyed preferential status: their distress was evidence of "ruinous competition." This final point was critical to the argument against Standard Oil, because while economists might take market dominance as evidence of anticompetitive practices, federal courts required proof of aggressive behavior.[79]

As in Pennsylvania and New York, government action against Standard Oil first took place in the state arena, and other states readily joined in. In defense of their own small businessmen, Ohio and Texas, among other states, attacked vertical integration through quo warranto proceedings, alleging that Standard Oil and others had exceeded the terms of their charters by engaging in more than one segment of industrial activity or by merging with firms that did business outside the chartering state. Ohio attorney general David K. Watson led with an ultra vires suit against Standard Oil on May 8, 1890. His principal argument was that the charter of Standard of Ohio had not contained specific permission for that firm to combine with others outside the state. The Ohio supreme court's decision, rendered nearly two years later, ordered the Standard Oil Company of Ohio to withdraw from the Standard Oil Trust, without stipulation of a date by which it had to do so, but it did not revoke its charter. Compliance followed in less than one month, with the stock of companies in the trust transferred to the Standard Oil Company of New Jersey (SONJ); the stock of the remaining companies was divided into 972,500 parts — of which John D. Rockefeller held 256,785, corresponding to the number of trust certificates, with vesting of proportionate ownership in the twenty firms. Sixteen other large trust certificate holders also converted them to shares, but the failure of small holders to do so left the trust in a "state of perpetual liquidation."[80] The initial antitrust broadside against Standard Oil prompted reorganization, but it did not succeed in dismantling the industrial giant. The perpetually hostile *New York World* took the occasion of the settlement of the Ohio suit to remind its readers that "the grip of the Standard Octopus will still be laid upon the necessities, though the Standard name may cease to exist."[81]

While states took legal action against Standard Oil, antitrust legislation clearly aimed at the company's activities emerged at the federal level. The legislative action in Ohio, instigated by George Rice and other of Standard Oil's competitors, was part of a campaign that they also waged in Congress. In 1888, they dominated a federal investigation of trusts by the House Committee on Manufactures. Thomas W. Phillips, president of the Oil Producers' Protective Association of America; Lewis Emery Jr.; Joseph D. Potts; and George Rice testified. They also succeeded in inserting favorable extracts from their Pennsylvania court cases.[82]

One year after the House investigation, United States senator John Sherman of Ohio, under pressure from independent producers and refiners — led by George Rice in his home state and by oilman Lewis Emery Jr. in Pennsylvania — introduced the first federal antitrust legislation, in the form of a bill that barred both vertical and horizontal integration. Above all, Sherman's measure was crafted to give Standard Oil's opponents the right to sue in federal courts, as they had done in state courts.[83] He saw his bill through the Senate Finance Committee, which he chaired. However, Sherman, "no great lawyer or expert in legal and constitutional affairs," left what colleagues considered major defects in the bill, which, they argued, might lead to its rejection by federal courts. Consequently, the bill was referred to the Judiciary Committee and largely rewritten there by Senator George F. Edmunds of Vermont, the Senate's expert on constitutional law. Two weeks after referral, the bill reached the floor of the Senate containing the traditional common-law concepts regarding restraint of trade. With the revised Sherman bill on the floor, the Senate rejected a stronger bill by Senator John H. Reagan of Texas, patterned after a Texas statute. The Senate, and later the House of Representatives, thus rejected a specific and restrictive alternative in favor of a more flexible mode of barring restraint of trade; the underlying assumption of the legislation was that competition always existed in the absence of a monopoly and survival of competition meant the survival of small businessmen.[84] This position made it necessary to assume that monopoly could be achieved only through inherently unfair competitive practices: "Undesirable consequences result from pernicious conduct."[85] Grandly circular and deductive, this process of reasoning created the presumption of guilt whenever one producer dominated a market, making defense all but irrelevant.[86]

The breadth of the concepts of common law embedded in the federal statute left specifics, regarding asset transfer trusts, for example, in the hands of the state courts, thus accepting a second tier of antitrust legislation. What was broad and nebulous at the federal level would be defined more narrowly in state statutes. Though this was an ingenious solution to the constitutional problem of federalism and to the contention of various parties for control of the issue, it left major problems unresolved, to be clarified by federal courts because of significantly different theoretical approaches of the two interfaced levels of legislation. There was a limit to the extent that the ambiguity of the Sherman Antitrust Act might be used by state courts to defend local interests.[87]

If the legal effects of the Sherman Act were slow in achieving definition because of the act's broadness, its immediate consequences were almost certainly quite contrary to the expectations of its initiators. By barring loose associations, the act forced firms to merge, thus cutting the number of competitors. Rate pools, for example, were clearly illegal, but horizontal integration was not clearly so. It took five years for the first related case to reach the United States Supreme Court, during which time mergers proceeded apace. In the meantime, such antitrust enforcement as there was occurred at the state level. By the time of the Court's initial decision, most of the trusts had reorganized as holding companies, under enabling legislation passed by New Jersey and other states. By 1899, this form of organization was used successfully to evade ultra vires prosecutions by state courts, a situation that held until 1904, when the Supreme Court ruled against the Northern Securities Company, a holding company.[88] In effect the Sherman Act is the first example of federal legislation, drafted at least in part to affect the petroleum industry, that had an effect opposite that intended.

Conflicting attitudes toward large corporations held by federal regulatory officials meant federal activity got off to a slow start. In general, federal officials responded to ambiguity in the law by expanding their roles in its execution. Referring to antitrust, John Garfield, George Rice's erstwhile correspondent and head of the Corporation Commission in 1904, expressed the view of the liberal economists and the regulators: "It is aimed not at the restraint of combination as such, or the maintenance of competition, but at regulating the methods of competition." His successor extended the Progressive regulatory ideal: "We must recognize concentration, supervise it, and regulate it. We must do this positively, through an active federal agency, and not merely by the negative prohibition of penal law. We must have cooperation with corporate interests as far as possible. We must have, of course, effective penal laws against specific forms of unfair competition and the misuse of monopoly powers." Thus, the Progressive regulators reframed enforcement of the Sherman Act, "from an anti-trust act into an act relating to the legal control of competitive methods." The goal was not prohibition of big business but control of it.[89]

Litigators and legal theorists found a useful methodology to facilitate this redirection from common law to the regulative state in the emphasis on the case history approach promoted in German scholarly circles. The new method emphasized broad, holistic approaches, structured historically and supported by research that was more archival than legal. In large part, the new approach was promoted by scholars, primarily in economics and history; the new reform-oriented American Economics Association, for example, supported it aggressively, and it was championed by the U.S. Bureau of Corporations. The most significant aspect of this case history approach for the oil industry was that it made the development of an exemplary antitrust case against Standard Oil the most cogent way to support broader assertions about society and economics and to enhance the social standing of the academic experts who were trained in the new methodology.[90]

The development of theory and practice relating to antitrust, thus, tended to favor two specific groups that advocated it, the competitors of large corporations, like Standard Oil, and the new academics and regulators. The former were the immediate beneficiaries of the case method because they had already succeeded in shaping the historical record to support their interests. They also proved to be adept at using the concept of predatory pricing.[91] As George Rice discovered, when Standard Oil cut prices in local markets to maintain its position, commonly in the face of his entry and price-cutting, he could define its action as predatory, because Standard Oil did not simultaneously establish the same prices in all its markets. Indeed, so elastic was the concept of predatory pricing that it applied to either raised or lowered prices. Critics of Standard Oil could never agree as to whether high or low prices were more destructive. Following this price argument, high prices were acceptable in the interests of preserving competition, and low prices were predatory. Other critics, primarily in the Bureau of Corporations and its predecessor, the Corporation Commission, emphasized the social value of low prices and alleged that Standard Oil gouged consumers, a charge that stuck against the industry ever after. In the end, it was difficult to determine the political economy of prices.

Though the question of the impacts of prices was murky intellectually and the term "predatory pricing" reeked with "implications of unfairness," prices assumed a highly important place in public discourse, and various estimates of Standard Oil's profits, as contrived by competitors and adversaries, were always newsworthy.[92] Economists and attorneys who argued that Standard Oil and other large companies produced social benefits through mass-produced goods recast the debate over prices. Policy makers attempted to work through these contradictory positions by shifting the price issue into the academic and public policy discourse relating to public utilities. In this context, the most important question was whether the market dominators were achieving an unreasonably high return on investment. Though this fixation on "fair return" was ultimately fruitless in terms of legal and regulatory effect, as Thomas K. McCraw has observed, it gave economists an important role in the public debate.[93] As the masters of quantitative techniques and the new case study method, they assumed prominence in the discourse relating to public policy on antitrust and the operation of that policy.

Standard Oil's competition regrouped in 1892, when Lewis Emery Jr., Thomas W. Phillips, and associates formed the Producers' and Refiners' Oil Company. Now most of the Pennsylvania independents of significant size were in a single corporation. Their attorney, Roger Sherman, advertised their continued opposition to Standard Oil in "The Standard Oil Trust: The Gospel of Greed," an article published in New York by the *Forum* in July 1892. Among other complaints, Sherman objected to the "backward" integration of Standard Oil by paying preemptive prices for producing oil leases in Indiana, Ohio, and West Virginia. He also rehearsed the independents' version of the history of the South Improvement Company, bringing Simon Sterne in as a supporter of his position and referring to the legislative investigations that independents had controlled. But, above all, he advanced a moral condemnation of Standard Oil. In a phrase that he made memorable, Sherman wrote: "It [Standard Oil] was founded in injustice and built up by an enormous wrong secretly done." Accordingly, Sherman drew battle lines on moral — and gendered — terms: "Let the spirit of manhood, moral sentiment, and religious conviction unite with the active business interests of the land in protest against the gospel of greed."[94]

*Believing the Worst*

Most of the major elements of discourse relating to the petroleum industry and Standard Oil were firmly established in public discourse by 1890. Once the competitors of the Standard Oil interests developed the theme of monopoly as a defensive and competitive device, the manipulation of discourse to affect public policies thus became a central element of business strategy for the remainder of the nineteenth century and well into the twentieth. Major themes in public discourse included the moral superiority of small businessmen, the predatory nature of big businessmen, and the erosion of democracy, civic virtue, and social harmony by new concentrations of wealth and power. Writers continued to target Standard Oil as the most significant example of the new perils to America, by repeating the well-known tales of its malfeasance, thus enhancing the credibility of their accusations, through repetition, and by borrowing material from one another to expand their stories. As clamor mounted over Standard Oil, it became politically vulnerable, a liability that would ultimately lead to its dissolution in 1911.

A combination of traditional attitudes, class interest, and conservative moral crusading forms the core of *Wealth against Commonwealth*, the most widely read book-length attack on Standard Oil. The heart of Henry Demarest Lloyd's book, appearing in 1894, consists of half a dozen chapters devoted to George Rice, drawn from Rice's pleadings in court and before the Interstate Commerce Commission; other materials supplied by Rice and Roger Sherman; the Buffalo refinery case; and one of the favorite devices of the mass press, the cheated widow story. From these elements Lloyd put together an extended, essentially moral condemnation of Standard Oil and John D. Rockefeller, going well beyond his 1881 *Atlantic Monthly* article.

The widow Backus story, which Lloyd took from the *Scofield* suit of 1880, is perhaps the best example of Lloyd's construction of a moral indictment of Rockefeller. With it Lloyd presented a verbal portrait of Rockefeller as a cynical hypocrite, a churchman who consistently "does unto others." The widow was "weak with grief" when Standard Oil purchased her refinery, but she was obsessed with the "gallant task of paying her husband's debts" and maintaining her family. Rockefeller, in turn, was the conniving deacon, a man lacking in all chivalrous feeling, and hence in manliness, who took advantage of her distraction to pay a small fraction of the real worth of her plant. In peroration, Lloyd quoted her missive to Rockefeller, condemning "professing Christians [who] do as you have done by me." After the tale of the widow Backus, no one could conclude that Rockefeller was a good man, or even that he operated anywhere within the confines of generally recognized ethics. And, as one of its founders, he stood for Standard Oil.[1]

Beefing up his recitation of Standard Oil's iniquities, Lloyd gave three chapters over to the Buffalo refinery case. Here Lloyd aimed to disclose the menace of Standard Oil's power and the lack of moral constraint in its activity. As is appropriate to moral tales, he developed sharp contrasts between contending sides. Describing the refinery partners as "capable men [who] showed great business sense in their arrangements," Lloyd even brought the virtuous "thrifty wife" into the story, as one who helped her husband save several thousand dollars to launch the venture as well as the cheerful guardian of the hearth who tried to keep her husband from drink. He thus enlisted domesticity into the battle against Standard Oil in a conventional gendered context. On Vacuum's side, he depicted managers who were so inept as to confess their planned arson to an opposing attorney in advance, and he even had one of them tell the Buffalo refining partners, "We have ways of making money you know nothing about." The famous phrase, originally attributed to John D. Rockefeller in the Hepburn committee hearings, was now borrowed for further use. Lloyd even made playful note of the repetition by observing that the Standard Oil man was "using singularly enough, the phraseology employed by a greater man in the interview with another would-be competitor," thereby using the apparent implausibility of the repetition to enhance his tale. For the most part, Lloyd's material came from the opening speech of the prosecutor of the case, supplemented by the testimony of the Buffalo refinery partners as reported in the mass press; he borrowed heavily from the *New York World* for interpretation of various aspects of the trial. Lloyd's notion that the judicial process misfired when the Vacuum managers were convicted only of hiring away a Buffalo refinery employee and not of attempting to sabotage their installation, for example, picked up the *World*'s observation that the ruling was handed down because the judge was "unfit to be a judge." In short, virtue was overcome by Standard Oil and allied forces of corruption.[2]

*Wealth against Commonwealth*, though focused tightly on Standard Oil, established common ground with reformers and the "yellow" press with its powerful exposure of the immorality of the first and most dangerous trust, Standard Oil. According to Lloyd, Standard Oil was not an isolated case: "Monopoly anywhere must be monopoly everywhere." To support his notion that Standard Oil's financial success had created a massive menace to competition and liberty, Lloyd appended an eight-page list of "Combinations or Trusts."[3] By broadening the issue to trusts, Lloyd obtained support from a widely varied constituency, including liberal economists, populists, moral reformers, and the new mass-circulation newspapers. Certainly the *New York World* approved. Not only had Lloyd borrowed its perspectives, but its publisher, Joseph Pulitzer, had already singled out John D. Rockefeller for his "genius at money getting" and consistently carried the octopus image of Standard Oil, the *World*'s centerpiece in its attack on trusts.[4]

Like the later muckrakers, Lloyd buttressed the credibility of his judgments with apparently accurate evidence, which he "quarried out of official records." In doing so, he commonly assumed that claims against Standard Oil were true if the company did not contest them at the time, though there had usually been no opportunity for Standard Oil's rebuttal or cross-examination in the proceedings. Moreover, his assumption that when Standard Oil won in court it was through bribery or judicial error drew on the reformers' belief in the pervasiveness of corruption and corporate bribery. Lloyd liked one device that numerous muckrakers would follow—the representative biographical sketch. In addition to George Rice, he profiled Franklin B. Gowen, offering a dramatic account of his death in Washington during a fight with Standard Oil. Similarly, Lloyd described Samuel Van Syckle, a rival pipeliner, as "one of the type of country-bred, hard-working American manhood of the last generation. . . . His clear gray-blue eyes, tall strong frame, firm mouth, large features and limbs, eager face, fit the facts of his career." But American manhood was in peril because, as Lloyd told it, Van Syckle's career ended tragically when he was forced out of business by Standard Oil.[5] In short, for Lloyd, as for the muckrakers, the world was filled with heroes and villains, and the villains, for all stalwart men fought valiantly, gave no quarter.

Lloyd's work drew a mixed reaction. Since his book fitted perfectly with the views presented in the mass press, newspapers tended to applaud it. B. O. Flowers, editor of the Populist *Arena*, saluted it for its "masterly and conclusive expositions of the menace of corporate greed to a republic."[6] *Outlook*, the

*New England Magazine*, the *Review of Reviews*, and Congregational religious publications tended to support this generous assessment.[7] Moral reformers found much to admire in his work. William D. P. Bliss, a Christian Socialist and editor of *The Encyclopedia of Social Reform*, characterized Lloyd's writings as "intensely moral" and observed that Lloyd and his wife lived "for the cause of the sorrowing and the oppressed."[8] Bliss's encyclopedia drew its entry for "Standard Oil Monopoly" from Lloyd's book as well as from M. W. Howard's *The American Plutocracy*.[9] George D. Herron, the leading Congregationalist Christian Socialist, blamed the current depression, the Homestead strike, and all industrial violence on "the centralization of wealth . . . in the hands of the cunning and the strong." John D. Rockefeller and the other vastly rich industrialists were the current agents of evil: "Commercial tyranny and social caste are at war against God."[10]

Standard Oil's competitors, with whom Lloyd invested,[11] also found the book to be highly acceptable and for good reason: Roger Sherman contributed data and sources to the work and scrutinized working drafts of the manuscript. When the work appeared in print, he wrote Lloyd, "Your book should be the *Uncle Tom's Cabin* of this Era, and I pray that it may be." Lewis Emery Jr., Sherman's client, bought one hundred copies and distributed them to congressmen and other "influential Americans."[12] Though Sherman's hope for Lloyd's book was unfulfilled, his comparison of it to Harriet Beecher Stowe's novel was apt: like Lloyd, Stowe successfully avoided relatively complex issues, in that instance of federalism–states' rights and the economics of slavery, by advancing an ethical argument, ably supported by the conventions of the sentimental novel. *Wealth against Commonwealth* was no novel, but it certainly aimed to rouse sentiment with its moral tales of Standard Oil's wickedness. Lloyd's skillful development of moral discourse similarly skirted issues relating to the economies of mass production and the equity and efficacy of legal limitations on corporate conduct by focusing on the allegedly immoral behavior of John D. Rockefeller and his associates. Both authors reduced complex public issues to moral discourse any citizen could understand.

Still, Lloyd's unbridled vehemence about Standard Oil offended some reviewers. The *Nation*, aghast at Lloyd's socialism, described the volume as "over 500 octavo pages of the wildest rant," based on "questionable evidence . . . calculated to arouse incredulity in the mind of any reader who understands the nature of evidence." The new professional journals either ignored Lloyd's book or dismissed it as unscholarly. In the second edition of *Trusts: or Industrial Combinations and Coalitions in the United States*, Ernest von Halle dismissed it as "an acrimonious pamphlet," with no attempt "to present the facts on both sides."[13] In many instances, however, economists who found abundant fault with Lloyd's book continued to advance its circulation; Richard T.

Ely, E. Benjamin Andrews, and John Bates Clark, among others, recommended the book to their students.[14] Thereafter, it enjoyed an immortality on college and university reading lists that belied its professional assessment.

Lloyd's work found ready adaptation in Alabama congressman M. W. Howard's *The American Plutocracy*, published the following year. In common with many observers, Howard identified Standard Oil as the real cause of the trust problem because it demonstrated the market control that could be established through trusts. The result of Standard Oil's success as a monopolist was the appearance of trusts in most aspects of American business. To support his allegation, Howard appended a twenty-page list that included some well-known trusts, such as oil, and some that were not matters of common knowledge, including candles, honey, vitriol, fish oil, pitch, whips, doors, snaths, buttons, chopping bowls, sanitary ware, eggs, butter dishes, piano covers, castor oil, ergot, job printing, and athletic clubs.[15] Standard Oil also "sowed the seeds of political and legislative corruption which have germinated and form such a fruitful crop" and was responsible for the spread of urban poverty. Money power, as established by Standard Oil, had grown to the point that it controlled most of the American press—except the *New York World*, from which Howard quoted freely. Carrying on an intra–Democratic Party feud, Howard charged, "The greatest tools of plutocracy in this country today are Grover Cleveland, President of the United States, and his trust champion, Attorney-General Richard Olney."[16]

One year after Lloyd's book was released, F. F. Murray, an editor in Titusville, Pennsylvania, also made considerable use of it to broaden the relevance of the crusade against Standard Oil. In his *The Middle Ten*, he identified Lewis Emery Jr.'s Producers' and Refiners' Oil Company with the interests of the American middle class, which was caught in between "the modern pirates of the land and sea—the monopolies and trusts" and the "Lower Ten: the class seemingly imbued with the conviction of its having been created especially for the purpose of sticking closely to the beer centers, loafing, breeding and railing at all existing governments and institutions." Producers' and Refiners' represented "the faithful remnant" of America's besieged middle class. That remnant should defend itself, Murray argued, by organizing consumer boycotts of Standard Oil and of the newspapers and journals that supported it. Picking up on George Rice's pamphlet, the author reminded his readers of the "turn another screw" episode, in the song that concluded the book: "And so long as man thinks little as now of his fellow worm of the dust, Ourselves or others will turn the screws, And sing the Song of the Trust."[17]

The linkage Lloyd and other writers made with Standard Oil's competitors was significant to the development of the case against the trust, but the conventional discourse of academic economics, focused by public concern over trusts, also reinforced the more popular arguments. In 1897, for example, Arthur T. Hadley linked academic and public discourse through application of moral discourse in "The Good and Evil of Industrial Combination." Moreover, he advanced the case of the anti–Standard Oil men by offering a technical definition of "predatory pricing," heretofore taken as synonymous with selling for less than a competitor. In a refinement of the conventional definition, Hadley argued that predatory pricing amounted to charging a price that covered variable costs but not fixed costs, a distinction that was increasingly accepted in business circles. Presumably, Hadley's definition could be used to identify the practice precisely, once Standard Oil and its competitors made the relevant data on costs available. Hadley's definition reduced social and economic problems to manageable dimensions by making solution an exercise in the gathering and analysis of data. He thereby made it the domain of the academic economist, extending professional hegemony over the major issue of the time.[18]

Following Hadley's argument, other influential economists advanced related theories. Edward W. Bemis accepted the common definition of predatory business activity, which he labeled "evil" and "culpable," but he also recognized the centrality of fixed costs in the setting of prices and in competitive relationships, arguing that by 1899 capital had become a principal barrier to the entry of new competitors in capital-intense industries. Bemis's remedy for predatory pricing seemed simple: if a trust cut a price in one locale, it would be required by law to charge that price everywhere it operated — an approach that would have enabled George Rice to have put Standard Oil out of business.[19] All the discussion occurred in a vacuum because there was no reliable data on pricing and consumer behavior. During the following decade, Harold Hotelling, of Stanford University, discovered that buyers rarely changed stores even when prices fluctuated, that detected shifts were gradual, and that vendors tended to enjoy local and regional monopolies for the most part: "The difference between the Standard Oil Company in its prime and the little corner grocery is quantitative rather than qualitative."[20]

While academics theorized on predatory pricing and its presumed impact on Standard Oil's competitors, George Rice continued to harass the company. Rice bombarded federal officials with anti–Standard Oil correspondence, urging U.S. Attorney General John W. Gruppe to bring suit in 1898 against "the most gigantic and unlawful combination the world has known." He followed up with a sixteen-page reminder and a twenty-page complaint of inactivity. Thereafter, he sent seven more letters during the McKinley administration, including one to the president.[21] During 1898 Rice instigated another investigation of Standard Oil, this one by the Ohio Senate. Appearing as the second witness, Rice read his testimony to the Fiftieth U.S. Congress investigation into the Ohio record, recapitulated his Interstate Commerce Commission (ICC) testimony, and reiterated his allegation that he was being driven out of business by the continuation of railroad rebates to Standard Oil. Finally, Rice played the theme that appealed to the anti–Standard Oil press, that the company had raised the prices of petroleum products from 50 to 100 percent above what they would be in the presence of strong competition.[22] Not all yellow press readers might grasp monopoly as a concept, but they could understand that prices seemed too high.

Rice was followed on the stand by John Teagle, a Cleveland refiner, who repeated his earlier courtroom allegations that Standard Oil tried to drive him out of business by competing for sales of cheap grades of kerosene and that it tried to bribe his clerk to obtain production data. An associate, B. W. Browne, claimed that Scofield, Schurmer, and Teagle could no longer compete with Standard Oil in Iowa because Standard Oil gained a three-fourths of one cent per gallon advantage from the proximity of its Whiting refinery to the market, because it had vast bulk storage facilities, and because it had somehow forced railroads to raise rates for competitors from Cleveland to Illinois, despite the ICC.[23] Peter Shull complained that Standard Oil got customers to cancel orders by offering lower prices. Finally, Mrs. G. C. Butts, George Rice's daughter, claimed that her company "would have built up a large and successful industry" were it not for Standard Oil's secret rebates. In short, the same claims were advanced by many of the same people, to the same effect: repetition established veracity.[24] Newspaper coverage of the hearings was extensive, especially in New York City. The *New York Daily Tribune*, not recently hostile to Standard Oil, ran more than a dozen stories on the investigation and a subsequent lawsuit of the State of Ohio against Standard Oil, as did the *New York Times*, the *World*, and the *American*.[25]

Shortly after the Ohio investigation was concluded, Standard Oil submitted to another investigation, this one conducted by the United States Industrial Commission, as authorized by Congress in June 1898. The commission, a mixture of members of Congress, academic experts, and members of the public, was dominated by Standard Oil's opponents. Its vice chairman, former congressman Thomas W. Phillips of Pennsylvania, was an officer of the Pure Oil Corporation; and in the usual absence of the chairman, Phillips controlled proceedings. Control them he did: the main witnesses were from Pure, except for Ohio attorney general Frank S. Monnett. James W. Lee and Lewis Emery Jr. of Pure testified, along with George Rice and Theodore Westgate.[26] Henry Demarest Lloyd appeared to read a part of *Wealth against Commonwealth* into the record. Marcus L. Lockwood repeated his testimony to the Hepburn committee from 1879; he also read a part of Lloyd's book into the record. Lockwood went on to claim that Standard Oil had bribed the judge in the

Buffalo refinery case, an unsupported allegation to which three committee members objected but that Phillips insisted on including in the record. Roger Sherman entered his version of the undocumented testimony of Alexander Cassatt into the record, as he had done in 1888 before a U.S. House committee. The effect of this sleight of hand was that his interpretation of Cassatt's testimony was read into the record as if it had been Cassatt's testimony to Congress in 1888.[27] Phillips, who controlled the writing of the report, edited his own testimony with that of Emery, Lee, Monnett, and Rice to make them seem more reasonable. The report, not surprisingly, obscured significant manipulation of the record and proceedings. Thus, Lee supported his claims regarding the number of refineries that Standard Oil had closed by referring generically to the 1888 investigation, though he was actually referring to Lewis Emery Jr.'s testimony. Once again, Standard Oil's opponents controlled the record and, through it, discourse.[28]

While the Industrial Commission was still newsworthy, Standard Oil was hauled into the court of public opinion again, this time by Ohio attorney general Monnett. Monnett was, in fact, familiar with the attacks on Standard Oil before he entered office: his father was president of Bucyrus Gas, which competed with a Standard Oil subsidiary in Ohio. He entered the lists against Standard Oil when George Rice brought him charges that Standard Oil had failed to comply with the 1892 Ohio decision that mandated the dissolution of the Standard Oil Trust. Using new enabling legislation, Monnett filed quo warranto proceedings in November 1898 and in January 1899. The related investigations, carried on in New York, where John D. Rockefeller and other Standard Oil executives testified, made front-page news and reinforced the charges of the company's competitors. When Standard Oil refused to open all its records to the Ohio investigators, the *New York World* depicted Rockefeller as a clam. After proceedings had dragged along without progress for some days, Monnett claimed that Standard Oil had incinerated incriminating documents; his source was a Bohemian rabbi who claimed to have overheard an Irish teamster in a Manhattan barroom.[29] Coverage of the trial and testimony was extensive and, in New York, anti–Standard Oil.[30] While the Ohio Supreme Court was considering Standard Oil's appeal of lower-court verdicts in the Monnett cases, Monnett, now out of office, broadened his attack by claiming that the company had tried to bribe his successor.[31] Though this charge against Standard Oil was dismissed for lack of competent evidence, Monnett continued to repeat the allegation in articles and in personal appearances sponsored by Standard Oil's opponents and adversaries. Repetition accomplished what "the incorruptible Attorney General of Ohio" was unable to do in court.[32]

Just as Standard Oil and its success raised a variety of problems relating to the role and scope of big business in the United States, the public persona of

John D. Rockefeller projected a variety of contradictory values. As Howard Horowitz has reminded readers recently, John D. Rockefeller might have been seen as the personification of nineteenth-century individualism and Standard Oil as its reification. Writing in 1883, Henry James restated the Emersonian creed: "We measure the greatness of an individual by the push he gave to what he undertook." Had he considered John D. Rockefeller, James might well have added: "and by the enemies he made." Rockefeller's competitors and other opponents transformed him from potential hero to ominous villain largely by translating business rivalries into moral confrontations. As Horowitz notes, Ida M. Tarbell and others "demystified" Rockefeller's standing and achievement by emphasizing his presumed moral flaws. These flaws—materialism, cheating, and close dealing—offset his admittedly admirable characteristics, diligence and intelligence.[33]

This perception of Rockefeller was significant for the entire petroleum industry because both Standard Oil and the industry were increasingly seen, in Emerson's phrase, as "the lengthened shadow of one man." As the *Nation* stated the situation, because other leaders of Standard Oil were not often visible to the public, "Standard Oil and Rockefeller have become synonymous." Admittedly, some of the attacks on Rockefeller seem at first glance to be narrowly targeted, hardly implying a judgment on the industry. Thus, a popular early-twentieth-century play, *The Vanderbilt Cup*, depicted him as "a man who has one eye on the money market and the other on a failing digestion." Common depictions of Rockefeller in cartoons as an avaricious, dyspeptic, and hairless old fogy would also seem to be largely ad hominum. On closer examination, however, these images conveyed additional information.[34] Depictions of Rockefeller's appearance tended to make him both unmanly and anomalous. Bald and abnormally gaunt, in an age of stout manly girth, shown with near skeletal limbs, Rockefeller contrasted with the current image of virile manhood. Recurrent references to his status as America's first billionaire reinforced the anomaly.[35] This approach, as cultural anthropologists have pointed out, is commonly used to reinforce threatened moral codes.

Condemnation through anomalization was applied even more viciously to his son, John D. Rockefeller Jr. The *Baltimore Sun* described him as "a religious fanatic with a weak stomach and an indifferent set of brains," "a semi-invalid." The *New York American* frequently described him as sickly and unstable.[36] Like physical weakness, the younger Rockefeller's commitment to religion was used by journalists to detract from manliness. Such a linkage of piety and effeminacy may reflect a variety of attitudes. As Barbara Welter has pointed out, piety was one of the most important virtues of the ideal nineteenth-century woman.[37] But piety was also identified by the religiously skeptical with hypocrisy when its practice was obtrusive; the working press tended to be cynical about religiously involved people. As for the clergy, whose piety was a demand of profession, their status had diminished and with it their image of manly calling.[38] In any event, the popular press made much of John Jr.'s unmanly religion. The *New York World* thus told readers that his Bible class was to be served a "pink tea" by the ladies of the church; so much for "muscular Christianity"! The Pulitzer paper also reported that John Jr. was thin skinned. After a professor at Union Theological Seminary criticized him for praising Joseph's cornering of the grain market, he was supposed to have stayed home and pouted, skipping his class.[39] In *The Vanderbilt Cup*, he was lampooned as a "goody-goody."[40] His wealth was also described in anomalous terms: thus, he was wealthier than everyone in Oregon, Nevada, and California, according to Hearst's *New York Evening Journal*; his infant son, with a net worth of three hundred million dollars, could pay the wages of 560,000 workmen for one year, as the *Brooklyn Eagle* pointed out on a front-page article.[41] The adversarial press was also fond of stories that emphasized that John Jr., with his hundreds of millions, was parsimonious. Hearst's New York paper favored stories about John Jr. beating down railroad porters and delivery boys on their tips. Here he failed to perform a basic duty entailed by riches, to be generous and open handed.[42]

In short, neither John D. nor John Jr. qualified as a Christian gentleman, a role that required generosity and empathy; neither embodied "manliness with muscles," a nonreligious variant on the gender model. Thus, gender signals strengthened the classification of anomaly and served as a correlative support to the moral code and those whose interests were served by it. Such a use of gender to separate someone from his fellow businessmen and make him a moral pariah was not new with late-nineteenth-century journalists. As Toby Ditz has demonstrated, eighteenth-century Philadelphia merchants described failed or unscrupulous colleagues as "feminized or ambiguously gendered figures," the "negative counterparts" of manly, honest traders. More to the point, Ditz has shown that giving a man ambiguous gender underscored his position as an outcast. That was precisely what critics of the Rockefellers did.[43]

But what happened when the Rockefellers were generous, when they gave millions away? The various benefactions of the Rockefellers, to the University of Chicago, the Rockefeller Institute, and religious bodies, were all used against them, often opportunistically, sometimes idealistically, and usually incorrectly. Illinois governor John Peter Altgeld, a friend of Henry Demarest Lloyd's, used rumors about the gifts to the University of Chicago, "the Rockefeller University," to secure sizable appropriations for the University of Illinois. It was rumored at the end of the nineteenth century that John D. Rockefeller, having given to Brown University, prevailed on its trustees to fire E. Benjamin Andrews, the university's president, and used his influence to undermine professors E. W. Bemis at the University of Chicago and John R. Commons at Syracuse University because of their criticism of Standard Oil in particular and trusts in general.[44]

In some measure, the financial success of Standard Oil created the preconditions for renewed attacks on it and its founder. Its profitability—a matter of contention for more than twenty years—made it conspicuous, especially during the financial downturns of the 1890s and the first decade of the twentieth century. Moreover, when its leading stockholders regularly used Standard Oil dividends to invest outside the company, they also attracted unfavorable notice, lending apparent credibility to the "money power" theme in the press. In 1900, for example, the *New York World* publicized the acquisition of the Carnegie Trust by a group of Standard Oil directors. Pulitzer's newspaper claimed that the Standard Oil group controlled bank deposits equal to one-fifth of the U.S. money in circulation, that it had taken over the Bank of Hong Kong, and that it had thereafter taken over banking in former Spanish possessions. The *New York Tribune* opposed the creation of a national bank on the grounds that Standard Oil would control it if one were created.[45] Not to be outdone, the *New York Journal* carried frequent stories about the attempt by the Standard Oil group to control world copper, at the expense of consumers. Its companion Hearst newspaper, the *New York American*, was also fond of plots and money power stories. Thus it claimed that the Rockefellers and J. P. Morgan controlled prices on Wall Street and that the Rockefellers, Nobels, and Rothschilds had formed an international oil trust. The latter story was headed by a photo of a cadaverous John D. Rockefeller, who looked for all the world like a sinister death's head.[46]

To counter such attacks during the 1890s, following a conventional practice of the age, Standard Oil hired a literary bureau as well as journalists to place favorable stories in newspapers. It paid the commercial rate for the space, though sponsorship was not indicated. Samuel C. T. Dodd, Standard Oil's attorney, burst into print in the 1880s and 1890s with his defense of the trust.[47] Moreover, Standard Oil, like its critics and competitors, also invested in newspapers in Cleveland and other places, presumably to obtain favorable notice. In other instances, it subsidized periodicals through overpayment of subscriptions. Such was the case with George Gunton's economics journal.[48] In 1903, as Ida Tarbell's articles were appearing, the company supported Gilbert Holland Montague's *History of Standard Oil*, a portrait with no blemishes. The directors of Standard Oil were aware of the need to improve their image, but they were outnumbered and outgunned. In 1906, in an effort to mount a better-organized campaign, Standard Oil hired J. I. C. Clark as its press agent. Clark had been editor of the *New York Herald*, which was relatively uncritical of the company. The rival *New York World* claimed that Clark's job was to "try to change in some degree the practically unanimous newspaper denunciation of the Oil Trust throughout the country." The *New York Times* reported that Clark would be paid twenty thousand dollars per annum, a princely salary for a working journalist, and added, "That [Standard] should pay any such amount to gain publicity is almost staggering." Both the *World* and the *Times* saw the hiring as an admission "that its policy of silence was ineffective," as the *Times* put it. Quotation of Clark's salary was also an additional reminder of Standard Oil's immense wealth.[49]

No doubt one of the important considerations in the hiring of Clark was the hope of offsetting the highly negative press campaign of the first decade of the twentieth century. Following the numerous shots from politicians, editors, economists, and competitors, Standard Oil took a broadside attack in the pages of *McClure's Magazine*, beginning in November 1902, when Ida Tarbell's series of articles began; the bombardment continued through the final chapter, which was published in October 1904 and was amplified later in that year when the articles were published in book form.

Tarbell's earlier efforts, biographies of Napoleon and Abraham Lincoln, had given the magazine much-needed boosts in circulation, and publisher Frank McClure looked for the same outcome from the series on Standard Oil.[50] Tarbell set about the daunting task of chronicling Standard Oil with characteristic industry. She reread Henry Demarest Lloyd's history and began an extensive correspondence with him; worked through pages of the *New York World*; gathered up the various histories written by Standard Oil's adversaries in the Oil Region; traveled through Pennsylvania collecting affidavits from attorneys who had represented Standard Oil's enemies; and, finally, worked through the collection of congressional hearings, ICC cases, and Corporation Commission reports in the Library of Congress. While she was in the capital city, she renewed her friendship with Senator George Frisbie Hoar, fierce antitrust advocate and outspoken critic of Standard Oil. As she conducted research and wrote up her findings, she conferred frequently with other *McClure's* writers, including Lincoln Steffens and Ray Stannard Baker, and she hired John R. Commons, labor historian and student of Richard T. Ely, to read her manuscript. She often conferred with her brother Will, an officer of the Pure Oil Company, who kept alive her father's hatred for Standard Oil and perpetuated the family myth, that Rockefeller had ruined Frank Tarbell. She also conducted extensive interviews with H. H. Rogers, a Standard Oil director, whom she met through Mark Twain, but she did not get to see John D. Rockefeller, who sustained his well-known reputation for reluctance to meet the press.[51]

When the articles began to appear in the early months of 1902, public reaction was quick and favorable. As David Freeman Hawke, a recent biographer of John D. Rockefeller, described it: "Her audience swelled with each month's chapter, and soon a large segment of America was awaiting the next installment." When the book appeared, in 1904, it was "an immediate success."[52] Hawke correctly attributed its success to timing—appearing at the end of two decades of highly publicized attacks, numerous lawsuits, and a series of highly publicized investigations of the company.[53] Peter Lyon, historian of *McClure's*, describes public reception even more strongly: "Month after month Miss Tarbell's history was studded with . . . evidence of dishonesty and her readers followed her with the absorption they would ordinarily have accorded the most suspenseful detective story."[54] McClure's expectations were fully realized as circulation grew with each of Tarbell's Standard Oil articles.[55]

The preface to the *History of the Standard Oil Company* clearly establishes the theme that McClure set for Tarbell, to expose the domination of Standard Oil "in dealing with other industries and political institutions," as a way of illustrating the process of trust formation. Tarbell's assignment embodied the assumption, a commonplace by 1902, that Standard Oil set both the legal form of the trust and the methods by which all trusts operated. From the beginning, the articles and book were also intended to be more than mere exposé; they were intended to seem meticulous and impartial. Thus Tarbell emphasized that she worked from "trustworthy documents," including nineteen volumes of investigations under McKinley's administration alone.[56]

With all her apparent impartiality, Tarbell demonstrated two basic truths familiar to historiographers: that the way a historian tells a story expresses a point of view and that every historian has a point of view. Tarbell's was profoundly shaped by family history; her father had gone from farming and teaching school in Iowa to building tanks in the Pennsylvania oil fields. He prospered for a time and then failed, a painful experience for him and his family. Were she to come to terms with her father's past, Tarbell could see him as inept and misguided in business or as caught up in broader developments beyond his control, developments presumably affecting hundreds of small businessmen like him. Instead, as she would indicate in her later autobiography, *All in the Day's Work*, she saw her father as one of the many victims of "the great oil trust" that "turned the men of the Oil Region into hired men" and that was responsible for "bitterness and unhappiness and incalculable ethical deterioration for the country at large." No doubt Tarbell had a personal ax to grind, and she would succeed in wielding it against Standard Oil.[57]

Mixing fantasy with material from Wright's *Petrolia*, Tarbell began her history with the oil industry before Standard Oil, a world with thousands of independent producers and refiners. Her small refiners were quick to apply the latest technology to their operations; her romanticized wildcatters "loved the game, and every man of them would stake his last dollar on the chance of striking oil." Here were sturdy, self-reliant republican yeomen in the oil fields, and "life ran swift and ruddy and joyous in these men." Tarbell presented an orderly industry, with careful and systematic development of production, pipelines, and refineries. The Oil Region before Standard Oil amounted to a petrolial Eden. It had its seekers for quick riches, its overproduction, its fights between pipeline owners and producers, its speculation, bogus stocks, and swindles, even its rebates to producers from railroads — "the open rate was enforced only on the innocent and the weak" — but it was still a wonderful world dominated by pioneers. And these pioneers "came in shoals, young, vigorous, resourceful, indifferent to difficulties, greedy for a chance, and with each year they forced more light and wealth from the new product."[58]

Thus, with a mixture of melodrama and morality typical of muckraking style, Tarbell set the stage for what was to come; into this Eden a serpent would appear. Using a novelist's page-turning technique, she ended her first chapter by putting her happy independents in danger: "But suddenly . . . a big hand reached out from nobody knew where, to steal their conquest and throttle their future. The suddenness and blackness of the assault on their business stirred to the bottom of their manhood and their sense of fair play, and the whole region arose in revolt which is scarcely paralleled in the commercial history of the United States." Tarbell then shifted the scene from the Oil Region to the refineries of Cleveland, with their "heavy and evil-smelling burdens." She thus took the reader from Eden to Hades and coyly introduced her devil, his presence being discovered as his "chief competitors began to suspect something." Suspected but did not know: like any talented writer of melodrama, Tarbell did not want to reveal her plot — or that of her villain — all at once.[59]

Tarbell's John D. Rockefeller offered a striking contrast to her open, manly, venturesome oil producers. Thus, he was "too pious," "a man to suspect or fear," "a secretive man," "brooding, cautious, secretive," and lacking "a sense of humor." Overall, Tarbell gave Rockefeller an assortment of feminine qualities: piety, secrecy, timidity. He was "low-voiced, soft-footed, humble," a man of "vast patience" and "placid demeanor," a self-effacing man who, at one gathering, "sat in a rocking-chair, softly swinging back and forth, his hands over his face." Tarbell created a Rockefeller who was a womanly man, and she underscored his anomaly and moral unacceptability.[60]

Not content to draw Rockefeller as an effeminate social outsider, Tarbell added to her portrait details emphasizing his immorality and his power, ending with a near demonic figure and reference to "the cloven foot." Putting him in a class with Jay Gould and Jim Fisk, she made him supremely avaricious, never tiring "until he got his wares at the lowest possible figure." In the end, his characteristics inspired "a terrible popular dread"; men began to dread him and to invest him with mysterious qualities. Oilmen were up against "a power verging on the super human — a power carrying concealed weapons, fighting in the dark, and endowed with an altogether diabolical cleverness." By demonizing Rockefeller, Tarbell explained how he succeeded when others failed.[61]

When Tarbell was not vesting Rockefeller with near supernatural powers, she saw him in terms of secular might. Drawing from her earlier biographical work, she likened him to Napoleon. Thus, she visualized "Mr. Rockefeller again bent over a map of the refining interests of the United States. Here was a world he sighed to conquer . . . [like] the first Napoleon." In her concluding chapter, she returned to the theme: "He saw strategic points like a Napoleon, and he swooped on them with the suddenness of a Napoleon."[62] Though the metaphor was not original to Tarbell, having been employed by Jeremiah Jenks in an earlier book, she used it more dramatically and effectively.[63]

Casting Rockefeller as Napoleon made for dramatic unity in Tarbell's long work, but it swore at the reality of committee governance at Standard Oil, which was noted repeatedly in the cases and investigations she consulted and which she acknowledged. With no discomfort at apparent inconsistency, in other sections of her work and especially the final chapter, "The Legitimate Greatness of Standard Oil," she credited Standard Oil's superiority to "energy . . . intelligence . . . dauntless," "minute economies," and "the men who formed it." Rockefeller the lone titan was more exciting a subject than Rockefeller as chairman of a board.[64]

Using literary devices to construct a Rockefeller as diabolical Napoleon of American business was scarcely presenting the history of Standard Oil from an impartial point of view, and Tarbell's recent biographers have recognized that both the McClure's articles and the book present a view of the company that is sometimes biased and lacking in factual foundation. Isabelle Sheifer, for example, suggests that in some instances Tarbell worked from innuendo rather than evidence. Mary E. Tompkins dismisses related articles as "hatchet jobs," and Kathleen Brady sees them as lacking "judiciousness and objectivity."[65] Yet for all the legions of historians who have drawn from Tarbell to comment on Standard Oil, there has been remarkably little critical examination of either Tarbell's use of sources or the internal consistency of her presentation. Such examination is necessary, both because citing voluminous and often authoritative-sounding sources was an important part of Tarbell's strategy to establish credibility and because her strategy worked. In the end, many readers believed Tarbell did offer an impartial and accurate history. Unlike Lloyd, she would not be criticized for want of facts.

One element in Tarbell's discursive strategy was her apparent willingness to listen to what the company's own leaders said about their organization. She talked to H. H. Rogers and Henry Flagler; she was allowed to work with company documents at 26 Broadway in Manhattan. Company officials could not, therefore, dismiss Tarbell's work on the grounds that she had not bothered to look at their side of the story. When it came time to tell the story, however, in episode after episode Tarbell relied on what was said by Standard Oil's bitter enemies, those with a personal interest in damning the company. Thus, in the first major dramatic segment in her story, the fight over the South Improvement Company (sic), Tarbell relied above all on the *History of the Rise and Fall of the South Improvement Company*, distributed by Oil Region refiners.[66] When she looked at the Hepburn committee investigation, she passed over the actual hearings to base her conclusions on Simon Sterne's report: no wonder she concluded that Rockefeller had "an entirely false standard of values." Taking up the New York state senate investigation of 1888, she used the counsel's opening statement—in effect the indictment of the company—slighting what the senate heard thereafter. Understandably, in her account of Pure Oil's contest with Standard Oil, she turned to brother Will and friends Roger Sherman and Lewis Emery Jr. for materials.[67]

Reliance on hearsay was also an important part of Tarbell's approach; she worked quite happily with second- or thirdhand information. In the sic episode, for example, when she presented what was supposed to have been a confidential remark of John D. Rockefeller's made to one of his opponents, she quoted an anti–Standard Oil newspaper's quotation of what the oilman said Rockefeller had said. In other episodes she used secondhand accounts of Standard Oil's operations presented in civil suits against railroads to which the company was not a party and thus had no opportunity for rebuttal. Tarbell's most significant use of secondhand information, important because many subsequent historians saw it as hard evidence of Standard Oil's profit from railroad rebates, came with her treatment of the testimony of Alexander Cassatt. From the historian's point of view, the difficulty with this evidence is that there is no record of the testimony outside of highly biased sources. Thus, at a later time, Standard Oil could deny that Cassatt ever offered such data. Tarbell resolved the problem by relying on the unverified version of the testimony in Sterne's Hepburn committee report.[68]

What she could not find Tarbell concocted, a practice within journalistic convention of the time that could greatly enhance the dramatic impact of a story. Chunks of imaginary dialogue with Rockefeller began, "He could tell them . . ." Taking the testimony of William H. Harkness, she had Harkness blame Standard Oil for his misfortunes—which he had not done—and indulge in imagined reflection. Coming to the Hepburn committee investigations, she devised an imaginary monologue condemning Rockefeller: "You have taken deliberate advantage of the iniquitous practices of the railroads to build up a monopoly. . . . You are guilty of plotting against the prosperity of an industry."[69] As reading, these fictional bits spiced up Tarbell's text, and after the reader plowed through hundreds of pages of apparent fact, their fiction might not be obtrusive.

Far less gripping was the abundance of statistical information Tarbell offered, but here she resorted to some literary as well as statistical sleight of hand. For example, she implied regional price gouging by asking, "Why should Colorado pay an average of 16.90 cents for oil per gallon and California 14.60 cents, when the freight from Whiting differs but one-tenth of one cent?" The answer to this question was obvious to one who read trade journals: California produced, refined, and sold petroleum in a Pacific Coast micromarket in which there was excess capacity. California did not need products from Whiting, whereas Colorado, with only modest production and refining capacity, did — sustaining higher prices. Again with prices and profits in mind, Tarbell offered tables of crude oil and refined product prices to show Standard Oil's profit margins growing; she omitted reference to cyclical variations in demand and contrasted times of lowest prices, such as 1873–75 or 1893–94, to times of greatest prosperity.[70]

Failing other devices, when Tarbell wanted to drive home her case against Standard Oil, she simply abandoned internal logic and consistency. This is especially apparent in her treatment of railroad rebates, the means by which Tarbell saw Rockefeller gaining the whip hand over the oil industry, unfair advantage gotten by illegal and immoral means. Did his competitors get rebates? At the beginning of her story, Tarbell mentioned one receiving a 37.5 percent rebate, another a 10 percent drawback; Rockefeller was paying from $1.20 to $1.60 to ship a barrel of crude as compared with a competitor's $1.33. Later she made it seem as though competitors were not taking rebates by saying that Oil Region refiners were flatly opposed to rebates of any kind. If receiving rebates was the critical element in Standard Oil's success, it would certainly be an aid to its prosperity if no one else enjoyed what Tarbell called "the vicious system of rebates." Yet by Tarbell's account, the New York firm of Ayres, Lombard and Company got the same break on freight rates as Standard Oil. Tarbell reported this item but omitted comment. Standard Oil's rebates, by contrast, showed its directors were deaf to "the most obvious principles of justice . . . unhampered, then, by any ethical consideration."[71]

How can one account for Tarbell's inconsistencies on so important an issue in her story as rebates? To be credible, Tarbell had to seem accurate. Rebates were a fact of oil industry life, whether for Standard Oil or George Rice. Tarbell knew it, and so did the industry participants who would read her book. What Tarbell did was minimize their importance to Standard Oil's competitors — a passing reference sufficed — and dramatize their use by Standard Oil. Through moralistic condemnation, she made the company anomalous for doing what its competitors also did. What did not merit comment when done by Standard Oil's competition was "vicious" on the part of Standard Oil. For that matter, in similar fashion, normal business prudence about disclosing information about operations became a sinister cloak of secrecy when Tarbell described Standard Oil. Things that were normal when others did them became anomalous when Standard Oil did them — because the company stood for anomaly, a target of condemnation.

Notwithstanding her vigorous condemnation of Standard Oil's use of rebates, Tarbell aimed to be restrained and moderate in tone, especially as compared with Henry Demarest Lloyd; no one was going to shelve her work as a "wild rant." For that reason, though Tarbell included many of the ripping yarns from Lloyd — the tale of the widow Backus, the Buffalo refinery fire, the saga of George Rice — she toned them down considerably. She greatly condensed the widow's tale, using it to show Rockefeller having his way with his competition because of his advantageous railroad rates. She dubbed George Rice "irrepressible" and, like Lloyd, cast him as an individualistic fighter for free markets and liberty. With her usual selection of anti–Standard Oil sources, she presented Rice's version of the "Turn Another Screw" episode, ignoring the positions of both Standard Oil and the Louisville and Nashville Railroad. Though she quoted none of them, Tarbell noted that Rice showed her hundreds of letters relating to Standard Oil's predatory pricing tactics. As usual, when the company cut prices, it was predatory, whereas when Rice and other Standard Oil competitors cut them to gain market share, as she acknowledged they did, they were being competitive. Moreover, the device of referring to hundreds of Rice's letters suggested that they were correct, though she omitted quotation and analysis of them.[72]

To support the highly important allegation that Standard Oil had corrupted the political system, Tarbell, like others before her, told the tale of the supposed involvement of Standard Oil in the election of H. B. Payne to the United States Senate. Though she noted that no one in Ohio and Washington found evidence to support the charge, Tarbell cited various documents that "contain the evidence of bribery, collected by the Ohio Legislature and the majority and minority reports of the Senate committee,"[73] and then concluded with the minority report, thus reading into the record the judgment of her friend Senator G. B. Hoar, "The Standard Oil Company undoubtedly exerted its influence against all trust investigation and legislation," to support the more general allegation of corruption. Similarly, in the absence of evidence, she quoted another member of the minority on the issue, who warned of "a power which controls business, railroads, men and things, [and] shall also control here." By introducing the unsupported allegation Tarbell enhanced the credibility of the overall charge against Standard Oil. Again, on the defeat of the

Billingsley antitrust bill, Tarbell pointed out legal and practical difficulties in the proposed legislation and noted that charges that Standard Oil bribed members of Congress were never proved. But by introducing the possibility of bribery, she sustained the theme of corporate power and political corruption. Dubious evidence was equally useful. Recounting Mark Hanna's alleged attempt to secure dismissal of Ohio's quo warranto suit against Standard Oil, she drew all her material from the *New York World*'s version of August 11, 1897, which was based on an altered letter. In short, in Tarbell's approach, all that was necessary to convict Standard Oil of corruption was repetition of allegations made against it.[74]

Lest the reader have missed the moral message in Tarbell's history, she ended her long chronicle with a grandly simplified homily on what had taken so many pages to describe:

> As for the ethical side, there is no cure but in an increasing scorn of unfair play — an increasing sense that a thing won by breaking the rules of the game is not worth the winning. When the business man who fights to secure special privileges, to crowd his competitor off the track by other than fair competitive methods, receives the same summary disdainful ostracism by his fellows that the doctor or lawyer who is "unprofessional," or the athlete who abuses the rules receives, we shall have gone a long way toward making commerce a fit pursuit for our young men.[75]

Standard Oil, as Tarbell presented it, had broken the rules of the game. It threatened not only the economics of an industry but the virtue of the Republic. Unless its activities were condemned, virtuous republican manhood was at risk. In short, behind apparently economic questions were moral anxieties. That was the alarm Tarbell wanted to sound. For that reason, her *History* was from its beginning the case against Standard Oil, and Tarbell as prosecutor constructed it with only one morally acceptable outcome, whatever the evidence — or lack of it.

Not everyone thought Tarbell made her case. Negative notices appeared in *Gunton's Magazine*, the *Nation*, and the *New York Times*, though the *Times* noted that "nearly a third of the volume is given over to documents and records on which are based the facts of her story." George Gunton reviewed the articles before they were reprinted in the book and dismissed them as a mere repetition of Henry Demarest Lloyd's earlier work, as "little more than a re-hash of Lloyd's *Wealth Against Commonwealth*, which was the most inflamed, unfair book that was ever published by a respectable house." More specifically, he attacked Tarbell for including obviously biased materials, particularly from the investigation of the Industrial Commission, which was presided over by the head of the Pure Oil Company. Gunton also pointed out the peculiar position of Standard Oil regarding the issues of prices: "The Standard gets damned whichever way the price goes; if it goes up, it's damned by the public; if it goes down, it is damned by the small competitor." *Gunton's Magazine*, which enjoyed strong circulation in a geographically limited area, thus mounted both a defense of Standard Oil and an attack on Tarbell.[76]

The *Nation* used the occasion to defend some of its views, including the basic fairness of discrimination between long-haul and short-haul shippers, but also focused on the wider characteristics of Tarbell's work, which it characterized as "a railing accusation." The elements of bias and melodrama did not go undetected: "In impassioned, if turgid, language, a desperate struggle is described between the powers of evil incarnate in the Standard Oil Company and the powers of goodness appearing in a metaphysical entity called the 'Oil Region.'" As to motives, the *Nation* was no more charitable: "This book seems to have been written for the purpose of intensifying the popular hatred. The writer has either a vague conception of the nature of proof, or she is willing to blacken the character of Mr. John D. Rockefeller by insinuation and detraction."[77]

Gilbert Holland Montague, a young scholar at Harvard University and author of an article and a Standard Oil–sponsored book-to-come on Standard Oil, was equally negative. He claimed that Tarbell "combined fact, rumor, common reputation and current fiction" in her history, detailing his objections in a sixteen-page essay in the *North American Review*. He made a special point of Tarbell's tendency to ignore inconvenient facts contained in her appended documents relating to the history of Standard Oil's involvement in SIC and successor organizations; he chided her for including seven pages of bribery gossip before she acknowledged that no evidence existed to corroborate allegations and for including a twenty-three-page rehearsal of the Buffalo refinery case although she acknowledged at the end that Standard Oil officials were not convicted. Montague was especially critical of Tarbell's tendency to include George Rice's various charges, despite his failure to make most of them good when they had to be proved. For good measure, he included a judgment from the New York Court of Appeals in one suit that supported Standard Oil's allegation that Rice's intention was to be so much of a nuisance that the company would pay a highly inflated figures, $250,000 and later $500,000, for his operations. Montague's final judgment was that Tarbell's book was mainly useful as an example of "the psychological condition of that portion of the community which assigns mythical attributes to the Standard Oil Company."[78]

Though the reaction of hostile critics was sometimes on target, theirs were decidedly minority reports. The *Oil Investors' Journal*, published in Beaumont, Texas, reflected the verdict from the countryside: "There is enough in her story to convince an unprejudiced reader of the truth of the charges that have been iterated and reiterated against the oil trust."[79] Here was one element of

Tarbell's credibility: she offered additional support for what people already thought. The *Dial*, the *Arena*, and *American Magazine* all hailed publication.[80] *Outlook* told its readers that "the author has no thesis to sustain and is willing to let her readers draw their own conclusions." It concluded that Standard Oil "actually gained its ascendancy, as the public learned many years ago and as Miss Tarbell's record makes perfectly clear, through the enjoyment of railroad shipping rebates, denied to its rivals." This would become a commonplace in history textbooks. The reviewer placed Tarbell's history "among the few great historical undertakings of American authors of this generation."[81]

The exceptional popularity of the articles and book prompted Arthur Hornblow (Charles Klein) to write a play, *The Lion and the Mouse*, which enjoyed considerable success at New York's Lyceum Theatre, beginning in late November 1905. It ran for 686 performances, the longest continuous run of any play written in the United States to that time, and four separate road companies carried it across the nation.[82] The play, in turn, inspired a novel, *The Lion and the Mouse: A Story of American Life Novelized from the Play by Arthur Hornblow* by Charles Klein. Published by a major New York house, copies of the play and of the book eventually circulated throughout the country.[83]

The new political expert professionals, including those at the Bureau of Corporations, also found Tarbell useful. By advancing a case ostensibly on the basis of a rigorous analysis, she promoted the professionalization of economics in the realm of popular discourse. Though she incorporated a wider range of materials, she included the data of economists as the basis for authoritative opinion. In doing so, she offered a model that other studies would follow. Tarbell also incorporated the critical worldview of the Progressive Era academics, with a strong emphasis on the interconnectedness and interdependence of life. Thus, Standard Oil's threat to the economic order became a threat to the political system and republican virtues as well. This broad systemic view had something for everybody: the older idealists among political scientists could find applications of their school, while younger statists found warrant for regulation of economic behavior by experts, to restore "balance" between forces in society. In the end, her vagueness as far as political theory went made her acceptable to both schools of political scientists.[84]

Tarbell's persistent introduction of moral discourse did not impair her credibility among social scientists any more than it did among the clergy and the general public. Her persistent use of language from moral discourse — evil, diabolic, secret, sordid, piracy, innocent, weak — harmonized with the perspectives of the "ministers of reform," along with politicians, the mass press, fellow muckrakers, small producers, anti-Darwinists in religion, and cultural conservatives alarmed at change and expecting the worst. By carrying the discourse on trusts and Standard Oil from the technical and operational levels into moral discourse, the most general stratum of common discourse,[85] Tarbell's work offered a holistic view of the trust problem. It also offered a commonplace solution, the regulation of large and dominant corporations to restore conventional fair play in the marketplace, to end the threat of concentrated economic power to small producers and consumers, and, above all, to preserve the republican virtues. Standard Oil as symbol and reality of modernity could still be tamed and put to the service of a conservative worldview. As Tarbell framed the problem, one could eliminate the menace of a Standard Oil and, thereby, strengthen the capitalist system.

Her simplicity was compelling. Tarbell's analysis and remedies fended off a whole range of troublesome reflections: if the octopus could be regulated and tamed, then America could deliver itself from tyrannical power and the related threat of social disorder, using only its own institutional and cultural resources. One might appropriate Dorothy Ross's description of the grand effect of social science to describe this work of Tarbell's and account for its persistence in discourse. In effect, she "consistently constructed models of the world that [embodied] the values and follow[ed] the logic of the national ideology of American exceptionalism."[86] In line with this ideology, a remedy would be sought in legislatures, Congress, and the courts.

When it attracted the attention of Theodore Roosevelt, Standard Oil came up against a truly daunting adversary, expert in manipulation of discourse, the most successful manipulator of the press to occupy the White House to that time. Roosevelt recognized a useful target when he saw one, and, beginning in 1902 and continuing through the rest of his administration, he used moralistic attacks on Standard Oil to advance his personal political agenda and to effect resolution of inconsistent elements of Progressive economic policy.

Since the 1890s, academic economists had worked to reconcile their defense of mass producers, in the interests of efficiency and consumers, with their moral preferences for small producers. Roosevelt's rhetorical approach to this knotty problem was to separate the sheep from the goats, the "good trusts" from the "bad trusts." The former he never enumerated, but the latter included railroads, meat packers, and Standard Oil. Once he identified his villains, he warned his attorney general to be selective in antitrust prosecutions, and, at the same time, he unleashed repeated torrents of moral indignation against Standard Oil and other "bad trusts." To the end of documenting the abuses of the malefactors, Roosevelt established the Bureau of Corporations in the Department of Commerce and Labor in 1903. In Roosevelt's view, by gathering data from and about corporations and publishing what it found, the bureau would head off more aggressive antitrust legislation that might curtail mass production and disturb capital markets. As Roosevelt explained it, "Such publicity would by itself tend to cure the evils of such."[87] In pushing the creation of the bureau through Congress, Roosevelt made effective use of the specter of Standard Oil, by concocting a telegram he claimed Standard Oil sent to six senators, urging them to block the measure. The *New York World* gave Roosevelt's bogus telegram front-page coverage, as did the *Philadelphia North American*, whose headline screamed, "The Standard Oil Plot."[88] This reaction, in Roosevelt's opinion, secured the legislative success of his initiative.[89]

Attacking Standard Oil paid political dividends for Roosevelt, just as it had given Lewis Emery Jr., George Rice, and others at least temporary competitive advantages in business. Within the Republican Party, Roosevelt's attacks placated the increasingly numerous Progressives.[90] Beyond that, as George E. Mowry observed, "Roosevelt knew that to attack the trust problem was the one action calculated to win the admiration and support of middle-class America."[91] As the *New York Times* observed, "The Standard Oil Company has been chosen as the scapegoat of Mr. Roosevelt's administration. . . . Any attack upon it . . . is bound to be popular."[92] Roosevelt and other Republicans also believed that an attack on trusts was needed to defend Republican control of the federal government. Thus, in response to William Jennings Bryan's inclusion of trusts — along with the monetary issue and imperialism — as a central issue, Republican candidate William McKinley had criticized them in general terms in his inaugural address. One Republican insider, George Tichenor, wrote to Senator William Boyd Allison, "Something more has got to be done . . . or the Republican Party's hide will be on the fence."[93] Standard Oil, thus, became a doubly useful "goat" in national politics.

Tarbell's volume was still attracting comment when Standard Oil became embroiled in an acrimonious battle with independents again, this time in Kansas. The company had acquired properties before the state became a significant producer of crude oil. In 1895, it expanded its Kansas investments by building a small refinery in the producing region, with the intention of supplying refined products to the Waters-Pierce Oil Company for distribution in the south-central part of the country. Thereafter, Kansas production expanded, taking off in the early years of the twentieth century. Standard Oil responded by building an extensive pipeline system, by enlarging its first refinery, and by building a second one near Kansas City. By 1904, Standard Oil's operations were indispensable to producers in Kansas, though small refiners from Texas and Pennsylvania had also secured footholds in the state. Among the arrivals from Pennsylvania were three independents who had crossed swords with Standard Oil before: T. N. Barnsdall, C. D. Webster, and Marcus L. Lockwood.[94]

When Standard Oil expanded its refineries, tank farms, and pipeline systems to accommodate increased Kansas production, its relations with Kansas producers were generally harmonious, a rare situation and one that ended in February 1904, when Congress passed a bill permitting pipelines to be built from the Indian Territory to Standard Oil's refineries in Kansas, creating competition from Oklahoma crude for Kansas producers. Acting for home-state interests, a Kansas congressman attempted to block this measure with a stiff prohibition of such shipments. For its part, Standard Oil attempted to placate Kansas oilmen by providing an additional outlet for their oil, building a pipeline from its Sugar Creek refinery, near Kansas City, to its mammoth plant at Whiting, Indiana. It built additional tank farms and bought increasing amounts of crude, to the point of accumulating enough oil in storage to supply its Kansas refineries for about two years. As Kansas production mounted ever higher, however, Standard Oil urged Kansas oilmen to declare and enforce a moratorium on drilling to support the price of crude. When oilmen did not respond, it announced price cuts and classified oil into two grades, paying significantly higher prices for higher-gravity oil, which yielded more kerosene. The less useful "heavy" oil declined from $1.18 in 1904 to 17 cents in April 1905.[95]

Rather than acknowledge that their flush production swamped markets, Kansas oilmen blamed Standard Oil for the decline in prices, claiming that the company aimed to take over all producing properties by driving the independents out of business. In Kansas, oilmen followed the example of their counterparts in Pennsylvania and formed producers' associations, the most important of which was the Kansas Oil Producers Association, and attacked Standard Oil in the newspapers, in public meetings, in the state legislature, and in Congress — repeating the commonly successful combative tactics of Pennsylvania independents. In some ways, the Kansas independents outdid their Pennsylvania counterparts: they organized an extensive direct mail campaign, taking advantage of the efficient rural free delivery system. They also influenced local editors by purchasing thousands of copies of newspapers for mailing to all the state's farmers, if editors supported the KOPA. Finally, they organized a "literary bureau" to circulate their version of events within the state and across the country. The KOPA also distributed "That the People May Know," a four-page pamphlet, in every corner of the state and organized a massive lobbying campaign aimed at the state legislature.

Kansas independents, adopting a tactic commonly employed in Pennsylvania, staged a media event, a gigantic protest rally in Topeka. The KOPA brought in now famous Ida Tarbell to address them and to confer on strategy, while William Randolph Hearst sent Frank Monnett, whose bribery allegations he had publicized several years before. Marcus L. Lockwood, of Pennsylvania, participated as an adviser, aiding in the drafting of the resolutions of the KOPA meeting.[96] Using a technique that Pennsylvania independents, Theodore Roosevelt, and William Randolph Hearst used with success, they manufactured a bogus letter, this one from John D. Rockefeller himself. The *Topeka Daily Capital* claimed that it had seen the letter, in which Rockefeller said that the refinery business was so taxing that "after preparing myself to properly instruct my Sunday school class, I am able to clean up the beggarly sum of 50 million dollars per year out of it." Rockefeller is supposed to have continued by writing, "What the people need is poverty and contentment. The possession of wealth brings with it cares and woes."[97] A less likely pronouncement is hardly imaginable, but in 1905 no allegation against Rockefeller and Standard Oil was held to a reasonable standard of credibility.

The fully developed publicity program of the KOPA was brilliantly successful. New York–based periodicals, including the *Arena*, the *Independent*, the *Review of Reviews*, the *Literary Digest*, the *Outlook*, and the *World's Work*, all published articles highly favorable to the independents' crusade. Writing in the *Saturday Evening Post*, Philip Eastman described the battle in Kansas as "an open conflict between the Standard Oil Company on the one side and the people of the State on the other."[98] Eastern newspapers, including the *Philadelphia Ledger*, *New York Evening Post*, the *Journal of Commerce*, the *New York World*, and the *Evening American*, all reported the heroic resistance of Kansas independents to another move by the octopus.[99]

The public action campaign of the KOPA had three major objectives: the creation of a state-owned and convict-operated refinery; common carrier laws for Kansas; and the prompting of a major congressional investigation of Standard Oil. In the end the in-state objectives were not realized because state and federal courts rejected the association's refinery and common carrier measures, but Kansas congressman Charles F. Scott succeeded in obtaining support for a Corporation Commission investigation, and he introduced a federal common carrier bill in February 1905, shortly after New York congressman William Randolph Hearst promoted a similar measure. Scott's bill appeared too late in the session for passage, but it received sufficient publicity to attract the support of both House Speaker Joseph G. Cannon and President Roosevelt, and it won enthusiastic support from congressmen from the Oil Region of Pennsylvania. In April, the president dispatched John Garfield of the Corporation Commission to Kansas. Garfield interviewed independents and supported their arguments in well-attended press conferences. According to the commissioner, events in Kansas proved once again that "Standard's system of operation is a grand scheme of deception."[100]

In compliance with Roosevelt's orders, Garfield returned to Washington and organized a massive investigation of the affairs of the Standard Oil Company of New Jersey (SONJ). In 1905, in his annual message to Congress, Roosevelt used the bully pulpit at the onset of the major political campaign against the company. Few listeners could have doubted the specific identity of his target: "The kind of business prosperity that blunts the standard of honor, that puts an inordinate value on mere wealth, that makes a man ruthless and conscienceless in trade, and weak and cowardly in citizenship, is not a good thing at all, but a very bad thing, for the nation. This government stands for manhood first and for business only as an adjunct of manhood."[101] Created as the icon of "mere wealth," of ruthlessness, lack of conscience, unmanly competition, and poor citizenship, John D. Rockefeller was, once again, at war with manhood in Roosevelt's encoding of the voluminous attacks on the oilman. In this pronouncement as in so many others, Roosevelt enlisted gendered language and concepts to sustain his moral pronouncement.[102]

Above all, Roosevelt used the Bureau of Corporations, whose revelations he manipulated when they were disclosed to the media. His campaign received an invaluable boost from a widely heralded series of muckraking articles, "The Treason of the Senate," by David Graham Phillips, which underscored the political importance of the bureau's work. The fifth installment, which appeared in July 1906, tarred Joseph Weldon Bailey as Standard Oil's man on the Democratic side of the United States Senate. As Phillips concluded, "With leaders like this . . . 'the interests' grow and the people diminish."[103]

The following year, the bureau released the first part of a study of the position of Standard Oil in the oil industry. Like the previous report, this one was heavily documented, drawing on the report of the Hepburn committee (1879), one of the Rice cases before the ICC, the Ohio cases against Standard Oil, the New York Senate investigation in 1888, and the findings of the Industrial Commission, especially the testimony of Lewis Emery Jr. Both bureau reports maximized the advantages Standard Oil had in transportation, especially through discriminatory railroad rates, and minimized its economies of scale, managerial skill, and expansion through retention of earnings.[104]

Press reactions to Roosevelt's campaign against Standard Oil were generally, though not uniformly, positive. The company had its unpaid defenders. The *New York Times*, for one, asserted that the statistical basis of the bureau's publication "The Position of the Standard Oil Company in the Petroleum Industry" was shaky, inadequate to support the statements made against the company, but it concluded that Standard Oil had given the public "none of the benefit of its superior efficiency." In general, however, the reports offered journalists another opportunity to attack the company. Thus, after the release of the third part of the bureau's report, "Foreign Trade," the *New York American* concluded that the government had proved its claim of excessive profits at the expense of consumers and proved "the existence of a private power so gigantic and arbitrary and a political influence so insidious and corrupting as the Oil Trust."[105]

As it coped with attackers, Standard Oil also faced state-level prosecution of the Waters-Pierce Oil Company and the Republic Oil Company, its subsidiaries. Missouri's attorney general, Herbert S. Hadley, who won election to that office as a reformer in 1904, developed the cases. Through litigation that dragged through the courts and newspapers for several years, Hadley succeeded in ousting both companies from his state and in obtaining fifty-thousand-dollar fines against each of them. Newspapers and journals had a field day. The *Wichita (Kans.) Daily Eagle* created a letter in which an executive of SONJ boasted: "We are bigger than the government. Standard Oil is larger than the United States. We own the senate [*sic*] and the house [*sic*]. Rockefeller is a bigger man than Roosevelt." The letter was reprinted in Hearst's *New York American* on January 16, 1906, and, thereafter, by other periodicals.[106]

The *New York World* made especially effective use of the New York phase of Hadley's investigations. Before the sessions actually began, it ridiculed John D. Rockefeller by recounting his presumed dodging of a subpoena by hiding out. The *World* reported that it had reporters searching for him in three New Jersey towns as well as in Savannah, San Juan, St. Louis, and Oconomowoc, Wisconsin. In one issue, it ran a "missing-persons" description of Rockefeller with height and weight, adding, "When last seen Mr. Rockefeller was clean-shaven and bald. He may, however, wear a wig of brown and gray mixed. If he is wearing the wig he will also wear a pleased expression. He is suspicious toward strangers and if he talks at all it will be about charities and church and Sunday-school work." The *World* also reported that it had discovered a secret passageway between Rockefeller's house and that of his son-in-law, to be used, no doubt, to evade the law. Thus the *World* ridiculed Rockefeller, placing him in the company of ham actors, religious hypocrites, and law evaders at the same time that it called attention to his unmanly appearance.[107]

Not to be outdone at Standard Oil–baiting, William Randolph Hearst amplified the "money power" theme strongly in 1908 with the publication of letters from John Archbold to Senator Joseph Foraker, Representative Joseph Sibley, and other political figures, dated from 1898 to 1904. Hearst's first salvo came on February 14, 1908, with the publication of a telegram from Archbold to Senator Matthew Quay; on the basis of it, Hearst accused Standard Oil of using Quay to undermine enforcement of the antitrust law. Archbold refused comment on the allegations and the telegram, but other newspapers, including the *Philadelphia North American*, *Philadelphia Evening Telegraph*, *Pittsburgh Post*, *New Orleans Daily State*, *Providence News*, *Wichita (Kans.) Daily Beacon*, *Albany Argus*, and *Bridgeport Evening Post*, reported Hearst's allegation with strong approval.[108]

Hearst, who acquired some of Archbold's correspondence from a Standard Oil employee who stole it, began using Archbold items extensively in his *New*

*York American*, in connection with his attempt to launch his Independence Party. Hearst made a media event of reading the letters across the county, in Chicago, St. Louis, Memphis, Denver, El Paso, Los Angeles, and New York.[109] He demeaned John Archbold as the "Little King of Standard Oil," the man who specialized in the buying and selling of politicians just as Henry Flagler had made a specialty of securing rebates from railroads. Clippings from other newspapers in the "King Oil" files of the *New York American* show that the press generally supported Hearst's attack, though the *World*, equally antagonistic toward Standard Oil, made a point of quoting Oklahoma governor Charles N. Haskell's denunciation of Hearst as "a disappointed egotist, one who sneers at virtue in women and leers like a fiend while assassinating the characters of men — one who revels in inherited millions while I toil for a living."[110]

Undeterred by the *World*'s attacks and by critical coverage in *Collier's Magazine*, Hearst continued to publish items. In 1910, for example, he published a telegram from Archbold to Congressman Sibley, one that smelled like the proverbial smoking gun:

> I beg to enclose you herewith certificate of deposit to your favor for $5,000, sent you at the request of Mr. Griscom, the purpose of which you no doubt understand with him.
>
> Permit me to improve this opportunity also to express my high appreciation of your most courteous and efficient action in response to our request regarding the consideration of the subsidy matter with Mr. Griscom.
>
> Very truly yours,
> Jno. S. Archbold.

This communication, with Archbold's handwritten signature, was published in the *American* on May 31, 1910. As was common with Hearst creations, the letter was both less and more than it seemed. At the top of the page, the address for Archbold, 25 Broadway, has been erased, and a tracing of Archbold's signature has been pasted on the bottom; the styles of the two paragraphs are strikingly different, and the closing is improbable in a telegram. The first paragraph is written in the condensed form favored by senders who were charged by the word, but the second paragraph is more fulsome, typical of letters of the time. Such a closing would have been common in letters but not in telegrams. The likelihood of Archbold, or anyone else at Standard Oil, missing the famous 26 Broadway address is remote. These anomalies may be explained in one of two ways: either Archbold was having an uncommonly bad day, and so was his secretary, or the telegram is a fabrication, perhaps of parts of a telegram and parts of a letter, or perhaps it was entirely fabricated. In the end, these "revelations" prompted a congressional investigation and additional lessons in the dangers of corporate power; indeed, the controversy outlived the Standard Oil holding company.

As the government readied a comprehensive antitrust case against Standard Oil, it filed a barrage of lesser actions, intended primarily to keep the company in the news and on the defense. In 1907, the company was accused of accepting an illegal rate from the Chicago and Alton Railroad in Illinois. The case drew attention largely because it involved Standard Oil, but it became national news when United States district judge Kenesaw Mountain Landis fined the company $29,140,000 for 1,462 violations of the Elkins Act. Judge Landis exploited the situation for publicity, even claiming at one point that he might place Standard Oil in receivership because it might not have cash available to pay the fine. The *New York World* jubilantly reported, "If paid in silver dollars, the $29,140,00 would make a weight of 1,327,500 pounds. Allowing that two horses could draw three tons, it would take three hundred and four double teams to haul the amount of the fine in silver from place to place. The amount of the fine is more than Jefferson paid for the Louisiana Purchase, with Alaska thrown in; more than the whole Philippine Archipelago cost the U.S. in money; greater than the net incomes of five independent monarchies bordering the Danube and the Mediterranean. Great wars have been waged on lesser sums."[111] When Landis's decision was reversed by the U.S. Court of Appeals one year later, Hearst's *New York Journal* reported that the result was "MOST SATISFACTORY TO MR. ROCKEFELLER AND WALL STREET."[112]

The Chicago and Alton case prompted motion picture producer Sigmund Lubin to bring Standard Oil and Rockefeller to the silver screen. In September 1907, he released *John D. —— and the Reporter*, in which the lead character, president of the Rancid Oil Company, evaded a legal summons until an ambitious young reporter trapped him and served it. At the conclusion of the film, John D. paid a twenty-nine-million-dollar fine but was undaunted because it was "squeezed from the poor consumers' pockets and John D. is happy again."[113] The case against Standard Oil was now so well embedded in public discourse that a moviemaker could assume that consumers of his product would pay to see a version of it.

Following the Landis decision, Standard Oil faced legal challenges in New York, Tennessee, Mississippi, Minnesota, Wyoming, Iowa, Ohio, Oklahoma, and Texas. In the last, the renewed prosecution of Waters-Pierce ran through 1908 and 1909, with both Henry Clay Pierce and Texas politicians showing their customary flair for self-advertising. At one point, the Texas attorney general garnisheed six million dollars in accounts receivable at Waters-Pierce so that he could be certain that the company would pay the fine he anticipated. Thereafter, the company was fined $1,623,000 and ousted from the state. For a follow-up, Pierce was indicted and tried for perjury in Austin, all of which was more noteworthy than his acquittal. It was clear to reporters at the time that the federal government was mounting a multifront campaign against Standard Oil and that hostile state governments were joining in the concerted assault.[114]

In the midst of it all, John D. Rockefeller released his chatty autobiographical essay, *Random Reminiscences of Men and Events*. Attempting to offset the Tarbell book and subsequent articles, Rockefeller emphasized the working of the committee system at Standard Oil, where decisions were usually compromises and ended in unanimity. There was, in fact, no new information in the book, but it did reveal his sorest spot—the allegation of hypocrisy; his version of the widow Backus story was far more elaborate and better told than the rest of the volume.[115] No doubt Rockefeller expected few plaudits from his opponents. Hearst's *New York American* critiqued the book editorially before it appeared, charging that Rockefeller's "tyrannical methods are too well known and too thoroughly authenticated to be smoothed over with cunning words." In the editor's opinion, Rockefeller "was indicted long ago by the grand jury of public opinion."[116] That was probably a realistic assessment.

The culmination of the muckrakers' exposés, the independents' crusades, and the government's onslaught came in 1909, when the United States attorney general filed suit against Standard Oil for violations of the Sherman Antitrust Act. The suit charged that John D. Rockefeller and six other directors of Standard Oil who were also officers or directors in thirty-seven subsidiary companies "conspired and confederated for the purpose of combining all the said companies in restraint of trade . . . for the purpose of monopolizing commerce." The case, filed in the court for the Western District of Missouri, was expedited under the act of February 11, 1903, so it was first heard by the four circuit judges of the Eighth Circuit in April 1909. After adverse judgment, Standard Oil's appeal came before the Supreme Court on the grounds that it was imperative to the national interest to do so and because the discussion in the case was "largely involved" in the American Tobacco case, which was also before the Court.[117]

The government's case rested on five major points: (1) Standard Oil had a strong position in ownership of production in Pennsylvania, the Lima-Indiana oil field, and Illinois, though it owned only one-ninth of total U.S. production; (2) the company controlled 89 percent of pipeline traffic to the East Coast; (3) it had near total control of refining in New England and handled 79.3 percent of the crude oil processed in the United States; (4) Standard Oil controlled 87.9 percent of the export trade in 1905; and (5) it secured "enormous profits," especially in domestic refining and marketing and in the operation of domestic pipelines.[118] A critical element of the government's case was its acceptance of the independents' definition of predatory practices, "price cutting in particular localities" and not at all points, their attacks on operation of a vertically integrated company, and allegations of industrial espionage.

Above all, the government argued a strongly historical case against Standard Oil to prove that market dominance had been won through illegal and unethical practices over a long period of time. The argument began with the South Improvement Company and the opposition of Oil Region refiners to it. In court, the government used Lewis Emery Jr. of Pure Oil as an important witness and had him read items from a scrapbook he brought, repeating his testimony from 1883, given again to the Corporation Commission in 1906; he read sections of Colonel Joseph D. Potts's *Brief History of Standard Oil* into the record as well.[119] James W. Lee and William W. Tarbell of Pure Oil testified, as did Robert D. Benson of Tidewater. They argued consistently that from its inception in 1870 Standard Oil operated to establish a monopoly.[120] Standard Oil's attorneys objected in vain that Emery was "giving us a lecture on the history of the early oil days."[121] Emery's testimony and the government's summation both emphasized strongly the elements of secrecy and conspiracy in the case.

The Supreme Court accepted the government's historical case that "from the beginning [Standard Oil] took its birth in a purpose to unlawfully acquire wealth by oppressing the public and destroying the just right of others, and that its entire career exemplified an inexorable carrying out of such wrongful intentions." The Court also accepted the argument against the company based on its size, which gave it "vast property and the possibilities of far-reaching control." Finally, the Court found that Standard Oil had violated the Sherman Act by operating to destroy "the potentiality of competition." Again, the Court tied its judgment to common-law tradition, which regarded the deliberate construction of barriers to entry of new competitors as monopolistic behavior. The Court, thus, defended traditional and conservative values when it found against Standard Oil. The Court's remedy was to break the holding company and its affiliates into twenty-six separate corporations, with pro rata distribution of shares in the new independent companies.[122] The Standard Oil Trust was thus dissolved, the octopus slain: morality triumphed over moneyed power in a decision that blended moral and economic considerations.[123]

Predictably, reactions varied. Louis Brandeis and Andrew Carnegie approved dissolution, but William Jennings Bryan, Robert M. La Follette, and the editor of the *Wall Street Journal* did not. U.S. Senator Atlee Pomerene, an Ohio Democrat, insisted that John D. Rockefeller and the other defendants should serve prison terms.[124] The Hearst newspapers found the decree wholly unacceptable: "The Standard Oil Company has suffered no set-back in its career of aggression and . . . its officers feel no discomfiting reproof." An editorial cartoon in the *New York American* showed a smiling John D. Rocke-feller happily clipping coupons. The major effect of the decision, according to the *New York Evening Journal*, was "to reduce the Sherman Anti-Trust Law to punk and putty." The only effective remedy was a new antitrust law from the Democratic Congress.[125]

While the legal staff of Standard Oil was busy working out compliance with the Supreme Court decree, the ICC, which by that time employed Frank Mon-nett, launched a highly publicized investigation aimed at bringing interstate pipelines under its control. As in past proceedings, independent producers and refiners prevailed at the hearings, which were held in several cities, and some of them were invited to the White House to discuss the situation with President William Howard Taft. With Taft's blessing and strong congressional support, the ICC declared interstate pipelines common carriers under its jurisdiction, an action that was upheld by the Supreme Court in 1914. It is clear that Standard Oil's opponents viewed dissolution as a first step, the second being divestiture of pipelines from other operations. The ICC, duly empowered, held no pro-ceedings on the subject after 1912 and heard no complaints until 1921–22.[126]

And was the Standard Oil monopoly still at work? Reformers continued to take alarm that it was. Attorney General George W. Wickersham called for another investigation of Standard Oil in 1913, in response to a complaint and suit by Henry Clay Pierce against John D. Rockefeller and others. Two years after the dissolution decision the *New York American* complained: "The power of one great monopoly to breed new monopolies is still unfettered."[127] Bio-graph, a major motion picture producer, kept the monopoly theme in the view of audiences in 1913, when it released *By Man's Law*. Built loosely on Ida Tarbell's articles, the film exhibits a ruthless oil baron building his company by crushing competitors and causing, thereby, desperate suffering among work-ers and their families. In the closing minutes of the film, only the fortuitous intervention of the hero saves an unemployed factory girl from prostitution.[128] By the beginning of the war in Europe, Rockefeller was the personification of villainy in all media, except, possibly, recorded music.

Standard Oil's attempts to cope with labor problems in a positive manner fared no better. Frank P. Walsh, chairman of the Industrial Commission and formerly a political ally of Missouri attorney general Hadley, attacked John D. Rockefeller and the Rockefeller Foundation for hiring Mackenzie King of Canada to undertake an investigation of management-labor relations. Guard-ing his agency's turf and spotlighting the report he would release the next day, Walsh cast the foundation as the inheritor of the monopoly's power: "The powers it is exercising are practically unlimited. Whether such powers can safely be permitted in the hands of any authority less than the Government is a question which forms an important part of the commission's forthcoming report."[129] The *World* never changed its tune: in 1915, along with the *Journal*, it lamented that Standard Oil companies had profited by $1,350,000,000 in four years. In 1916 the Justice Department blamed Standard Oil companies for a rise in the price of gasoline and threatened a new investigation.[130] The *Journal* observed that Standard Oil had been "buttressed" by dissolution and that a new suit should be prepared.[131] In the years that followed the Supreme Court's decree, then, a common negative attitude toward Standard Oil and dissolution remained the hallmark of progressivism and liberalism. Whenever there was talk of monopoly, partisan investigators of the company and its successors advanced the same arguments and examples, much in the spirit of Claude Rains's memorable concluding line in the film *Casablanca*: "Round up the usual suspects."

Thus it was that when the Federal Trade Commission was created, one of its first assignments was an investigation of gasoline price rises in 1915, with special attention to industry price-fixing and possible conspiracy on the part of Standard Oil companies. Its 1917 report made clear that commissioners did not assume that dissolution had changed the petroleum industry. Adopting what it called generally accepted usage, throughout the report the FTC referred to Standard Oil companies as opposed to "other companies" and explained that "the various companies called 'Standard' . . . are owned by bodies of stockholders . . . so similar in membership as to justify the common usage." It admitted that these companies had separate organizations, officers, and directors, as well as separate refining and marketing operations, but at the same time the FTC implied that the Standard Oil companies were not really separate industry entities, an implication that later observers took as fact. It noted the amount of crude Standard Oil companies held in storage, the number of pipelines Standard Oil companies controlled, the amount of gasoline Standard Oil companies refined, the market share of Standard Oil companies, and the exports of Standard Oil companies. In short, with respect to what it reported to the public the FTC spoke as though dissolution had not happened.[132]

When it came to explain the rising gasoline prices of 1915, the FTC decided that price movements and differences could not be explained either by variations in costs or by market conditions. Thus, it reasoned, the Standard Oil companies must be responsible for price differences because they had divided up the country into marketing territories in which there was "the absence of effective competition." *Cherchez le Standard*, in effect, when no ready explanation lay at hand. The FTC decided that Standard Oil companies "practically" fixed gasoline prices and had "considerable control" over crude prices as well. Indeed, it blamed Standard Oil of Indiana for *falling* gasoline prices that pressed other gasoline distributors in June 1915, though it admitted that, within Standard Oil of Indiana's marketing territory, "the 'independent' jobbers . . . have engaged in a competitive struggle that has sometimes involved price cutting."[133] Whether prices rose or fell, Standard was to blame.

The FTC's "remedies" for higher gasoline prices show that, despite its reliance on traditional monopoly-oriented themes of discourse, it was not unaware of newer conservationist arguments. To offset higher gasoline prices, there could be "economy in production and use of petroleum and its products." "Numerous wastes can be avoided." It also reminded readers that crude oil was "a limited natural resource which one day will be exhausted."[134] These reflections headed its list of remedies. But, having made them, the FTC returned to familiar channels of antimonopoly discourse about flawed competition and called for more Justice Department action, more antitrust litigation. To facilitate restoration of competition, the FTC urged that some branch of the federal government be charged with collecting petroleum industry statistics; as its United States Geological Survey (USGS) counterparts would do later, the FTC tied improvement to more work for bureaucrats scouting out facts.[135]

Quite illogically, and with no clear connection to its gasoline price investigation, the FTC also recommended that ownership of pipelines be divorced from other parts of the petroleum industry. Here it built upon its investigation of pipelines in 1916, a study begun by the Bureau of Corporations and given over to the new agency, which patterned its conclusions on the work of its predecessor. Thus, in 1916 it saw pipelines in the context of railroads, as essential to the petroleum industry "as the railroads [were] for agriculture," charging "high rates" and requiring "excessively large minimum shipments." It found that the cost of pipeline construction was so great that small refiners could not build them from fields to major markets and that they were thus "forced to build" plants near the oil fields rather than markets. Here the FTC described practical economics in terms of compulsion. The "Standard Oil group" had "exclusive use" of pipelines from Mid-Continent oil fields to markets, and its affiliate Prairie Oil and Gas controlled "extensive storage facilities," giving them competitive advantages of which the FTC disapproved. In 1916, however, the FTC was content to say that pipelines were not complying with the legal requirement to be common carriers and that they ought to—which would end much of the advantage of the Standard Oil companies.[136]

Five years later, in 1921, the FTC was again hot on the trail of Standard Oil, arguing that it controlled gasoline prices in the country as a whole, and on the Pacific Coast in particular. As before, in discussing Pacific Coast retailing, after an initial correct reference to the Standard Oil Company (California), it spoke of "the Standard Oil Company," and though it admitted that four other large companies in addition to Standard Oil (California) dominated the Pacific Coast industry, it steadily maintained that Standard Oil effectively controlled petroleum prices on the coast, as it did in the rest of the nation. It warned of the Standard Oil companies' "solidarity, arising . . . from an interlocking stock ownership resting largely in the hands of a few great capitalists and its great financial resources and credit" and took special alarm that such companies as Humble Oil and Refining and Midwest Refining had been absorbed by companies in the original Standard Oil group. Indeed, in a separate report of the same year, the FTC concluded that Standard Oil of Indiana's purchase of Midwest Refining stock had created "monopolistic conditions" throughout the entire Rocky Mountain area.[137]

Regardless of whether one saw dissolution as the end of a Standard Oil monopoly or as an ineffectual measure allowing monopoly to continue under camouflage, in terms of discourse, Standard Oil stood convicted. The victorious moral crusade that produced antitrust action both defined and implemented it with reference to Standard Oil. The victors reigned supreme in discourse; compared with what people such as Henry Demarest Lloyd, Ida Tarbell, and William Randolph Hearst did in shaping public discussion, Standard Oil's attempts to defend itself against moralistic antimonopoly attackers were paltry and ineffectual. So much for the reformers' nightmare of corrupt moneyed power's ability to control opposition. In discourse, Standard Oil largely failed to carry the day.

Out of the moral upheaval over Standard Oil, however, came two developments totally unforeseen by any of the principals in it. First, the debate over dissolution, subsequent recurrent investigations of pipelines, and recurrent controversy over competition and prices in effect generated a federal energy policy that was largely the extension of antitrust policy: priorities in public policy were the prevention of all anticompetitive and collusive practices and maintenance of a vigorously competitive domestic petroleum industry. The federal government's actions with respect to Standard Oil and its successors also shaped the business strategy of oilmen, as subsequent chapters will show. Albeit negative, incoherent, and at times virtually perverse, the federal penchant for prohibition and investigation affected the oil industry's operations. Second, in a way that Standard Oil's competitors, long adept at wielding discourse to further their competitive positions, could not have predicted, the onus they so successfully attached to Standard Oil gradually extended to cover all large oil companies and, ultimately, the entire oil industry.

## Running Out of Oil

In mid-March 1910, one more San Joaquin Valley wildcat, Lakeview No. 1, roared in at eighteen thousand barrels a day. It blew away the derrick, buried the engine house with sand, and made a crater of the well site. Lakeview No. 1, totally beyond control, ran wild for eighteen months. One torrent running from it was so strong and continuous that it was dubbed "the trout stream," while airborne crude misted down on people and property for miles around. Visitors nonetheless flocked to the scene of oil run amok, and among the many spectators whose clothing was gently freckled with petroleum was the arch-apostle of conservation of American natural resources, Gifford Pinchot.[1] Not coincidentally, he and other conservationists would broaden the scope of discourse on petroleum to focus on oil's use and, as they saw it, misuse. They would emphasize that the petroleum industry had not been a wise steward of national resources, and they raised the alarm that, unless something was done, the nation would run out of oil. It was not just Standard Oil acting counter to the public interest but a whole industry. Oilmen, great and small, were irresponsible plunderers of national wealth.

From hindsight, it seems paradoxical that early-twentieth-century conservationists should worry about exhaustion of the United States' petroleum at a time when prospectors opened area after of area to production, launching a golden age of oil discovery.[2] In 1899 most U.S. oil came from fields in Pennsylvania, New York, and West Virginia—the Appalachian region—as well as Ohio and Indiana. By the end of the first decade of the twentieth century, great oil fields had opened up in California, the Mid-Continent (chiefly Kansas and Oklahoma), the Texas and Louisiana Gulf Coast, and Illinois. There were promising indications of future bonanzas in Wyoming, New Mexico, and Arkansas. With these discoveries, new oil companies proliferated, among them such major companies as the Texas Company (Texaco) and Gulf (now Chevron). The first filling stations appeared, retail outlets especially for gasoline and related petroleum products, as opposed to the general store with a pump out front. Every sector of the petroleum industry grew explosively, and that was, in effect, what bothered some of the early conservationists. Notwithstanding discoveries, the United States was producing and using its oil at a dizzying, unprecedented rate.[3]

Thus the direction of public discourse on petroleum began to shift from questioning competition to questions of conservation and national energy self-sufficiency even before the dissolution of Standard Oil. New concerns intertwined with old. Political leaders such as Theodore Roosevelt and Robert M. La Follette, for example, waved the banner of conservation while carrying on the antimonopoly cry; indeed, for many who shared a negative perception of the modern economic order, the waste of natural resources and the exploitive character of big business went hand in hand. As W. J. McGee told the Mississippi Valley Historical Association in 1910, abundant resources "opened the way to monopoly, and the resources passed under monopolistic control with a rapidity never before seen in all the world's history."[4] But there were many conservationists who accepted and approved of the new order and who saw conservation as essential to continued prosperity and long-run national economic strength. On the score of conservation, both groups faulted the petroleum industry.

Conservationists could agree that petroleum was essential to the modern age, but beyond that, their perspectives diverged. Because petroleum was an essential but nonrenewable resource, should as much of it as possible be kept in government hands and under the ground? Certainly, this would prevent monopolists from growing rich on it, but it would also keep consumers from benefiting from many of the conveniences inexpensive petroleum products would provide. Should Americans use their oil, but use it "wisely," prioritizing uses and ruling out some in order to make whatever they had last as long as possible? That course required some agency — presumably the federal government — to set priorities and impose them on industry and consumers, a greater degree of economic regulation than Americans had experienced in peacetime. Should Americans save their own oil by acquiring oil reserves overseas and, in effect, pump other nations dry first? Here, too, a greater role for government would probably be necessary, this time in foreign economic affairs, and all those industry participants, from prospectors to distributors, who depended on domestic oil production would lose opportunities. All these questions implied an enhanced role for government in petroleum industry affairs—on this conservationists generally agreed. They also implied that if the nation was not yet running low on oil, it would do so in the future; most conservationists agreed that shortage was not far away.

The question of whether the United States would run out of oil surfaced within twenty years of the birth of the petroleum industry.[5] At the time of Drake's discovery petroleum had been the subject of relatively little scientific scrutiny, and there was no geological consensus on how petroleum deposits formed or accumulated. Some scientists speculated that petroleum might constantly be in formation, in effect precipitating from rock. But as observers of Pennsylvania oil fields saw gushers dwindle into production of only a few barrels a day with no sign of revival, the replenishment view lost credibility. Instead, geologists began to think that petroleum normally appeared in structures of limited production potential; as one produced from a structure, gas production would give way to oil and then to salt water, at which point the petroleum production ended. By 1890 American geological opinion inclined to the latter, pessimistic view of petroleum reserves.[6]

Chief among the pessimists were Pennsylvania state geologist Peter Lesley and his subordinate, John F. Carll. Having seen many small fields come and go, throughout the 1880s Lesley and Carll sounded the alarm that not only were petroleum reserves exhaustible but within a matter of a few years Pennsylvania crude oil production would be finished. There was no reason to believe that old fields would revive or be significantly extended, that drilling deeper to untested rock strata would have positive results, or that future discoveries would bring in anything more than minor pools. Natural gas would be likely to run out before crude oil. In short, in Lesley's words, Pennsylvania petroleum production would turn out to be "not only geologically but historically, a temporary and vanishing phenomenon." That meant the next generation would be out of oil: "Our children will merely, and with difficulty, drain the dregs."[7] Similarly, the first issue of the *American Geologist* contained a bleak forecast of natural gas depletion by E. W. Claypole, who compared gas production to drawing down on a bank account; eventually there would be insufficient funds to cover drafts. The following year, the editor of the *American Geologist* endorsed Claypole's view, adding that petroleum production was "destined to fail and before very long."[8] In short, if one believed the experts, by 1890 one could fear that the United States was on the brink of an oil famine.

The experts agreed not only that petroleum was running out but also that the way Americans had produced their petroleum left much to be desired. Carll, for example, felt that Pennsylvania oil fields had been wastefully depleted as the state supplied the world with cheap light. Lesley went beyond geology to morality, condemning the "gambling spirit" that brought so many persons to scramble for oil profits in "so unmanly and thriftless a manner." Here he echoed earlier observers on speculation and resorted to traditional gender imagery to condemn it. The next generation, Lesley warned, would condemn what the present one had done.[9]

For Progressives, the issue of conservation linked anxiety about exhaustion and waste of natural resources with concern about monopolistic wealth. By attacking waste of resources, conservation could head off future shortages of them; by keeping resources still in public hands from falling into private possession of powerful business interests it, in effect, preserved what Progressives saw as a public patrimony.[10] Both conservation stratagems aimed at protecting consumers from a future in which monopolistic wealth could control resources in short supply and gouge consumers; just as with antitrust, in advancing conservation, the federal government would take the role of defender of the public interest against irresponsible private power. Conservation, in sum, had great political utility as an expression of Progressive belief. But it presumed future shortage of resources, and therefore, to be credible ideologically, let alone to emerge as an acceptable alternative to economic individualism, some empirical evidence had to demonstrate that, within a foreseeable future, there would be an end of abundance. Some source without direct economic interest in natural resources had to assess what and where America's publicly owned resources were and offer facts and figures justifying keeping them out of private hands in the interest of the nation. The more empirical ammunition a conservation campaign could draw from science, the better.

As Theodore Roosevelt took up the conservation cause, the scientists of the United States Geological Survey (USGS), headed by George Otis Smith, thus became very useful to him. Closely allied with Forestry Service head Gifford Pinchot; James R. Garfield, who moved from Commerce and Labor to interior secretary; and Reclamation Bureau chief Frederick H. Newell, the last a former USGS expert, Smith and his subordinates shared professional associations and policy perspectives with Roosevelt's archconservationists. Their work was obviously essential to any conservation program, not only to justify it in general terms but to do anything about its implementation. With conservation in mind, national management of public lands meant distinguishing land with valuable mineral resources, to be leased but not alienated, from land suitable only for agriculture, which could be given over to homesteaders' private ownership. Only the USGS had the staff of experts ready to take on such a classification.[11] The stage was set for a mutually supportive interaction of politicians and scientists.

It was also set for political conflict. While there were many who saw Roosevelt's suggestion to lease rather than alienate public lands bearing coal as a radical attack on private enterprise, there were others for whom Roosevelt did not nearly go far enough. Progressive senator Robert M. La Follette, for example, argued that all coal-, oil-, and lignite-bearing public lands should be kept from private hands. Caught in congressional crossfire, Roosevelt had the USGS work on identifying public lands bearing substantial deposits of coal, petroleum, and phosphates; once identified, such lands might be kept from private ownership.[12]

When they worked to carry out Roosevelt's policy by classifying public lands in terms of petroleum potential, the USGS experts faced a number of formidable problems. First, they could determine the likely presence of reserves only if they could find surface structures suggesting it or if there was production adjacent to public tracts. Second, going beyond mere possibility to confirming the presence of petroleum reserves and hence justifying continued public possession required drilling. Since mineral exploitation of public lands was covered by placer law, a wildcatter on public land had to show production before he could file for ownership. In the time he put up a derrick and drilled, a competitor could file for his tract as a homestead and gain title to it, thus setting at naught the wildcatter's investment and effort. Legal reality, therefore, made wildcatting on public land, especially in the most promising areas, so risky as to be economically unjustifiable. If the USGS was to establish the presence and extent of reserves on federal land, it could not expect wildcatters' help unless land policy changed, and if it was going to lock away resources from private exploitation, it could hardly expect that help on any terms. These problems became particularly pressing after 1905 as California oil fields opened up on private tracts adjacent to or near public land in the western San Joaquin Valley. Oilmen asked the General Land Office to withdraw likely tracts for wildcatting from homesteader entry, and the Land Office responded with temporary withdrawals.[13] But, as the USGS experts surveyed public lands near the San Joaquin Valley oil fields, George Otis Smith and his subordinate in the field, Ralph Arnold, pushed the interior secretary for withdrawal of large amounts of public land not only from agricultural (i.e., homesteader) entry but also from oil development.[14]

Smith advanced two arguments. With respect to the California lands, he drew Garfield's attention to a recent Department of Commerce report stressing the value of petroleum as steamship fuel, something the British government recognized in naval planning. Smith argued that the naval value of petroleum justified keeping public lands for their reserves. Once privately owned, such lands would produce oil the United States would have to buy back in order to fuel its ships, so it was surely more sensible to keep the lands and their oil against future need. Here was the germ of a federal naval petroleum reserves policy. Smith's other argument focused on petroleum waste. Looking at

California and Louisiana, Smith told Garfield, "There is no record in the history of the United States of such wanton waste with absolutely no effort to check it." Obviously public lands ought be kept from unbridled exploitation. As committed to conservation as he was, however, Garfield balked at permanent withdrawal of mineral lands. He and others like him would need more convincing.[15] But, in spreading the conservationist message, Smith and his USGS subordinates were assured of active presidential support; indeed, during 1908 Roosevelt called two national governors' conferences on conservation and established a national conservation commission. With the election of William Howard Taft, however, conservationists such as George Otis Smith found themselves in far less congenial political surroundings.[16]

One recourse lay in turning to the public, and, to judge from media response to Roosevelt's conferences, a receptive audience awaited. The *New York Times*, for example, had called Roosevelt's address to the first conference the most important of his administration, and its editors had used the occasion to decry "barbaric prodigality" and "criminal waste" in American use of resources.[17] Surely this kind of sentiment outside Washington could work to conservationists' advantage. At any rate, among the materials the experts at the USGS had at hand was staff member David Talbot Day's report for the conservation commission. Day had prepared an extensive study of the location and extent of the nation's known and probable oil fields, with estimates of their longevity based on production rates, the fruit of twenty years' data gathering for the survey's Mineral Resources division. The *American Review of Reviews* published his findings for the conservation commission in January 1909, and the survey followed suit the next month. As political scientists have recognized in other instances, Progressive administrators commonly emerged as policy and administrative entrepreneurs by promoting their specialties and agencies through public discourse.[18] Day's report was well suited to capture public attention, for its message, conveyed with rhetorical restraint and an abundance of statistics, was simple and dire: the United States would soon be out of oil.

Day's forecast, an excellent example of ideologically driven mathematics, involved estimating how much oil could be retrieved from known fields, something he admitted was guesswork.[19] His calculations relied primarily on the amount of production already obtained and the rate of increase of production of oil taking place from year to year, in effect an adaptation of the bank draft perspective on oil production. These statistics, a record of oil that reached market, gave Day hard data. Beyond this, Day's estimates required considerable creativity, for the information he needed was as yet outside the scope of existing oil field engineering and technology. Public discourse defined his task but offered no means to carry it out. Day resorted to averages — that a cubic foot of pay sand of average porosity would yield a gallon of oil, that on average where pay sands were five feet thick, 5,000 barrels of oil per acre would be recovered, that various oil fields had pay sands of an average thickness: such computations gave Day figures that sounded authoritative even if they would not stand up to close scrutiny. Day revised estimates thus obtained downward, seeing the rapid rate at which production in some fields dwindled from early highs, and argued that in most fields, recovery per acre would be closer to 1,000 than 5,000 barrels. But, given these assumptions, Day concluded that the United States probably had somewhere between 10 and 24.5 billion barrels of oil left in known fields.[20]

Day proceeded to set this estimate against what was happening to oil production. Looking at the half century since 1859, Day pointed out that in every nine-year period as much oil was produced as had been in all previous years. Up to 1907 total production amounted to 1.8 billion barrels; by 1916 it was logical to expect that total production would be 3.6 billion barrels. In other words, in the next nine years the United States would use as much oil as it had in the previous fifty, a prospect Day found alarming. Worse yet, by that time, production in older fields such as those in Pennsylvania, New York, West Virginia, Ohio, and Indiana would be exhausted, production in the Gulf Coast and Mid-Continent known fields very much depleted, and only fields in California likely to continue to supply much oil. Prices to consumers would soar, and industrial need for petroleum would go unmet. By 1935 U.S. oil production would be virtually finished.[21]

Part of what made Day's forecast a worst-case scenario was the realistic assumption that more and more oil would be produced in response to demand, but Day made no attempt to analyze demand or see either demand or supply as responsive to price. He simply looked at past increases in production rates and reasoned from them. He projected a world without market forces. Equally important was his unrealistic projection, an outgrowth of the bank draft perspective, that there would be no additions to reserves in known oil fields or outside them. Like his Pennsylvania predecessors, Day dismissed those who would cavil: their "reliance upon unknown sources of supply after a few decades seems to be the characteristic attitude, as if these new fields of great size were a foregone conclusion."[22] Day's forecast rested exclusively on existing reserves being produced, used up. That made it worst case indeed.

If only the rate of oil production stopped increasing, Day thought existing fields might be good for nine more decades of production: but what could bring this about? A chemist by training, Day suggested that more scientific research might uncover alternatives to petroleum — cheap alcohols to serve as illuminants or power sources, or even artificial petroleum produced from animal and vegetable wastes.[23] Subsequent chemists commenting on possible petroleum shortage would echo his faith in the marvels of modern science;

ideas for manufactured alternatives to petroleum would become a staple in energy-related discourse.

It was more important, however, to stop waste and prioritize uses. Day called all exports of oil "the most profligate waste," and all use of oil rather than coal as boiler fuel was wasteful. Unlike natural gas, Day did not see loss of oil from runoffs or evaporation as a serious problem in the United States, a contrast to what he saw in other nations. But oil was produced in excess of what was needed to meet essential uses such as lubrication and lighting, and the way to attack this was through public lands policy: "Every acre of oil-bearing public land should be withdrawn from every form of entry and be subjected to a suitable and fair system of lease." To do this, Congress would have to make legal provision for leasing. More to the point, public land likely to overlie oil reserves had to be identified, requiring more knowledge of the geological features of oil fields and more mapping and survey work.[24] That, of course, was the direction in which the USGS operations headed, so, in one sense, Day offered a splendid rationale for what his agency was doing, accompanied by ominous implications for the future should its work cease.

Day's study was excellent propaganda not only for the usefulness of his own agency but also for the conservationist cause in general, as National Conservation Conference participant and University of Wisconsin president Charles Van Hise recognized. In September 1909, Van Hise gave Day's arguments additional exposure in an article titled "Patriotism and Waste" for *Collier's* magazine. Advancing the general thesis that natural resources were "extremely limited" given probable future needs of the United States, Van Hise argued that there had been "wanton waste" of oil and gas, producing data drawn from David Day's reports.[25] The following year, Van Hise produced what amounted to his own reiteration of the findings of the National Conservation Commission, *The Conservation of Natural Resources in the United States*.

On the subject of petroleum, a part of his first chapter, he repeated David Day's figures and conclusions, but with more rhetorical embellishment in the form of the language of luxury and overspending, as he would in his general conclusions: "The change from an apparent plethora of natural resources, free to anyone, to paucity, has come upon us so suddenly that the people find themselves in a position similar to that of the youth who, bequeathed a fortune, believes it far beyond his needs and draws heavily upon it . . . until one day the bank refuses to cash his check." What the American people had done with resources — comparable to money in a bank — was, in effect, the story of Coal Oil Johnny writ large. The answer lay not only in a return to frugality but, more important, restriction of private enterprise. And, with respect to petroleum, as much oil as possible should be left in the ground, "in order that succeeding generations may have the advantage of this most valuable product."[26]

While outsiders such as Van Hise publicized it, in his own agency Day's report launched the USGS on a course of public pronouncements it would follow into the twenties, and through which it established many of the main themes in conservationist discourse about oil. From 1908 to 1914, Day was responsible for the petroleum section of the USGS's annual survey, *Mineral Resources of the United States*, continually repeating alarm over mounting production and consumption. In the 1915 report, Day's successor, John D. Northrup, continued to sound the alarm by stating that it was "at variance with natural laws" to expect petroleum resources to be adequate to meet demand indefinitely: "Farsighted men within the industry itself already feel that the crest in production is close at hand."[27]

Northrup's USGS colleague Max W. Ball was not willing to go that far. Nonetheless, Ball argued that failure of domestic supply was inevitable for, like many another expert, Ball discounted oilmen's chances to continue to discover large fields. Unlike Day, however, Ball charged the oil industry with wasting what was produced, particularly in faulty storage through which many valuable petroleum fractions evaporated. Looking at California alone, Ball estimated loss from evaporation to amount to 25 percent of the value of oil production, though he did not offer evidence to support his estimate. He also singled out legal obstacles to conservation, particularly the law of capture, which gave ownership of oil to the person who produced it, even if he drained adjoining properties. As Ball and others pointed out, this principle obliged an oilman to drill quickly and to produce at top capacity, either to protect his well from those on adjoining tracts or to pump oil from under his neighbors' properties.[28] These themes of conservationist argument would be developed at great length in succeeding decades.

In all, the experts at USGS did more than anyone else to establish the basic themes of oil-related conservationist discourse. As they did so, their cry for conservation found a bureaucratic echo in another branch of the Department of the Interior, the newly created Bureau of Mines. From its beginning in 1910 to his death in 1915, the bureau's head was Joseph Austin Holmes, former USGS employee and member of Roosevelt's pro-conservation circle. Holmes saw his department's mission as promoting mining safety and increasing efficiency in producing and using mineral products, and he had no doubt that there was "great waste in the production and handling of petroleum and natural gas." Indeed, he wrote on the waste of petroleum for the *Annals of the American Academy of Political and Social Science* the year before his Bureau of Mines appointment.[29] Accordingly, under his leadership, the bureau began a series of technical papers and bulletins on petroleum technology intended to stress oil field safety and prevention of waste. To press for the latter objective, bureau experts had to come up with statistics.

Doing so was easier said than done, for as with the USGS survey of oil reserves, discourse drove data rather than vice versa. In December 1912, the bureau assigned former Indiana state geologist Raymond S. Blatchley to investigate waste of petroleum, both past and present, in the Mid-Continent of Kansas, Oklahoma, and North Texas, a daunting task. As in other early-twentieth-century oil fields, in the fields Blatchley visited it was easy to see that substantial amounts of oil and gas brought from underground were put to no use whatever. Operators flared casinghead gas, gas produced with oil, for which they had no local market; wells roared out millions of cubic feet of gas no one expected or wanted as operators hoped gas flow would give way to oil; lacking storage and separation technology, operators let crude flow into earthen reservoirs from which lighter petroleum fractions evaporated; operators permitted residue of separation to flow over the ground and into nearby streams and creeks. But for Blatchley such familiar visual evidence of what could be called waste was not enough: he had to quantify what he saw and try to use his figures to estimate what had been lost in the past.

Had Blatchley been able to work with reliable production data for the thousands of wells covered by his assignment, he still would have had an enormous undertaking. Many wells, however, had no accurate production records, much less refinements such as gauges to measure flow of oil and gas. In the absence of reliable technology, estimates of variations in reservoir pressure or productive capacity were more a matter of guesswork than measurement or engineering. Indeed, petroleum engineering was in its infancy, and the petroleum industry had not yet developed the kind of operational ability or, in effect, operational discourse to describe what Blatchley was supposed to do. Driven to disclaimer, Blatchley ultimately admitted, "It will never be possible to estimate the amount of either [past or present] loss with accuracy because of the lack of information and the extent of development. The estimates that are given must necessarily be mere approximations."[30] Nevertheless, by relying heavily on an earlier report on gas production done by Kansas state geologist Erasmus Haworth and by interviewing individuals personally familiar with fields he had to study, Blatchley came up with figures dramatic enough to please Washington conservationists.

Of course, his figures had considerable elasticity. Of waste in the Cleveland, Oklahoma, region, for example, Blatchley reported, "Very few of the Cleveland wells were gauged, but it is thought that each well produced from 3,000,000 to 30,000,000 cubic feet of gas daily."[31] Blatchley drew even more astounding conclusions: enough gas had been wasted in Mid-Continent fields to meet that region's demand for gas at present rates for twelve years; enough gas had been wasted in Oklahoma to meet state demand for sixty years; the "grand total" of gas wasted in the Mid-Continent amounted to 425 trillion cubic feet. Of the last figure, Blatchley asserted, "The writer is confident that this is a conservative estimate."[32] For a man who admitted his figures were "mere approximations," Blatchley's self-confidence was undeniable.

For the most part, Blatchley concentrated on the waste of natural gas, and when he offered figures on gas wasted, he usually referred to casinghead gas or uncapped gas wells. His interpretation of waste, however, was broader than gas put to no constructive use. Waste also included using gas for industrial fuel, selling gas at bargain rates, and, compounding error, giving free gas to industries as an inducement for their location in communities. Here Blatchley allowed himself some righteous indignation: "The greedy exploitation of the extensive gas fields of Kansas and Oklahoma by the large pipe line companies is one of the most deplorable features in the waste of natural gas."[33] Blatchley overlooked the possibility that large pipeline companies might make possible the marketing of gas otherwise put to no use, or that they might purchase gas from small producers; the very presence of large companies roused his suspicion and condemnation. In fact, this passage, ostensibly technological discourse, is a good example of a technocratic use of traditional political and moral discourse.

Blatchley's remedy for waste was "a campaign of investigation and education" that would show the industry that it paid to conserve oil and gas.[34] How fortuitous it was that this was precisely what the Bureau of Mines said it was doing. Three months after publication of Blatchley's report in 1913, the bureau held a conference of industry-related associations in Pittsburgh to promote founding a national petroleum society, a group that would promote standardization of petroleum products and cooperate with the bureau in the study of geological, technological, business, and conservation questions relating to petroleum. At the same time, the bureau continued its reports, turning out six technical papers on petroleum conservation–related topics in 1913 alone; Ralph Arnold of the USGS collaborated on three of them.[35]

While the USGS and Mines experts generated conservationist warnings, George Otis Smith successfully wooed Taft's interior secretary, Richard A. Ballinger, to accept his position on public land withdrawal. Not only did Ballinger buy the argument that current high rates of petroleum production would not last long, but he was especially receptive to the idea that future naval need for petroleum dictated setting aside for naval use West Coast public lands that might contain petroleum. For that matter, he used Smith's statements word for word, albeit without attribution, in arguing to Taft for withdrawals. At the end of September 1909, Taft responded by withdrawing 2.87 million acres of public land in California and 170,000 acres in Wyoming from all forms of entry or exploitation pending public land legislation. Then, in 1912, following Smith's recommendations on location, Taft created the California naval reserves at Elk Hills and Buena Vista Hills.[36] Given the unpromising political position from which Smith started, this was a major conservationist coup. Smith had similar success with Taft's successor, Woodrow Wilson; three years later, Wilson created the Teapot Dome, Wyoming, naval reserve.

For oilmen in areas affected by the withdrawals and creations of the naval reserves, the conservationist victory meant a serious disruption of operations. In the face of uncertainty, an operator had the choice of stopping all operations, thus fully complying with whatever course the government would take, and giving up his investment; or he could go ahead with his operations, in the hope either that federal withdrawals would be reversed by Congress or the courts or that he would simply get away with drilling and production. Many operators chose the latter option and, in the face of uncertainty, developed what they had at an all-out rate, cheerfully draining the tracts of those who had opted to be law-abiding. But that led to another problem, for purchasers such as Standard Oil of California were extremely reluctant to buy oil with a dubious title. In short order, the California crude market saw more sellers than buyers, prices fell, and producers had a great deal to be angry about.[37]

In all, there was much less to the conservationist triumph than met the eye. Ostensibly the federal government had locked away in the ground a supply of cheap oil for its battleships, to be tapped as needed. But had it been possible to say that these reserved lands contained all the oil the navy would need, or even some of it, in order to use the oil, the federal government would have to drill wells, construct gathering lines and storage, build pipelines, and set up refineries — or pay someone else to do all these things. Under ideal circumstances, producing such reserves would take months or years, something of a strategic liability. More to the point, seeing those tracts of public land supposedly containing ample petroleum as akin to a private oil barrel from which the government could draw as needed was a totally misleading and unworkable vision of oil field operation. For it to have been at all functional, the tracts would have had to encompass the whole of three reservoirs; otherwise, drilling on tracts adjoining government acreage would drain oil from government reserves. When the Teapot Dome reserve was created in 1915, no one could be sure it completely encompassed a reservoir — later geology showed it did not. But one could be sure that the California tracts certainly did not ensure ownership of reservoirs because acreage in them was checkerboarded between federal and private ownership. What made perfect logic in conservationist discourse did not translate into operational coherence.

Thus, while federal officials thought to ensure the navy's future supply of oil by keeping Elk Hills and Buena Vista federal acreage unentered and undrilled, they in fact guaranteed that operators on adjoining sections could drain Uncle Sam's oil as well as their own. True, in 1912 Smith and his colleagues could argue that little of the private acreage had been drilled. It was said that the Southern Pacific was waiting for higher oil prices before opening it up to operators, and to help along the conservationist cause, the Justice Department began a campaign in 1910 to cancel the Southern Pacific's patents, a campaign that would end unsuccessfully in 1919.[38] The fact remained that it was virtually impossible to keep private operators from drilling near federal reserves. When they did so, they made it impossible for the federal government to keep its intended hoard of oil. Conservationist imagery notwithstanding, oil in the ground did not behave like money in a bank vault. Unlike money, it moved.

Apart from conservationists, only a person profoundly suspicious of business in general and the oil industry in particular could see real economic justification for a naval underground oil hoard. Such a person was Woodrow Wilson's secretary of the navy, Josephus Daniels. A self-proclaimed Jeffersonian Progressive from North Carolina who preached against concentrations of wealth and railroad conspiracies, Daniels was more antibusiness than he was committed to conservation. He clearly would have liked a navy virtually independent of contractors for its supplies, and in his first report he argued that the navy should be spared expense by the federal government's taking over manufacture of its armor plate and munitions. But what Daniels most wanted was a navy with its own private oil supply, a navy that would no longer "fatten the pockets of a few oil companies." Rather than purchase oil at "exorbitant and ever increasing prices," he wanted the navy to produce and refine its own oil, and, to that end, the naval reserves were a step in the right direction. In short, he wanted the federal government to go into the oil business, a desire shared by the Minnesota legislature, which decided it would like to see federal ownership of the entire oil industry.[39] Daniels eagerly took up the campaign to safeguard California reserves by taking back the Southern Pacific Railroad's patents, arguing that it had gotten them through fraud. Here he had an ally in Attorney General Thomas W. Gregory, an old trustbuster and railroad foe, and his assistant, E. J. Justice, a personal friend from back home who was more avidly antitrust than his superior.

Justice's position in what began as a conservationist issue offers a useful example of older antibusiness discourse brought to bear in a new context. For Justice, who took a leading role in Senate hearings on oil leasing of public lands, the California opposition to public land withdrawals once again demonstrated an unholy alliance of railroads and oil; the Southern Pacific Railroad's acreage throughout California naval oil reserves represented an attempt to defraud the federal government because at the time it got the land — well before any oil discovery, let alone the creation of federal naval reserves — it "foresaw there would be developed enormous oil fields." The oil companies pressing to drill on federal land, albeit they included firms, such as Caribou Oil and Monte Christo Oil, that did not appear to be industry titans, were in fact representing big oil interests, including Standard Oil, all "associated together in fixing prices for the consumer and producer." Justice used the familiar discursive elements of uncontrolled power, conspiracy, fraud, and harmful size against oilmen whom he condemned as a group. Like Daniels, he did not limit industry malfeasance to companies with Standard Oil in their names.[40]

Throughout the decade from 1910 to 1920, congressional committees wrangled over how to settle differences between strident preservationist conservationists and antitrusters on the one hand and angry oilmen on the other. In the long run, this political combat culminated in the Teapot Dome scandal, whose antecedents and political history have been so thoroughly examined by such scholars as J. Leonard Bates and Burl Noggle as to make repetition unnecessary.[41] In the short run, as World War I pushed national security questions to the foreground, the controversy over naval reserves generated a new, imperialist theme in petroleum-related discourse, and the person perhaps most responsible for advancing it was California oilman and self-styled petroleum engineer Mark L. Requa.

A onetime copper mine promoter, Mark Requa became an oilman in the Coalinga oil boom of 1908. He became a director of Independent Oil Producers Agency. Though the agency had originally been a coalition against Standard Oil and the Southern Pacific Railroad, Taft's land withdrawals put old adversaries on the same side of the political fence. A skeptic about the merits of Standard Oil's dissolution, Requa saw no benefits to independents from the breakup; it merely made the men from Standard Oil gun-shy of cooperation on the problem of too much oil and too few purchasers. By his own telling, he began to rethink industry problems and questions of national energy future.[42] When he finished, Requa had written an article on petroleum resources that California senator James D. Phelan found so useful he had it reprinted as a Senate document in 1916. And no wonder: what Requa did was take established elements of conservationist discourse and put them together in a new way that stood the argument of Daniels and the naval reserve proponents on its head.

Relying on David Talbot Day's 1909 USGS report, a document he termed of "terrific significance" and which he quoted at length, Requa brought forward familiar USGS arguments — that, if consumption kept increasing, the United States had only enough oil left for a few decades; that many older producing regions were close to exhausted; that the California, Gulf Coast, and Mid-Continent regions were the only hope of sustained production; and that one could not assume new discoveries would be so great as to meet need. But where Day had simply suggested that some uses of oil amounted to waste,

Requa used the rhetoric of extravagance with poetic verve worthy of Holly-wood: "Our way of prosperity makes us careless of the future; we feast and revel while the handwriting blazes on the wall in letters of fire. . . . As a nation we are wasteful, apathetic, and forgetful. We waste our natural resources with shameful prodigality; we are apathetic of the future, and we forget that our reserves of natural wealth are by no means inexhaustible." In but a few years, Americans would face, "subdued and chastened, the real truth"—that their oil was used up. When that happened, there would be "commercial chaos or commercial subjugation by the nation or nations that control the future source of supply from which petroleum will be derived."[43] Extravagance would lead to national weakness, a traditional gender-tied image.

Construed on Requa's terms, national security might better be served, not by setting aside oil on public lands for future naval use, but by not having the navy burn oil at all. Requa reminded readers that when battleships burn oil, "they are consuming the very lubricants without which war vessels must lie as idle as a painted ship upon a painted ocean." At the very least, Requa argued, ships ought to burn refinery residuum rather than crude oil. But, if domestic reserves were rapidly depleting, how could the nation's future security be tied to them? Surely the wisest policy was to acquire oil reserves in foreign coun-tries, particularly in Latin America. Other powers, especially Britain, were moving to acquire overseas reserves; in fact, they had even picked up reserves in the United States and were closing in on Mexico. If the United States did not act soon, "when it is too late we will awake to the fact that the oil resources of the world are in foreign hands, and that, so far as its lubricants are con-cerned, the United States has become the vassal of some foreign power."[44] In other words, it would be folly to think future national security could be guar-anteed merely by setting aside oil reserves for the navy. Requa countered the naval reserves argument with a call for an aggressive economic imperialism that would drain other nations first. Mexico, in particular, was an attractive target.[45]

In the meantime, Requa followed the lead of the Bureau of Mines and urged that something be done about waste of petroleum. He condemned production in excess of market demand and storage in earthen reservoirs that allowed seepage and evaporation, as well as the use of crude oil for boiler fuel. As chiefly responsible for such waste, Requa singled out small oil producers, often "utterly unfamiliar with the business," who produced crude regardless of market demand and sold it, unrefined, to anyone who would buy it. By con-trast, his praise went to Standard Oil—he spoke as though it had not seen dissolution—because the company always refined oil products from its crude and sold only residuum for fuel. Comparing Standard Oil with small operators, Requa saw "the one conserving the Nation's resources, the other destroying them"; when it came to conservation, Standard Oil was on the side of the angels.[46]

If Requa's article was far from the last word in the naval reserves controversy, it did offer an alternative channel in conservationist discourse, one distinct from the antitrust position. It envisioned a cooperation between government and industry in resource management, and, indeed, economic imperialism, rather than laissez-faire. The article also established his credentials as an oil industry expert. With the United States' entry into war in 1917, another of Requa's friends, Herbert Hoover, gave him a place in the Food Administration, but in January 1918 he moved to the directorship of the Oil Division of the Fuel Administration, where he served for a year and a half.[47] His bureaucratic title was further confirmation of his claim to expertise.

Once in Washington, Requa did not forget his California friends; he put many of them in Oil Division places, though their qualifications were often dubious. When posts were filled with non-Californians, they tended to go to small refiners, whose interests would be well heeded by Requa's division. The head of the Bureau of Conservation, for example, was W. Champlin Robinson of Baltimore, whose efforts concentrated more on economical use of natural gas than anything else.[48] No Standard Oil personnel appeared on the division's staff, a reflection both of Requa's sense of obligation to friends and of practical politics.

In his division position, Requa also supported California oil interests on issues. As J. Leonard Bates has noted, he urged opening withdrawn public lands to drilling and development, arguing that the Pacific Coast would run short of oil if this did not take place. Above all, many California oilmen, E. L. Doheny in particular, wanted the Buena Vista reserve opened up; they pointed out that its checkerboarded acreage meant it was already being drained. Naval reserve proponents countered that before this was done, all exports of California petroleum should end.[49] In effect, both camps used themes in existing conservationist discourse, with Requa and friends arguing that the Pacific Coast, rather than the nation, would run out of oil, and navy advocates claiming that ending waste in the form of exports would head off shortage problems. The political result was continued deadlock over oil development and naval reserves.

Following Requa's ideas about the cooperation of government and business, his Oil Division worked closely with the oil industry's National Petroleum War Services Committee. Prices of crude pushed upward by wartime demand stirred up familiar public anxieties about excessive profits and price gouging, and, because Requa's division had no power to control prices, he depended on the committee's help to achieve a lid on them in August 1918. As appropriate to a discourse shaped by Roger Sherman, Henry Demarest Lloyd, and Ida M. Tarbell, the division was solicitous of the interests of small businessmen in the industry, especially small refiners; it took pains to see that they were allocated crude to refine, and the lid on crude price rises was a further boost to them. Adopting David Day's ideas on conservation, the division stressed prioritizing the use of petroleum, especially natural gas, and avoiding waste; Requa called this the division's "missionary work." What the majority of Americans would have noticed most about the division's work, however, were the gasless Sundays it mandated from August to October 1918. Consumers were urged not to use automobiles, motorcycles, or motorboats for pleasure on the one day they were likeliest to do so. Requa, in effect, imitated the frugality policy of his friend Hoover's Food Administration, which had put America on wheatless Mondays and meatless Tuesdays: "joy riding" became unpatriotic.[50]

At the war's end, Requa offered some guidelines for future national energy policy. For all his faith in government-industry cooperation and his belief that industry should be "free and untrammeled," he had chafed under the limitations of his division's power, especially with respect to control of petroleum prices but in other ways as well. Thus, in some future conflict, a fuel administration should be able to control all petroleum and products. It should have full supervision and control over all phases of the petroleum industry, even though it would leave the actual operation of the industry to its peacetime participants; it should have "sole authority over the entire petroleum problem [*sic*]."[51] This, of course, assumed that supplying petroleum in wartime was a problem, one beyond private capacity to resolve. It also assumed that government could supervise industry functions more capably than the industry itself. Later, Harold Ickes would reach the same conclusions.

Requa also repeated his call for economic imperialism directed toward acquiring petroleum reserves overseas. Once again, he sounded the alarm of future shortage. Petroleum and plenty of it was vital to the United States' future, and encouraging oil companies to acquire overseas reserves should be a national priority. If oil companies had to cooperate with one another to do this—here was tacit recognition of the old bogey of monopoly—so be it; Washington should not only permit that cooperation but give it "most hearty and sympathetic support."[52]

Certainly Requa's position received hearty support from USGS director George Otis Smith and Bureau of Mines director Van H. Manning, for both built the argument for importing more oil into their positions on conservation. In frequent public appearances both also continued to warn Americans about running out of oil. Smith, in particular, sought out professional meetings such as those of the American Institute of Mining and Metallurgical

Engineers (AIME) and the Iron and Steel Institute to preach conservation. Thus, using the rhetoric of luxury and extravagance, he told the latter group in May 1920, "We are living beyond our means. . . . Where will our children get the oil they need?" Extending Requa's condemnation of using automobiles for fun to a peacetime environment, Smith admonished his hearers, "The use of gasoline to serve our pleasure cannot go on unchecked—the joy ride is not the kind of 'pursuit of happiness' regarded as an 'unalienable right' by our revolutionary fathers."[53] Statements of this sort by Smith and Manning commonly made the pages of newspapers such as the *New York Times*. Among periodicals, *Scientific American* gave them extensive coverage. A general reader interested in "scientific" opinion on the subject of the country' oil reserves could thus have concluded that the United States was running dry.[54]

To the dire forecasts of shortage the petroleum industry had no unanimous response. On conservation issues, as on other policy questions, what oilmen said was more likely to reflect their individual business strategies rather than a coherent industry perspective; this was as true after the founding of the American Petroleum Institute (API) in 1919 as before. In one respect, oil producers and royalty owners appeared to have benefited from the idea that oil was running out at a dangerously rapid rate, for in 1913 they first received a tax break in the form of an oil depletion allowance of 5 percent of gross value of annual production. The depletion allowance was modified a number of times before it was set at a rate of 27.5 percent in 1926, and its discussion seems to have generated remarkably little controversy on Capitol Hill, perhaps because those who advanced it said it would benefit small independent producers—which it did, just as it also benefited large companies. Similarly, in 1916, prospectors first got the tax incentive of expensing intangible drilling costs. Both tax breaks could be taken as incentives to find and produce more oil, something that might at least keep up the nation's energy supply.[55]

Whether or not oilmen benefited from the cry that the United States was running out of oil, discourse in industry journals and statements of oilmen indicate that few of them saw potential national oil shortage as a problem to be taken seriously. Between 1912 and 1917, the editors of the *Oil and Gas Journal* were far more concerned with airing industry complaints about California public land withdrawals than with whether the United States would run out of oil. As they told their readers, fear of running out of oil was groundless, one of the "many objectionable matters [that might] be charged up to the conservationist movement." Forecasts of shortage were "fanciful guess work," and as for oil running out in several decades, "practical oil men" knew better.[56]

Indeed, from time to time, the *Oil and Gas Journal* published what industry participants said, and those quoted almost always dismissed the danger of "oil famine." Thus, in 1920 California oilman and former Fuel Administration member Thomas A. O'Donnell, first president of the API, regretted that "the public has been frequently alarmed by statements of well-meaning and learned scientists, predicting an early exhaustion of our petroleum resources." There had been many such predictions in the oil industry's history, wrong every time; one could expect important domestic discoveries and domestic production to continue "long after the time limit set for exhaustion by some of our experts."[57] The following year Harry Sinclair told the API's annual meeting, "There is plenty of petroleum and always will be. Exhaustion of the world's supply is a bugaboo. In my opinion, it has no place in practical discussion."[58] By far the most eloquent in this vein was H. G. James, the president of the Missouri Oil Jobbers Association. As he put it in an *Oil and Gas Journal* of 1920, "I am wholly out of sympathy with those croakers who are constantly keeping the public mind inflamed with dismal predictions of declining production and nearby exhaustion of the supply of petroleum. The surprising thing is that some oil men engage in the same sort of bunk or are persuaded to approve what is being said by others."[59]

Some industry observers developed creative explanations for belief in shortage. Thus, during the war, the *Oil and Gas Journal* decided that some rumors of shortage were the work of "the untiring German propagandist," out to create producer and consumer discontent. After the war, H. G. James decided that coal producers were most responsible for keeping the shortage idea going, doing so to scare consumers away from switching to fuel oil. By contrast, a group of Kansas oilmen saw predictions of oil famine as "pernicious propaganda" circulated by large oil companies to encourage too much activity among producers and subsequent overproduction that would drive oil prices down. Oil promoters, however, learned that tales of oil famine could drive up the prices of shares they were selling.[60]

In general, even if they did not believe the United States would run out of oil, oilmen could use elements of conservationist discourse to advance their own business strategies. In particular, after 1918 some adapted Mark Requa's arguments into a demand that the United States government help American oilmen acquire reserves overseas. Their argument for foreign reserves tied conservationist apprehensions to nationalist rivalry and antimonopoly sentiment by stressing that if American oilmen did not receive government support the British would grab foreign reserves and dominate world oil. Britons would let America supply them with petroleum until supplies ran out and then offer supplies from Latin America and the Middle East to Americans at exorbitant prices: here was Mark Requa's commercial subjugation in action. This current in discourse received a tremendous boost in 1919 when British oil promoter Sir Edward MacKay Edgar, seeking to pump up stock sales in his Venezuelan Oil Concessions, told readers of *Sperling's Magazine* that the United States had wasted its oil and would soon be dependent on British oil companies for supply: an adroit use of conservationist discourse in business strategy, Edgar's remarks also served American oilmen who wanted State Department assistance in dealing with foreign powers.[61]

Among the oilmen most prominent in arguing for government support in getting foreign reserves were Standard Oil of New Jersey (SONJ) executives A. C. Bedford and Walter C. Teagle. Only two months after Armistice the *Oil and Gas Journal* reported that both felt it vital to conserve American oil while maintaining control of foreign markets — a position requiring reserves abroad. Using the USGS estimate that more than 40 percent of U.S. oil had been produced, Bedford said, "Our position in this most essential industry is not nearly so secure as it ought to be." Echoing Mark Requa, he concluded, "I particularly hope that public opinion will demand cooperative effort [of government and business] looking to the extension of our holdings of oil lest we be caught in the position of a petitioner for oil in foreign markets."[62] Teagle told the API that the United States was spending its petroleum wealth for the world's benefit and that it was imperative to develop oil resources in foreign lands. If this sounded like Mark Requa, that was not surprising. Teagle was a friend of Requa's and, while serving on the National Petroleum War Service Committee, shared Requa's Fuel Administration office. Like Bedford, Teagle complained that Americans were treated unfairly overseas; while foreigners were free to exploit American oil fields, they barred Americans from sharing foreign supplies in their hands. His company felt it could no longer depend on domestic wildcatters for its future supply. He did not explain that his company, left crude-short at dissolution in 1911, had not in its brief history been able to acquire domestic reserves adequate to its projected needs in the expanding American market. Holding market share would require that SONJ increase its reserves through foreign investments.[63]

Both the editors of the *Oil and Gas Journal* and API president Thomas A. O'Donnell echoed the SONJ executives' call for foreign reserves. In frequent editorials the journal warned of the British oil menace.[64] It told its readers that large oil companies — small ones could not compete — should pick up oil supplies all over the world; without support from the United States government, Americans would not have entry to foreign oil fields. If government did not support the efforts to pick up reserves overseas, the United States would be "left to deal with foreign oil monopolists who have planned to control the industry throughout the world and to eventually bring the oil consumers of America to accept such supplies as may be vouchsafed to them and at prices that may be fixed." Here, of course, was traditional antimonopoly discourse given an international context. O'Donnell also called for cooperative action of government and industry in acquiring foreign reserves. He noted, like Requa, that Americans needed to "abandon that indifference to the morrow which has hitherto characterized their attitude toward the petroleum industry and its problems at home or abroad." Working from conservation discourse, O'Donnell made it seem unpatriotic not to support oilmen in every way possible, certainly a perspective the industry could accept.[65]

As industry voices argued for foreign reserves and drew from the discourse of the prophets of shortage, one theme was conspicuously and ironically absent from their pronouncements: the idea that America was running dry. Indeed, the editors of the *Oil and Gas Journal* and the API's O'Donnell went to some lengths to reject the idea, though the editors always gave respectful consideration to SONJ's campaign for acquisition of foreign reserves. One can show, as we have done elsewhere, that oilmen could use the argument for foreign reserves to advance business strategy without accepting its corollary that the United States was running out of oil. To the extent that foreign policy makers believed the alarm, they may well have been encouraged to lend their support to oilmen eager for overseas support. But for oilmen, there was good reason to dispute the idea's currency among other federal officials. If the situation was as grave as people such as George Otis Smith indicated, that might warrant strict regulation of the industry on all levels. Few, if any, industry participants could accept such a development. Indeed, taken to its logical extreme, the idea might justify a complete shutdown of domestic production on the lines of what the government tried in a very limited way in the naval reserves. Nobody in the industry wanted that.[66]

In a little more than a decade, the main elements of a conservationist channel of discourse on oil emerged. Antimonopolists certainly helped shape those elements, but their principal originators were federal bureaucrats, scientific experts at the USGS, Bureau of Mines, and, in the case of Mark Requa, the wartime Fuel Administration. Arguably, those who developed the ideas that the United States was running out of oil, that it was wasting its petroleum, that it needed to change consumption patterns of petroleum, and that it should consider using the oil of other nations before using its own did so in part from personal conviction. Certainly they had no hesitation in falling back on moral discourse to support their condemnation of what they saw as waste and to speak for what they saw as the interests of the nation and its future citizens. But the shapers of oil-related conservationist discourse in federal offices were supporting their own positions as well as moral conviction; the latter readily fit the former. As "ministers of reform" the bureaucratic conservationists advanced the Progressive cause of government by experts in the public interest: and they were the experts who would be in charge. Shaped by these experts, conservationist discourse could ultimately be brought to support the idea of an oil industry essentially run from Washington. In the next decade maverick oilman Henry L. Doherty would use it to do just that.

When they developed conservationist discourse on oil, the federal bureaucrats such as George Otis Smith who did so also dominated public discourse on the subject. Their ideas received far more attention and repetition than those of industry participants on the same subjects, in part because, as we have seen, oilmen inclined to dismiss notions like an impending oil famine. Notions like an oil famine or an oil barrel for Uncle Sam or superior/inferior uses of petroleum did not translate readily into the economic and operational discourse oilmen actually used. By the end of the First World War some industry members had begun to see how elements in conservationist discourse could be turned to advance their goals, and thus some industry leaders picked up Requa's arguments to push for government help in gaining control of foreign reserves. A former oilman, Requa gave them something they could use. For the most part, however, industry participants were more attuned to older antimonopolist discourse. There were still many independents ready to complain of malfeasance on the part of large (usually former Standard Oil) companies, just as there were large (usually former Standard Oil) companies anxious to avoid federal investigation on such grounds. Moreover, when oilmen looked at their industry at the end of 1919, they saw it in the throes of the most widespread and intense boom it had experienced. That made it hard to take either the prophets of doom or the issues they raised seriously. Excepting those caught up in the struggle over oil on public lands, to oilmen conservationists were more annoying than threatening. Such complacency would vanish in the next decade.

## *A Wasting Asset*

At the same time the idea that the United States was running out of oil became part of conservationist orthodoxy, the automobile became a symbol of consumer prosperity, something many people bought before their homes had central heating and indoor plumbing. In 1921 there were 10.5 million motor vehicles on the road, and that figure jumped to 26.5 million by the end of the decade.[1] At the other end of the petroleum industry spectrum, quite contrary to conservationist pessimism, prolific discoveries took place in rapid succession, particularly in California, Texas, and Oklahoma. In some instances, these discoveries took place within city limits — in California's Huntington Beach, Signal Hill, and Santa Fe Springs, in North Texas's Burkburnett Townsite, and in Oklahoma City. That meant derricks clustered on tiny tracts and intensive drilling as oilmen tried to get their own oil before neighbors got it. The hectic pace and apparent turmoil of a town lot boom made excellent copy for journalists; as one reporter reflected in August 1923, "Oranges may be out of season in Los Angeles just now, but oil is very ripe." All a Los Angeles reporter had to do to cover oil field action was hop a bus from downtown to the suburbs.[2]

Not only did the number of giant oil fields in production multiply, but so did the number of oil companies exploiting them. The twenties saw the ranks of large integrated companies grow rapidly. The companies once part of the old Standard Oil and now large freestanding operations were joined by Shell, the Texas Company, Gulf, Union, Phillips, Continental, and Skelly. Other companies, such as Humble, Magnolia, and Sinclair, were affiliated with Standard Oil companies but had large-scale integrated regional operations. The multiplication of middle- and small-sized oil companies was far more phenomenal; every year, from 1917 through 1924, hundreds of new companies appeared. Thus, in terms of firms in the industry, there were literally thousands of legitimate players in oil by the end of the twenties.[3]

And there were, in oil ranks, the not-so-legitimate players as well. The twenties saw hundreds of thousands of Americans of all classes and income brackets succumb to the urge to get rich quickly by investing in "black gold." To some extent these investors may have been nudged toward oil by talk of scarcity and higher oil prices; more often they responded to the glittering promises of promotional literature, embroidered upon by traveling stock floggers. Hapless investors and predatory promoters also made prime copy for journalists, allowing them to dwell on the traditional theme of oil as undesirably speculative, adding the newer twist of oil exploiting gullible poor people. Even the better educated and more affluent investors, however, lost money in crooked oil promotions, giving many individuals a personal reason for seeing the industry in a negative light.[4]

Even if one did not follow the journalists' accounts of oil fraud, in 1923 it would have been virtually impossible for any newspaper or magazine reader not to encounter write-ups of the unfolding Teapot Dome scandal. Interior Secretary Albert B. Fall resigned and eventually went to prison. In what must surely be one of the least astute assessments made before the scandal's full extent was evident, the editors of the *Oil and Gas Journal* remarked that Fall's resignation was "a distinct loss"; and they hoped a man of "similar characteristics" would take his place.[5]

Navy Secretary Edwin Denby and Attorney General Harry Daugherty eventually had to resign; oilmen E. L. Doheny and Harry Sinclair went to trial on bribery charges and were acquitted, but Sinclair did jail time for contempt of court and Senate; and the reputations of many others — most notably Franklin K. Lane, Thomas W. Gregory, William G. McAdoo, and Standard Oil of Indiana chairman Robert W. Stewart — were tarred with the onus of oil-generated corruption. In all, as several scholars have noted, it was the greatest twentieth-century political scandal before Watergate.[6]

The investigation of Teapot Dome, like the original creation of the naval reserves, could be seen as another victory for Progressive conservationists; after all, Gifford Pinchot and his friend Harry Slattery had done a great deal to whip up the charges against Fall, and Senator Robert M. La Follette had taken a highly prominent part in the assault against leasing the reserves in the Senate, a part taken with a view to his own presidential candidacy. Pinchot used the pages of the *Saturday Evening Post* to construct what had happened as a battle over conservation, one in a long war, in which the righteous had been Theodore Roosevelt, Robert La Follette, and, by implication—with reference to forests—himself, and the forces of darkness people such as Albert Fall. Striking a suitably moral note by calling his article "Ships, Oil, and the Ten Commandments," Pinchot effectively cast himself as Moses by reminding readers that oil on public lands had to be kept in the ground against future need, both of the navy and consumers: "We are going to be short of oil before long, and then these lands will supply our needs and help to keep down the price of gasoline."[7]

Most commentators, however, did not see Teapot Dome as primarily about conservation, or even, surprisingly, as about the oil industry, as much as about traditional political bribery and corruption: surprisingly, because the theme of oil corrupting politics had been well established by the opponents of Standard Oil. Looking at the extensive newspaper coverage of Teapot Dome, one can find journalists and politicians alike pursuing the old theme of corrupt money-eyed interest at work. The *New York American*, for example, thought the oil industry ought to be subject to more regulation, for "the oil business, unorganized and suffering from savage and wasteful competition, need[ed] more and not less contact with the federal government." But Teapot Dome did not mean the industry itself was "rotten to the core." Instead, the scandal showed the "almost universal rascality and plundering," the "vast fabric of bribery and plunder and corrupt politics." *Literary Digest* and the newspapers it quoted shared this traditional perspective. By contrast, the editors of the *New Republic* rather quaintly observed that the scandal could not have happened in the Roosevelt or Wilson administrations, and as vigilance relaxed, corruption reappeared. They apparently had not listened when E. L. Doheny said whom he paid off.[8]

Because most commentators presented Teapot Dome as primarily about politics and incidentally about oil, the Teapot Dome scandal was a less important element in the evolution of industry-related discourse than one might expect. By the same token, exactly how important it was is harder to assess. Because congressional investigations were lengthy; because, in the end, so many politicians were involved in revelations; because Fall, Doheny, and Sinclair were not tried together or on the same charges; because revelations concerning the Teapot Dome leases, the Continental Trading Company, and Robert Stewart emerged later, in 1928, the scandal was in and out of the news for the better part of the 1920s. Allowing that corruption was the focus of journalistic concern, any informed reader knew that the bribers had been oilmen and their object had been getting control of oil reserves. In terms of public image, it is probably safe to say that none of this was to the industry's benefit.

Apart from the Teapot Dome scandal, when oil was the subject of commentary, whether on the part of journalists, economists, bureaucrats, or other nonindustry observers, conservationist discourse was the main channel of discussion, with antimonopoly discourse assuming an important but distinctly secondary role. Within conservationist discourse, the most important themes were the familiar ones of running out of oil, the need for foreign reserves, and waste, with a distinct emphasis on the last. As the *Outlook* put it in 1918, there was one "superlative danger" to oil resources: "That is not the danger of monopoly. It is the danger of waste."[9]

One reason familiar themes in discourse continued into the twenties was that the same people continued to speak out and authors drew on the same sources for information. George Otis Smith and the United States Geological Survey (USGS), for example, continued to warn Americans that they were running out of oil, and both newspapers and periodicals carried their statements.[10] Mark L. Requa wrote a series of lengthy articles for the *Saturday Evening Post* between late August and the end of October 1920, in which he repeated his call for the United States to pick up oil reserves overseas.[11]

It was easy to move from Requa's recommendation of foreign reserves to a jingoistic advocacy of them, and some journalists continued to single out Britain as the United States' great rival in this regard, responding to Edward McKay Edgar's boasts and reacting much like the editors of the *Oil and Gas Journal*. Gregory Mason, who made his perspective on reserves clear by titling an *Outlook* article "America's Empty Oil Barrel," thought that unless the United States acted at once, Britain would have "oil mastery of the world" in a few years. Similarly, Edward G. Acheson repeated Edgar's boast that the United States would be buying oil from Britain in a decade and warned that that would be disastrous to national interests. The *New Republic*'s Edward Mead Earle, however, dismissed Anglophobia as a fiction circulated by Albert Fall; he thought the idea that the United States was running out of oil was propaganda that had been made up by "the oil interests themselves."[12]

Most commentators who believed the country was running dry were less inclined to blame the British than their fellow citizens in general and, increasingly, the oil industry in particular. They talked about "huge wastes," "wanton waste," "criminal waste," and "reckless use."[13] Probably the most stinging indictment of the industry on the score of waste came from Walter N. Polakov, who told the *New Republic*'s readers that oil was a "grotesque industry sapping the crust of our earth of tens of thousands of years' accumulation of solar energy within less than one century." True, consumers of petroleum products were guilty of "the same reckless, unintelligent squandering of our rapidly vanishing resources" when they used gasoline and fuel oil. But on the part of the industry, waste was "criminal waste"; national wealth was deliberately "sacrificed by private interests" for quick profits. Wildcatting, in particular, was "trying to get something for nothing," "blind, ignorant speculation."

Here Polakov fell back on a familiar theme of moral condemnation, that of speculative wealth.[14]

Perhaps most interesting about Polakov's charges of waste was that he saw the American petroleum industry as symbolic of a more basic moral shortfall in national life. It embodied the American inclination to gamble and speculate; it was technologically progressive and "socially criminal." For Polakov the symbolic role once played by Standard Oil with respect to the evils of big business had been enlarged to let the entire oil industry stand for what Polakov saw wrong with the whole of American society. For Polakov, as for the earlier critics of Standard Oil, the situation called for "public intervention." Precisely what this would be was unspecified, but faithful *New Republic* readers could guess; as early as 1916 the magazine's editors had agreed with Robert La Follette that vital national resources ought to be publicly owned.[15]

It was one thing to use a normative term like "waste" to beat the American capitalist system about the ears; it was quite another to construe it in a way that made economic sense, that could translate somehow into units or dollars. What, after all, was the economic meaning of "waste"? To take the example of the wild gas well, was waste something that took place when gas, which could conceivably be used, was not used? What if putting it to use involved expense in excess of return? In that event, would not use of such gas amount to waste? How, in short, did one translate conservationist rhetoric into sound economic theory and practice? That was the task taken on in 1918 by economists Richard T. Ely and Ralph H. Hess, both at the University of Wisconsin, whose president, it will be recalled, was conservationist Charles Van Hise. Working with colleague Charles Leith and Harvard political economist Thomas Nixon Carver, they produced a collection of essays, *The Foundations of National Prosperity*, that attempted to develop a theoretical basis for conservationist ideas, some "principles of universal application," as Ely put it.[16]

The principles Ely developed, notwithstanding his insistence that conservation was "in large part economics," had far more to do with normative ideas than quantifiable relationships. Ely's definition of "conservation" involved keeping resources in "unimpaired efficiency" or allowing "wise exhaustion" of them. "Wise" conservation involved "wise property relations," which turned out to mean a "fair return" to labor and capital with any surplus put to public use. Not only did all this beg the sticky question of whether resources should be preserved or utilized, but it also relied on a tangle of normative terms— "unimpaired," "fair," "wise." Ely came somewhat closer to the quantifiable when he got to waste; if it cost more to use something than not to, use was wasteful. So far so good, but he then had to confront the conservationist's question of whether cost computation meant taking into account the needs only of the present or those of the future as well. Ely admitted that conserving resources meant "a sacrifice of the present generation to future generations," but how did one quantify sacrifice? And how did one quantify the needs of the future?[17]

The further Ely went with his attempt to define waste, the more reliant he became on normative thinking in the form of old ideas about frugality and luxury. Thus, there was "absolute waste"—destroying something with no measurable gain; "waste plus," where something was put to use but in a socially harmful way; and "relative waste," where something was used in a way "disproportionate with need." In traditional terms, what Ely's first two wastes amounted to was extravagance, while "relative waste" was another way of saying "luxury."[18]

But who was to decide what was socially harmful or disproportionate to need? Surely, as Ely argued, this decision could not be made by ignorant, profit-seeking individuals. Laissez-faire, unregulated competition was incompatible with conserving both natural and human resources. Particularly with respect to minerals, destructive competition ought to yield to some form of public control and ownership. At the very least, minerals on public lands had to be kept from private exploitation. Ralph Hess agreed and went further: privately owned resources ought to be under public supervision, and the "more exhaustible and highly essential" resources such as petroleum ought increasingly to be brought under public ownership.[19]

While Ely and Hess both raised questions about the benefits of unregulated competition, in the early 1920s members of Congress were more inclined to complain of its being lacking, at least with respect to oil, and to keep the old antimonopoly cry going. Some did so in the time-honored role of defenders of independent oilmen menaced by Standard Oil—no matter that the trust had been dissolved for a decade. In congressional circles, the man who worked harder than any other to keep the crusade against Standard Oil and monopoly alive was Robert M. La Follette. The Wisconsin senator seldom missed an opportunity to fulminate against Standard Oil, "the most brazen, the boldest, the most aggressive violator of the law that legislative bodies in any country ever had to contend with."[20] Years of debate on the leasing of public lands for petroleum development and the issue of naval reserves allowed La Follette to attack Standard Oil repeatedly and to defend the position taken by Josephus Daniels, that petroleum on public lands should not fall into the hands of monopoly: that is to say, Standard Oil. La Follette used familiar themes to assault Standard Oil; it had driven people out of business, crushed competition—in fact, La Follette told the public, "There is nothing reprehensible that the Standard Oil Company has not done." And, of course, the sinister presence of Standard Oil had penetrated the halls of Congress to influence debate on leasing public lands. La Follette's evidence for this, in 1919, was Standard Oil's *not* having objected to the leasing bill under consideration, proof positive that it had connived to get it.[21]

Not content with the opportunity offered by Teapot Dome to criticize oil, in 1922 La Follette used rising gasoline prices as pretext for a grandstanding Senate investigation of the petroleum industry, eminently suited as groundwork for the presidential campaign he planned for 1924. It is worth some consideration not only because such an investigation would have many future counterparts but also because the questions raised would resurface in the future. The investigation was also a splendid example of a phenomenon noted by Richard H. K. Vietor with reference to the period after 1945, that market disequilibriums giving rise to price changes precipitate policy issues and government initiatives or moves toward them.[22]

La Follette had his Senate subcommittee call up all the reports of the Federal Trade Commission (FTC) and send out questionnaires to 360 oil companies. In August it began hearing witnesses, questioned by La Follette's personal friend, attorney Gilbert Roe. La Follette's committee was ostensibly trying to determine if the petroleum industry was making excessive profits on gasoline, whether oil companies were fixing prices, and whether company ownership worked against effective competition, but the record of its proceedings makes clear that many of its members had convictions as firm as La Follette's about the answers to these questions. Industry spokesmen such as Robert L. Welch, of the fledgling American Petroleum Institute, found themselves confronted with questions that made sense ideologically but were meaningless in terms of industry operations. Thus Ellison Smith of South Carolina asked Welch how one could determine "the cost to a Standard concern of a barrel of crude oil" from season to season. In vain, Welch described how costs differed from well to well, from field to field, and prices from month to month in a competitive market; Smith responded, "We want to find out who it is that fixes the price of crude." Charles McNary of Oregon asked Welch how crude prices could fall "overnight, unless there is an understanding among the companies and all the producers that on a certain day crude prices shall go down or up." A discouraged Welch finally lamented, "I am trying to make myself clear, but I do not succeed."[23]

Welch and other industry witnesses found it equally frustrating to try to convince the senators that a monolithic Standard Oil no longer controlled American refining and retailing. He maintained that current conditions were entirely different than in earlier years. But several hours later, McNary proclaimed that there was no competition between Standard Oil companies: "The dissolution decree gave them a new name, but the interests are identical running right through." McNary and his colleagues much preferred the testimony of National Petroleum Marketers' Association president L. V. Nicholas, then under federal indictment for mail fraud, who told them that Standard Oil companies worked as a unit, of which, as he put it, "We are afraid." When questioned by Smith Brookhart of Iowa as to why independent marketers were "afraid," Nicholas said because Standard Oil cut gasoline prices; he and his organization were "morally certain" that when they could not make money it was because of the "manipulating and buying power of the Standard Oil Company."[24]

Not surprisingly, when La Follette presented his committee's report in March 1923, he emphasized, "The dominating fact in the oil industry today is its complete control by the Standard companies"—something the majority of the witnesses denied. According to La Follette, Standard Oil fixed crude, gasoline, and kerosene prices at all levels and controlled all pipelines. Making highly selective use of chunks of testimony by witnesses such as Welch, taken out of context, La Follette argued that his committee had found "intolerable conditions in the oil industry, the same conditions as existed prior to 1911 or worse. The industry as a whole, as well as the public, are more completely at the mercy of the Standard Oil interests now than they were . . . in 1911." The oil industry, especially "the great Standard companies," had made "excessive profits," "exorbitant profits," "fabulous profits," not by "economies"—they paid their directors "extravagant salaries" and ran their business in a "lavish and wasteful" manner—but by charging the public. If their control continued, Americans might soon be paying a dollar for a gallon of gasoline![25]

For these familiar grievances, La Follette had equally familiar remedies: there should be more action on the part of the Justice Department and grand juries against oil in general and Standard Oil companies in particular; freight rates paid by Mid-Continent refiners should be lowered; and pipelines should be divorced from other segments of the industry. With respect to this last, La Follette sounded an old theme of localism; ideally, all along common carrier pipelines, refineries would spring up to serve strictly local markets, in place of "transporting the product long distances from the refineries to the consuming public." Here was a quaint image of a vast network of little refineries located like railroad stations across America. Taking up the FTC's complaint, La Follette wanted a government agency to collect statistics on stocks of petroleum and products, and to make this easier, he wanted oil companies to be required to keep their books so "the reasonableness of the prices charged for any petroleum product can be ascertained on a cost basis"—so much for the testimony of Robert Welch! Overall, he was confident that action on these lines would "break the monopoly control of the business now existing."[26]

In sum, La Follette's investigation, like those of the FTC, amounted to reaffirmation of belief. In terms of discourse, if there was monopoly, there was an entity; if there was an entity, there was control; if there was control, there was collusion and coercion. This branch of discourse amounted to an onto-logical argument for the existence of Standard Oil; regardless of what the Supreme Court had done, regardless of the proliferation of oil companies great and small, there was monopoly and there was Standard Oil. They defined each other into existence, and they existed independently of operational dis-course, as offered by Robert Welch, who could indeed try to make himself clear but not succeed.

To the editors of the *Nation*, La Follette's findings came as no surprise; they had known all along that the dissolution of Standard Oil would not restore competition in the petroleum industry. They did not put much stock in the senator's proposed remedies. Instead, they told their readers, "When an indus-try becomes monopolized as is the oil industry today, the road lies forward into greater government control rather than backward into an outgrown com-petition." Readers of the *New Republic*, however, got a different message to-ward the end of the year from that ultimate oil industry authority Ida M. Tarbell; Tarbell agreed that dissolution in itself did not alter Standard Oil's control of both oil prices and the industry, but times had changed. Now there were signs that the Standard Oil companies might be competing with one another and thus no longer imposing uniform prices on the industry. She concluded, "It certainly looks very much as if the Standard Oil Company might be crumbling."[27]

Other readers of La Follette's report were less concerned with Standard Oil than with high gasoline prices. His sounding the alarm of one dollar a gallon gasoline encouraged anti-oil campaigns outside Washington. The American Automobile Association urged further congressional investigation and Justice Department action on gasoline prices. The National Conference of State At-torneys General passed a lengthy resolution in favor of action on many of La Follette's recommendations and asked for federal control of oil production. New York, Minnesota, and Wisconsin conducted their own oil industry inves-tigations, and the governor of Nebraska declared that if gasoline prices were not lowered, he would lead a campaign to nationalize the oil industry. The most drastic response came from the governor of South Dakota, who decided to challenge gasoline prices by opening state gasoline stations to undersell anyone offering gasoline for more than sixteen cents a gallon. In short, La Follette generated a real antimonopoly hue and cry.[28]

Faced with a revived antimonopoly crusade, industry spokesmen gave a variety of responses. The *Oil and Gas Journal* began to run articles with such titles as "Irresponsible Agitators Are Menacing the Oil Industry" and talk about political propaganda and demagoguery in Washington; they saw crit-icism of the industry on the part of politicians as a cheap maneuver for votes. As for La Follette, the *Oil and Gas Journal*'s M. C. Hill called him the "senior demagogue of the Senate," out to frighten the public into supporting government control of or interference with the oil industry. Such rhetoric showed the emergence of what would become a familiar channel of politically related discourse among industry participants, one whose chief hallmark was suspicion of politicians in Washington, also a traditional theme in more general popular political discourse. By October 1923, some industry writers had taken this suspicion to near paranoia; N. O. Fanning, for example, saw "powerful elements working together to the detriment of the future of the American oil industry," they being politicians, oil business failures, and speculators. Moreover, "propagandists" were conveying the idea that the industry was poorly led. This was an extreme example of defensiveness, but even the editors of the *Oil and Gas Journal* complained of the "never-ceasing stream of hostile propaganda against the oil business" and urged that old industry quarrels be forgotten in the face of "the common menace."[29]

By the time of this antimonopoly outbreak, some industry leaders recognized that the traditional Progressive anti–Standard Oil animus, so recently repeated by La Follette, had been generalized in the public mind to extend to the whole industry. As Charles Kern noted, "The hostility toward the Standard has given place to hostility toward the industry as a whole." Because some segments of the industry made money, the public saw the industry in general as guilty of "extortion." Then again, the public noticed the industry because of "the very largeness of the industry, and by the fact that some men have made large fortunes out of it. For these persons almost any report that would tend to besmirch the industry as a whole has caused a sympathetic echo from one end of the country to the other." The industry, not Standard Oil, was "the goat for the time being." Kern was certainly not the only person to realize that the industry as a whole had an image problem, but his was one of the more balanced views of it.[30]

Public animus against the whole industry, of course, made projects such as regulation of the whole industry on terms like those advanced by La Follette a much more likely possibility, and that raised the question of what should be done. The editors of the *Oil and Gas Journal* urged a united front among oilmen, suggesting that the industry might hold a conference in Washington to rally defense, "an indignation meeting"; in March 1924, some oilmen did hold a "Petroleum Week" demonstration in Chicago. Industry participants inclined to press for a public education campaign. Standard Oil of Indiana's Robert Stewart, for example, thought the industry had to deal with "the ill-formed public mind" and educate the public that what hurt the industry hurt people. But if educating was to be done, who would do it? Asked by an *Oil and Gas Journal* reporter whether there should be an industry conference on educating the public, Walter Teagle of the Standard Oil Company of New Jersey replied, "Who will call such a conference? Our company may not. If we did it would raise a hue and cry. . . . If the so-called Standard group of companies were to get together on a common campaign of defensive publicity it would bring down a storm upon us."[31] His was probably a realistic assessment of his company's position.

The *Oil and Gas Journal* also urged a grassroots campaign to get the industry's position to the public. It had in mind teaching gas station attendants about "the fundamentals" of the industry and having them distribute informative pamphlets to customers. Somewhat unpromisingly, when its reporters questioned Washington, D.C., service attendants on industry issues, they discovered that the attendants did not answer them in ways favorable to the industry. It nonetheless hoped to reach 1.5 million consumers with pamphlets and thought company advertising efforts would reach many millions more. Media leader Bruce Barton urged an advertising campaign on the American Petroleum Institute (API) at its Fort Worth meeting in December 1924, but he also warned, "Most industries when they advertise or start to 'tell their story' as they say, make a mess of it." Perhaps he was correct in this instance; there is no way to assess what the pamphlet campaign did to public attitudes.[32]

Beyond exploring public relations, industry members also began to respond to conservationist charges of waste and malfeasance in the early twenties; they began not only to arrive at their own understandings of waste but also to appropriate more conservationist themes to serve their own positions. In 1923, the *Oil and Gas Journal*'s Andrew M. Rowley acknowledged that the petroleum industry had "borne the stigma of wastefulness; of inefficiency in field operations; and a general disregard of the fundamental principles of conservation." The Texas Company's president, Amos L. Beaty, began to take the prominent role he would maintain for two decades in discourse on oil conservation by coming forward in 1922 to say that the industry suffered from "waste and extravagance" at many points — drilling too many wells, letting oil evaporate from earthen storage, and building too many gas stations. The following year he began to talk about the "economic waste" of producing more oil than markets demanded; this concept would come to dominate industry discussion several years later. Like earlier conservationists, Beaty stressed that oil was a natural resource that could not be replaced. Prominent Oklahoma oilman E. W. Marland, never one to mince words, decided that the industry "must admit to almost criminal waste" in the form of overproduction of crude, but he added this was not the fault of oil producers. They had only brought in as much oil as they had because they responded to the prophets of oil famine, the real parties to blame.[33]

In fact, these industry participants were responding to excess supply of crude oil on the market, forcing prices lower and signaling the beginning of the end of the great twenties boom. Ideally, if oil producers and purchasers could cooperate to limit the amount of oil on the market, prices would be shored up and downturn staved off. But what the industry liked to think of as "cooperation" was certainly in the eyes of La Follette and his fellow anti-monopolists "collusion," not to be tolerated on any terms, but least of all in the political climate of 1923. This basic political truth was blatantly obvious to industry leaders such as Walter Teagle. As he pointed out, "If a shut-down movement is inaugurated to prevent the production of unnecessary crude, the suggestion may not come from a Standard company because a sinister motive would be attributed to it. It must come from other operators."[34] Indeed, other operators could not have gotten away with such cooperation in the political climate La Follette's agitation created. It is one of the ironies of the industry's history that a period of all-out production in excess of demand could not be halted in the interest of conservation because the chief surviving Progressive leader made it impossible for the industry leaders to do so. Yet, more widely, such a conflict had always been inherent in the two streams of antimonopolist and conservationist discourse directed toward oil. If one worked to maintain competition and low consumer prices, one could not condone cooperation that restricted production. Keeping oil in the ground led predictably to higher consumer prices and greater industry profits, at least in the short run.

Of the two channels of discourse, however, the one apparently less threatening to the industry in 1923 was that of conservation. What industry leaders did was turn to the popular focus on waste, so apparent in both popular and trade periodicals and in the writing of academic economists, and identify waste not with abuses of consumption but production — specifically, levels of production that raised costs, pushed down prices, and narrowed profit margins. This was looking at waste in self-interested terms, but it was also approaching it from practical economics rather than the moral economy of Richard T. Ely. The oilman's waste translated into numbers and dollars. Once oilmen made this identification, they were free to draw on established conservationist rhetoric, as did the editors of the *Oil and Gas Journal* in 1923, in an editorial entitled "Waste or Conservation," to attack antitrust: "It is surely a sardonic example of the perversity of human wrong-headedness that today when our most valuable natural resource, petroleum, is being wasted by over-production through reckless exploitation, the mere rumor that the oil industry was considering methods to stop this exhaustion should have called forth the threat of instant prosecution for presumptive violation of Federal and State antitrust statutes."[35] Here the editors not only saw the conflict between two dominant channels of discourse but sought to exploit it for all it was worth and assume moral high ground while doing so. How could oilmen be held responsible for "reckless exploitation" if, as indeed was likely, they would be prosecuted under antitrust laws for trying to avoid it?[36] Surely it made rational sense to let the industry cooperate to solve its problems of soaring production, falling prices, and vanishing operating margins.

Oilmen also used elements of conservationist discourse to condemn low prices and, indirectly, high operating costs. Low product prices, fostered by production of too much crude, led to overstimulation of consumption, hastening "depletion of the already rapidly disappearing resources of the country." They would give "the unthinking public a joy ride for a few months to be repaid in famine prices when shortage looms." If that were the case, higher gasoline prices would be essential to conservation, a line of reasoning that would be common in the thirties. Turning to overproduction, seen as a result of the drilling of too many wells, oilmen began to lament the law of capture. E. W. Marland thought the industry and its lawyers needed to come up with a practical way to set it aside; he suggested that drilling fewer wells would lead to enhanced recovery of oil, surely an objective in harmony with conservation. In a more immediate context, however, drilling fewer wells to produce a lease also lowered costs of operators, an objective ever more desirable as crude prices were falling. Thus, by drawing on conservationist discourse, oilmen began to develop their own branch of conservation rhetoric, one that made sense in operational terms and could be used as defense against charges of industry irresponsibility levied by Progressive regulation-minded critics. As the editors of the *Oil and Gas Journal* would say in 1924, "Oil men do not want to waste their product any more than the most zealous conservationist. But they cannot save themselves because the laws will not let them." The menace of antitrust prosecution confronted those who might try to amend the law of capture or otherwise cooperate in the interest of keeping oil in the ground.[37]

Because controversy over the law (or rule) of capture assumed an important part in conservationist discourse in the later twenties and thirties, the concept is worth some discussion. It evolved out of the need to define ownership rights in petroleum. In common law, landowners owned minerals underlying their estate; if a landowner dug down to a vein of coal, for example, he or she owned all the coal under his or her land and could extract it or not. But when American jurists came to consider a landowner's rights in oil and gas, they confronted two problems: the lack of precedents in common law for a type of mineral production that was new and the fact that oil and gas were not solid minerals but moved out of well bores. Undisturbed under a landowner's holdings, petroleum unquestionably belonged to the surface owner. But once the drill tapped an oil pool and oil and gas flowed forth from it, how could one tell from which side of a property line this petroleum from under the ground originated? There was no way to see underground or to track oil droplets to their rock matrix. Searching for some analogy to oil in law, Pennsylvania jurists decided that oil's movement made it like wild animals or game; moving from place to place, game became owned when captured by someone, thus oil brought up out of the earth came into the ownership of the person "capturing" or producing it.[38]

By the time jurists worked out the law of capture, American oil producers had learned that production in oil pools came and went; fields and wells might come in with dazzling production and fizzle out in months or even weeks. The lesson in that was to get the oil out while the getting was good. Moreover, the man who held back drilling his leases might find that when he did drill, production was small; the logical assumption would be that those around him had produced the oil not only from under their property but his as well, since oil and gas in a tapped pool moved toward well bores. Such a producer could not accuse his neighbors of stealing his oil; once they produced oil, it was their property. In practical terms, then, because of the law of capture, according to which petroleum belongs to the person who produces it, there was compelling incentive for the individual producer to drill as many wells as he could on his tract; to match wells adjoining his on other leases; and to produce oil as fast as possible. In production, every producer raced against his neighbor to extract as much oil in the shortest time. That meant that with the discovery of a large pool, a great volume of oil soon swamped storage, pipelines, and markets, pushing crude oil prices downward. In this situation, however, there was no way the individual producer could hold back; he had to get his oil or others would get it. The only alternative producers had on their own, without aid from some public authority, was a voluntary agreement of all to cut back production. But such an agreement could be seen as combination in restraint of trade by the courts.

In the autumn of 1923, with the furor over gasoline prices at its height, with Senator La Follette apparently bent on major-scale oil industry regulation, and with the Teapot Dome scandal at a boil, an industry maverick, Henry L. Doherty, came forward with an idea that would influence thinking about petroleum conservation for the next decade. Doherty's idea was "unitization" of producing oil fields by federal regulation. From the point of view of many industry leaders, his timing could not have been worse.

Probably brilliant, definitely eccentric, and usually unorthodox, Henry L. Doherty had the kind of career one would sooner expect to encounter in late-nineteenth-century fiction than in the business world. Son of a civil engineer who died young, Doherty had to leave school at age twelve to help support his widowed mother and family, going to work as an office boy for the Columbus (Ohio) Gas Company. By 1905, the former office boy had become a successful utilities tycoon and founded Henry L. Doherty and Company to offer engineering and financial services to utilities. In 1910 he formed a holding company called Cities Service, which by 1913 had picked up fifty-three companies. His growing utilities empire led Doherty gradually into petroleum; in search of gas for his companies, he moved from manufactured to natural gas and from drilling for natural gas to drilling for oil. He loved novelty; not content to perfect a business strategy and stick with it, Doherty aimed to "acquire, re-build, merge, re-finance, infuse new life and new organization, and move on to the next field."[39] This swashbuckling approach to business took plenty of cash, which Doherty was shorter on than ideas. With mordant humor, his em-ployees joked that his initials, HLD, stood for "Henry Lacks Dough"; they were under considerable pressure to put their own dollars into Cities Service securities. Doherty's financial maneuvers certainly bore considerable resem-blance to those of the early-twentieth-century sharp oil promoter. Like many a promoter, moreover, Doherty loved the limelight, and he knew how to cap-ture newspaper headlines.[40] The issue of petroleum conservation allowed him to grab them.

Doherty's cause was unit operation, or "unitization," the working of an oil or gas field by one management rather than by many separate leaseholders. Under unitization, there would be none of the usual scramble to drill in order to capture as much oil as possible before it flowed into a neighbor's wells. Instead, one managing entity would take over development and production; it would decide when and exactly where to drill, how closely to space wells, and how fast to produce what it brought in. It could hold back production if market conditions were unfavorable or if there were transportation or storage problems. It could produce oil without haste or needless dissipation of gas pressure. It could thus make development and production not only a rational and scientific operation but also an economical one, because operating costs would be lower at the same time total recovery of petroleum would be greater. Unitization as a concept did not originate with Doherty. Bureau of Mines technologists William F. McMurray and James O. Lewis had suggested it in 1916, as had economists Chester Gilbert and Joseph Pogue two years later. But Doherty seized upon the idea as the answer to "the entire oil problem." The need for naval oil supply, greater oil recovery, gas conservation, lower industry costs, wise use of produced petroleum, provision for future supply — all would be achieved if all petroleum reservoirs were unitized.[41] What would make unitization a happy reality was legal reform ending the practice of letting oil belong to the individual who produced it, the old law of capture.

In the autumn of 1923, Doherty tried to interest the American Petroleum Institute's board of directors, of which he was a member, in unitization, only to be snubbed for his pains. A year later he tried once again to address them as they met in Fort Worth, to be rebuffed once more. His colleagues' reaction was understandable. It was not only that Doherty was unorthodox, but the last thing the API wanted to do was foster dissent in its own ranks by discussing a controversial idea. Moreover, some of Doherty's assertions simply could not be supported by industry experience: in contrast to what Doherty suggested, for example, science had not reached the point where oil could be located with reasonable certainty without drilling, and any oilman active in exploration knew it.[42] Many of Doherty's arguments were adaptations of arguments used by conservationists and others who criticized the industry, which did not make them more palatable. In short, there were few reasons to give this obnoxious maverick a platform. But when the API refused to hear Doherty, he parlayed that into press coverage and took his campaign to the public. He could, and did, pose as the lone patriotic oilman, shunned by his industry colleagues for the unpleasant truths he advanced, the courageous crusader for public interest facing down private greed. And whose aid would be more appropriate to enlist for his crusade than that of the president of the United States?

On August 1, 1924, Doherty wrote Calvin Coolidge a long letter laying out his criticism of the oil industry and his suggestions for change. The ideas he offered the president combined his own thinking on unitization with familiar themes in public discourse. Thus he picked up the cry that the United States was running out of oil. He used the theme of waste, advancing the emerging industry perception that oil left unrecovered in the ground was wasteful, identifying waste with dissipated unused natural gas and use of petroleum for fuel purposes coal could meet. With this he conjured up the image of oil as wealth wantonly spent. In terms of oil supply the United States was becoming a "pauper nation," "a bankrupt nation"; in later public appearances, he would use the phrase "Petroleum is a wasting asset."[43]

As appropriate in a letter to the commander in chief of the armed forces, however, Doherty stressed the military and strategic value of petroleum, whose importance made imperative government intervention in oil industry operations. Doherty pointed out that neither setting aside naval oil reserves nor pinning hopes of future oil to oil shale development was realistic. What Doherty promised the president, if his plan was adopted, was "large bodies of oil located and blocked out, and in event of war we can draw on these ground reserves very quickly to supply . . . our increased needs."[44] The core idea of this suggestion was still the underground strategic oil hoard, oil stashed away for a rainy day.

Legal change in ownership rights in oil reserves would make all this possible. The root of all evil was the legal concept that produced oil was subject to the law of capture: to this Doherty attributed not just depleted reserves but "practically every evil of the oil business." If laws could be changed so that petroleum could be treated like other minerals, as belonging to the property owner under whose land it lay, if states could pass laws so that oil pools could be handled like irrigation or drainage districts whose property was managed in common to the shared benefit of all, then oil could be produced slowly, methodically, and scientifically. At least double as much oil would be recovered, and there would be abundant gas for municipal use. No longer would unneeded oil flood the market. At one blow, if one believed Doherty, all conservation problems could be solved. Should states balk at necessary legal reform, Congress could take "jurisdiction" over all domestic oil production, ensuring universality of reform. In managing reform on the federal end, the secretaries of war, the navy, and the interior should work with the heads of the Bureau of Mines, United States Geological Survey, and Bureau of Standards. Doherty thus assumed that coercive federal power probably would be necessary to his plan.[45]

Doherty realized that unitization through federal coercion would not be popular with his industry colleagues, but that was a fate he accepted. He would be industry gadfly, crusader, even martyr, and he would not just get mad but get even: "If the public someday in the near future awakens to the fact that we have become a bankrupt nation so far as oil is concerned . . . I am sure they will blame both the men of the oil industry and the men who held public offices at the time conservation measures should have been adopted. I intend to make a record to which I can point whenever the inevitable time arrives when an indignant public asks for an explanation."[46] In other words, should the president fail to act, Doherty would point the finger at him as well.

But Coolidge did act. With the onus of Teapot Dome squarely on the GOP, it was expedient to make a gesture, at least, toward oil conservation. He did not go as far as Doherty wanted, but on December 10, 1924, he set up the Federal Oil Conservation Board (FOCB), chaired by Hubert Work, the secretary of the interior, and including the secretaries of war, the navy, and commerce, with George Otis Smith as chief adviser. His letter establishing the board mentioned waste, fear of future shortage, current use of cheap oil instead of coal, and the strategic danger of running short of oil. The FOCB's mission would be to find out if there was "an inexhaustible supply" of petroleum in the United States; whether the industry and government were squandering natural resources; whether consumption and production were economically regulated; whether there were substitutes for oil at "reasonable prices"; whether laws relating to public lands and pipelines should be changed; and whether consumption and production could eventually be cut back without disrupting the economy.[47] In short, it would investigate many of the questions posed in twentieth-century discourse on oil. Given the framing of the first two questions, there was not a great deal of doubt about what the board would find on some of these issues. In itself, that could give industry participants much anxiety.

As the FOCB got under way with its investigation by sending out questionnaires to more than three hundred oil companies and industry associations, the indefatigable Doherty took every opportunity available to preach unitization to the public. He addressed the National Petroleum Marketers' Association in November 1924; the Petroleum Division of the American Institute of Mining and Metallurgical Engineers in February 1925; the National Gas Association in May; and the American Gas Association in October 1925, getting full coverage of his appearances in the *New York Times* and other newspapers. With this much coverage of Doherty's conservationist campaign, a *Times* reader might well have been distracted from coverage of his various legal imbroglios — lawsuits against him by the Pierce Oil Company, the stockholders of Sinaloa Exploration and Development Company, and a group of investors including W. K. Vanderbilt.[48] Doherty the conservationist overshadowed Doherty the promoter.

Though Doherty charmed the *Times*, the same could not be said of his colleagues in the oil industry. Still, they were not uniformly hostile to him. Particularly in the ranks of the technologists, he found willing listeners; geologists Everett DeGolyer and F. Julius Fohs were receptive, as were present and former federal experts Ralph Arnold, Max Ball, and George Otis Smith. They could agree with Doherty's identification of many industry problems. But agreeing with his solution, compulsory unitization, was another matter; to most industry participants, Doherty's plan, which included wildcatting by permit only, distance limits on area development, and limits on drilling near peripheries of oil pools, was impractical. Education and technological progress might be better avenues to industry improvement, moreover, than sweeping legal reforms of the sort Doherty wanted. Even allowing that they disagreed with Doherty, however, oilmen began sustained discussion of his proposals.

While Henry Doherty preached for unitization, one of his former employees, now an aspiring economist, tried to bring together the two broad channels of antimonopolist and conservationist discourse on oil and merge all their elements into something that made sense. Son of a Clarendon, Texas, physician, George Ward Stocking graduated from the University of Texas in 1918 and went to work for the geological department of Henry Doherty's Empire Gas and Fuel Company of Fort Worth. For two years he did surface mapping and collected drilling and production data in North Central Texas, an area in all-out boom at the time. This experience not only let him later claim firsthand knowledge of the oil field but also directed him toward the subject he would write about, the consequences of competitive drilling and production. Stocking took an M.A. in economics at Columbia University in 1921, with a thesis titled "Waste in Oil Production," and a Ph.D. in economics in 1925, with a dissertation that became his most notable publication, *The Oil Industry and the Competitive System: A Study in Waste.* This study would assure him of a career in university teaching, at the University of Texas at Austin and Vanderbilt University, and make him a credible consultant for the New Deal. It would also become the most frequently cited work by writers on petroleum conservation and industry problems for the next half century.[49]

In *The Oil Industry and the Competitive System*, Stocking argued that the oil industry's chief flaw was not too little but too much competition, which resulted in the waste of both petroleum and capital. In an unstable fusion of antimonopoly themes, conservationist economics Ely-style, and popular discourse on conservation and waste, elements often at odds with one another, he argued that conservation of petroleum could only be achieved if usual all-out competition in drilling and production gave way to controlled oil field development. In this context, antimonopoly crusading like La Follette's was misguided: "While our law makers have been concerned with the matter of prices and profits and monopoly control, oil production in this country has been carried on in a wasteful and improvident fashion."[50]

Did he mean that either monopoly did not exist or, if it did, it was not socially destructive? Stocking straddled the line between challenging and accepting orthodoxy on oil by distinguishing between exploration and production on the one hand and everything else downstream on the other. Exploration and production were wastefully competitive, while transportation, refining, and marketing were wasteful but apparently less competitive. In short, one could believe most, but not all, of the traditional adversarial notions about oil, by saying in some places oil was monopolistic and in some that it was not.

Lest his readers doubt Stocking's general acceptance of orthodox condemnation of monopoly, he devoted the first third of his work to a historical survey of the American petroleum industry featuring the customary recital of the evils of Standard Oil in which he relied on Ida Tarbell and the Bureau of Corporations report. He had no trouble agreeing, as he put it, that "the Standard Oil Company had sinned against the competitive system." The question of whether light had really triumphed over darkness in 1911 was much tougher to answer, for here it was harder to locate orthodoxy. Stocking decided that dissolution had been "ostensible" and that Standard Oil continued to dominate the industry after 1911 "almost as effectively" as before. It had done so through overlapping groups of stockholders and "the forces of economic advantage and custom." However, that control was weakening, Stocking argued, for since 1911 "the Standard has lost ground." Here he echoed Ida Tarbell's recent assessment.[51]

Like almost everyone else worried about waste, Stocking relied on federal forecasts of waning oil and gas reserves. Working from David Talbot Day and others, Stocking assumed that petroleum was "a wasting asset," dwindling fast in the face of expanded consumption; domestic production could not be expected to keep up with growing demand. At current production rates, known reserves would be used up by 1931: only six years away! Stocking also agreed with Mark Requa that the United States would eventually be reduced to reliance on foreign oil unless something was done.[52] But where Requa used this point to argue for business imperialism by American oil companies, Stocking used it to justify his recommendations for dealing with what he saw as profligate waste.

Stocking sidestepped the thorny task of defining what "waste" was by offering examples of waste, often quite different in nature — parallel pipeline systems from oil fields to refining centers; too many gas stations on adjacent street corners; too many oil wells jammed together on town lot fields such as Burkburnett. This was waste of capital. Not making maximal use of petroleum drawn from the ground was also wasteful. Offenders included oil operators who let natural gas escape when there were no local markets available, and for figures on the amount of natural gas wasted, Stocking relied on Bureau of Mines experts Blatchley and Van Manning. There were also the small skimming plant operators who took only the highest petroleum fractions out, leaving residuum that could have yielded additional gasoline and lubricants for use as fuel oil. Such waste, like that of gas, primarily involved lost opportunity to retrieve products.[53]

The waste to which Stocking gave most attention, however, was that of oil left unrecovered in the ground, for that took him to drilling and production. Given the law of capture, out to get oil before his neighbor did, the producer covered his tract with as many wells as he could, sometimes letting derrick legs nearly touch. Wells were drilled as quick profit, rather than science, might dictate. The question of how far apart wells should ideally be, however, took Stocking to an area on which technologists disagreed. In a given field, wells wide apart seemed to have a greater per well output in the short term, but wells close together a greater cumulative output per acre over time, making it possible to argue for either close or distant well spacing. Though he did not admit it, such technologists' differences posed difficulties for Stocking's opinion that by applying modern engineering methods oil production could be made "to approach a scientific exactitude." Overall, however, he disapproved of drilling many wells, especially when small leases meant well proliferation. The Burkburnett Townsite boom, in North Texas, was his classic bad example, not only because of overdrilling but also because Burkburnett "experienced a wild orgy of stock promotion and lease speculation."[54] Here was the old bogey of quick riches, gambling, and social dislocation so familiar in American oil-related moral discourse; Stocking was seeing in North Texas what had been criticized in the earliest Pennsylvania fields.

Whether looking at town lot drilling or markets swamped with crude oil, what Stocking saw at the bottom of waste was competition: "Competition in oil production inevitably means waste . . . waste on a spectacular and magnificent scale." Condemning it in the language of frugality, Stocking said, "The provident producer is at the mercy of the improvident; and improvidence is nurtured by the competitive system." Less temperately, he asserted that, under free competition, oil production was "of necessity conducted on the principle of robbery."[55]

Stocking's solution to industry ills drew from the ideas of Ely and Doherty. He argued that public lands should not be open to oil development when there was an abundance of petroleum on the market; oil fields on public lands, moreover, should not be subject to the usual competitive scramble in development but should be developed as geological units. Moreover, a large measure of government involvement and control was needed, for the main difficulty, as Richard Ely had observed, was "the uncontrolled private ownership and operation of a community resource." There should be comprehensive federal regulation and control of petroleum production, managed through the USGS and Bureau of Mines, or an amalgam of both. Federal authorities would control the rate of oil field development, the location of wells, and well spacing. Presumably there would also be federal control of rates at which wells would be produced, and "the application of technique by private operators would be under the general guidance of the central [federal] bureau."[56] In short, what Stocking really wanted was federal management of all oil field development. Less than a decade later, Harold Ickes would try for it.

Overall, Stocking's contribution to petroleum-related discourse was not intellectual penetration but synthesis. He made no great effort to define difficult ideas such as waste or conservation, and he overlooked the conflict that his type of conservation would have with antimonopolist sentiment in the area of consumer interest. In terms of discourse what Stocking did was bring together so much of what had been said about oil within the confines of one volume, thus making it easy for industry observers to find support within it for a tremendous number of policy positions on oil, as we shall see. He took the many problems people had identified with the oil industry, and, like Ida Tarbell, suggested a solution to them that seemed to be relatively simple and did not involve total replacement of capitalism, albeit some mighty adjustments of it with respect to oil.

Stocking also took criticism of free competition much further than Ely and his friends and advanced an alternative to it. This gave him a ready audience, not only among Progressives but even among oilmen who watched markets collapse under flush production from new fields. Oilmen might not agree with Stocking's remedies, but in the later 1920s they could see more and more in his critique of competition. If one looks at Stocking's version of government control of production and development, and substitutes state for federal authority, one can see many of the practices that state conservation and regulatory bodies such as the Texas Railroad Commission began to experiment with during the 1930s. Put back federal control and one comes close to what Harold Ickes would attempt during the depression and World War II. Stocking's perspective on the oil industry was a harbinger of extensive discourse to come as well as a reflection of discourse past and present, a kind of landmark in discourse freely used by many a subsequent participant in debates over public policy. He was credible for so long because he repeated so much of what people had said before and what they continued to say thereafter.

By the mid-twenties, then, oilmen had begun to respond to conservationist discourse, generally on the defensive. They were right to be anxious about the direction of conservationist discourse, for whether it was used by an industry maverick like Henry Doherty or an ambitious young academic like George Ward Stocking, the whole industry was indicted as wasteful, and the remedy seemed to be extensive government involvement in industry operations. Were oilmen to think about abandoning free competition as both these critics suggested, however, they could be sure the political heirs of La Follette would charge them with designing monopoly. That undoubtedly limited their ability to respond to the charges against them.

# 7

## Talking Past One Another

As the newly created Federal Oil Conservation Board began its work early in 1925, many oilmen viewed it with apprehension. The editors of the *Oil and Gas Journal* told readers that the FOCB was better than meddling by "irresponsible politicians," but there had been so much talk of government control of oil that one might fear the worst. W. H. Gray, president of the National Association of Independent Oil Producers, shared anxiety that the petroleum industry had been singled out for "regulation and control." Journalist L. M. Fanning saw the appointment of the FOCB as the first positive federal move toward "a definite Government oil policy," one modeled on Henry L. Doherty's ideas: no comfort to those who differed from the maverick oilman. But no observers could foresee how thoroughly the incompatibilities of existing channels of discourse on oil would derail policy making and implementation. The appointment of the FOCB would precipitate a virtual showdown in discourse, a struggle for control that federal conservationists won but that resulted in functional gridlock. Thereafter, all parties talked past one another.

Certainly, to the directors of the American Petroleum Institute, the appointment of the FOCB was a matter for concern. Organized in 1919, the API was supposed to act as informational liaison between government and industry, a supplier of data and statistics offering a different perspective from that of the bureaucrats in the United States Geological Survey (USGS) and Bureau of Mines. Oilmen could hope that API information would improve the public's understanding of industry positions. With the appearance of the FOCB, the API directors decided to establish a public relations committee, and when the FOCB announced it would embark on fact gathering by sending out a questionnaire throughout the industry, the API appointed a committee to work up a lengthy set of answers.[1]

The kind of questions the FOCB asked invited such a response, for they amounted to a virtual litany of conservationist tenets, beginning with the first question: "To what extent, if any, do you consider a shortage of petroleum imminent in this country?" The FOCB brought up waste, prioritized uses, substituting coal for oil as fuel, increasing recovery, unitization, too many gas stations, and the need to acquire foreign reserves.[2] The API decided to confront the conservationist indictment of the industry head-on. In mid-1925, the API's volume, *American Petroleum Supply and Demand*, appeared, designed to reassure the public that the country was in no danger of imminent exhaustion of domestic reserves at the same time it aimed to refute charges that the industry was wasteful and incapable of conservation without extensive government regulation.

One way to challenge the idea that that United States was running dry — and thus that it needed government-run conservation — was to point out the shortcomings of past estimates. The API noted that David Talbot Day's much quoted statement had allowed for a 280 percent margin for error and that, by his forecast, Mid-Continent production would have been exhausted by 1923. This prediction was observably false in 1925, as was the USGS–American Association of Petroleum Geologists forecast of 1921 that California's reserves amounted to only 1.2 billion barrels; the current reserves estimate for California was almost twice as high. Of the last forecasting effort, which put total domestic reserves at 9,150,000,000 barrels, the API allowed that it was probably the best that could have been done at the time, but in only a year it was obvious that it was "grossly inaccurate."[3]

Then how much oil did the United States have? The API confined itself to figures based only on proven territory — producing wells, proven but undrilled acreage, and oil likely left in the ground after flowing and pumped production. It emphasized that most fields sustained flowing production for longer than predicted and that even in the oldest fields there was still some flowing production. From producing wells and proven acreage would come production of 5.3 billion barrels of crude, leaving some 26 billion barrels remaining in the ground, to be recovered by water flood, gas injection, or even mining, all depending on price levels. Should prices rise high enough, this 26 billion barrels would be targeted for recovery. Higher prices would also mean that 394 billion tons of oil shale might be exploited to yield more than 100 billion barrels of petroleum liquids; similarly, prices permitting, coal and lignite deposits might yield up to 600 billion barrels of liquids. All these resources were known, and only economics would determine whether they would be used. They stood as insurance against the United States running dry.[4]

The API pointed out, however, that before it would be necessary to look to oil shale or coal for petroleum liquids, it was reasonable to expect great additions to reserves from both deeper drilling in known fields and successful prospecting in likely regions. The most promising part of the exploratory picture was the sheer extent of territory that had promise for oil. Known oil fields covered but one-fortieth of land geologists thought had prospecting possibilities. If one eliminated the land apparently without sedimentary formations that might contain oil, that still left about 57 percent of the total land area of the United States, or 1,105,454,459 acres, where rock formations were like those in oil-producing areas and thus might contain oil. The API was confident that with extended search new supplies would be found — so confident that it called the billion unexplored but promising acres "the greatest of the national petroleum reserves," containing "vast quantities of oil." This kind of hyperbole was striking in what was supposed to be an argument from scientifically grounded statistics, and adversaries such as Doherty would seize upon it to discredit *American Petroleum Supply and Demand* as implausible and cynically self-interested.[5]

Having tried to prove that the United States had ample oil reserves, the API had to address the question of how long they would likely last. Here it sought to go beyond the crude projections of the conservationists, based on simple increments to demand, to analyzing components of demand. Those who put together the section on demand looked at future population projections, forecasts for growing use of automobile and other internal combustion engines, and engineering improvements likely to make engines more efficient, all considered over the next half century. They admitted that their projections were speculative, for technological change might radically shift energy demand.[6] In any event, given the premise that the United States still had vast untapped reserves of petroleum, not surprisingly the API found nothing alarming about rising demand. Nor was it necessary to think in terms of prioritizing consumer uses of petroleum as the conservationists did; to do so presumed scarcity, which the API denied.

While the API tried thus to refute conservationists and reassure the public, it also tried to defend the industry on the charge of irresponsible waste. The industry, it argued, did all it could to reduce actual physical loss of petroleum by spillage or evaporation; only 3 percent of all crude oil produced suffered such loss. As for loss of natural gas, some of that was virtually unavoidable; it took time to build gas lines from fields to markets, gas was not always readily marketable, and once produced gas could not be stored economically. But was much gas really wasted? Here the API played with the meaning of waste; surely gas was not wasted if, in being produced with oil, it helped force crude up through the wellhead. Anyway, the public impression of waste was a response to hectic development in booming oil fields. The petroleum produced was "safely sent to storage."[7]

To answer critics such as Doherty who complained about oil lost by being left underground, unrecoverable as the result of rapid competitive drilling, the API rejoinder took two approaches. It pointed out that most experts held that total oil recovery in a reservoir was higher when a great number of closely spaced wells were rapidly drilled than when a few widely spaced wells were drilled over an extended period. But even if slower noncompetitive drilling had a better recovery rate, it would represent "ideal but unattainable conditions."[8] Here the API carefully avoided Doherty's contention that these ideal conditions were attainable with a heavy measure of government control. Instead, it implied that all would be well if the government did not meddle in the industry.

*American Petroleum Supply and Demand* was the most substantial industry contribution to conservationist discourse in the 1920s. Unfortunately for the API, however, its reception fell flat. Within the industry, Henry Doherty castigated it, and geologists L. C. Snider and Everett DeGolyer lampooned it.[9] For the most part, the media outside the industry paid little attention to it. The *New York Times* noted its appearance and its general message that the United States was not running out of oil, but the paper was careful to report that estimates of future supply were "conjectural" and "speculative." Among popular periodicals, *Scientific American* told its readers that *American Petroleum Supply and Demand* would "lull the country and the government into an attitude of unjustifiable security as regards our future oil supply." Other periodicals ignored it; in effect, good news was no news. In short, the API failed to shift the orientation of conservation-related discourse from well-established channels.[10]

The API's aggressive rejoinder to the prophets of oil famine not only signaled an increasing tendency in industry circles to challenge and condemn predictions of shortage but also stung the believers in shortage to vigorous defense of their position. Thus in the American Institute of Mining and Metallurgical Engineers' Petroleum Division meeting in 1925, the USGS's David White said that forecasts of oil famine were needed to combat "dangerous complacency." The public had to be told oil was running out, and whatever the flaws of estimates, they were essential to "a campaign for the protection of the industry and of our domestic welfare, including our navy, our army, and our commerce." So much for objective fact finding. The experts had bamboozled the public but for its own good. The estimates had to be low, "for to have encouraged the expectation of a yield greater than might later have been realized would have been to court hazard of economic harm or possibly disaster."[11] Perhaps experts such as White and others had come to believe their own admittedly misleading predictions, but in any event, the end apparently justified the means.

White's point of view would be reaffirmed by the FOCB, whose chairman, Hubert Work, had been annoyed that *American Petroleum Supply and Demand* "barely mentioned conservation." In February 1926, Work convened a hearing by the FOCB by repeating familiar themes: United States oil was being produced too rapidly and in excess of real need, the industry had exploited petroleum on public lands counter to the national interest, and American use of petroleum could be compared to "the man who earns $100 a month and spends $100 a month." Conspicuously absent from Work's address were the reassurances of the API. Indeed, industry participants were cast more as defendants than witnesses, a situation reflected in the mildly sarcastic response of Texas Company president Amos L. Beaty: "We have grown so accustomed to being investigated that we may appear awkward for a while. We are familiar with the procedure when charges are brought, and speak the language of investigations. But this seems to be a study . . . and the novelty of our status may cause us some bewilderment."[12]

Industry critics showed no bewilderment, and their presentations were predictable. The star critic's role was Henry L. Doherty's, who repeated all his earlier charges, together with statements such as the "petroleum business has many evils, more than any other business I am familiar with." A common theme in critics' presentations was condemnation of prevailing competition within the industry and argument for a greater measure of government regulation. Even among those in the API camp there were doubts as to whether competition was compatible with conservation. Beaty of the Texas Company deplored "irrational drilling" spurred by intense competition. Engineering professor L. C. Uren offered a stronger indictment of competition and challenged antitrust orthodoxy by singling out the small producer as an economic misfit responsible for abuse; because small producers were inefficient, they produced unneeded oil, "demoralizing markets and compelling others to match [their] ill-advised efforts." Conservation would only be achieved through "consolidation of producing interests," unitization or, as it were, monopoly by another name.[13]

Most industry representatives took a conciliatory tone toward critics. For example, W. S. Farish, former president of Humble Oil and current API president, stressed that oil industry interests were perfectly compatible with those of the nation and its consumers; surely the oil industry did not want to see petroleum resources exhausted. Such industry defenders did not deny that the industry had peculiar problems, but they stressed that it was trying to work them out; government intervention on the state or national level was unnecessary. But there were those such as E. W. Marland who decided on a confrontational approach to conservation. As Marland put it, "I am not one of those who believe that there is any necessity for conservation, because I believe that there is a sufficient supply of oil to meet the demand for internal combustion fuel and lubrication for centuries to come, probably as long as our civilization and its necessities for petroleum shall exist." Those who predicted an early exhaustion of oil were "pseudo scientists."[14]

Marland singled out Doherty and the critics of waste for particular abuse: "Writers for petroleum magazines, writers in our press, on the subject of petroleum, have written about waste until they have fairly impressed the public with the idea that some enormous waste has been going on in the petroleum industry. I think that we ourselves and our press writers have had a confusion of ideas. We have been thinking about the wasting of an opportunity to make a profit. We have not really meant that we were wasting oil." Now, in large measure, lost profit—what the industry would eventually understand as "economic waste"—came to dominate the industry's understanding of waste in the late 1920s, but Marland's aggressive presentation conjured up the kind of exploitive, greedy, public-be-damned industry image present from the beginning in the conservationist discourse on oil, an image borrowed from the critics of Standard Oil. It was as though Marland was telling conservationists he was guilty and proud of it. To this Doherty mildly replied, "When I say we are wasting oil, I mean we are wasting oil. I do not mean we are wasting an opportunity to make a profit." That reinforced what industry critics had been saying all along.[15]

Had the FOCB genuinely hoped to be instructed by its hearing, it would have been left in some confusion at the end. The board had heard from industry participants who said there were serious industry problems and from those who denied them, as well as those whose opinions fell somewhere in between. But industry leaders had offered no unity of opinion, and the API could not feel that it had carried the day. It made a second effort at the end of May; this time it retained eminent jurist Charles Evans Hughes to argue its position before the FOCB. Hughes reiterated the API's arguments that the United States was not facing an oil shortage, that prices determined the amount of oil reaching the market, and that the oil industry was not wasteful; in effect, he read the main points of *American Petroleum Supply and Demand* into the record. But he also devoted considerable time to demonstrating that suggestions such as Doherty's for government regulation of oil would yield laws that were unconstitutional, and plans for voluntary limitation of production would be combinations in restraint of trade.[16] Whatever the industry's problems from competitive drilling, there was apparently no legal way to resolve them. Hughes's presentation prompted testy oral and written rejoinders from Doherty, who accused Hughes and the API of duplicity, of hiding the "real facts," pretending to cooperate with the president's call for conservation while actually trying to subvert it. Moreover, Doherty charged that API leaders had intimidated other industry members who knew the API position was false, so the "rank and file of the men of the industry" would not speak against it.[17]

On September 6, 1926, the FOCB produced its first report, written, according to Hubert Work, "with open minds, without conscious prejudice or thought of confirming theories already conceived." Nonetheless, the main themes of the report were what government bureaucrats had argued for the better part of two decades, with a few new and alarming additions such as the forecast that existing production probably would yield only six years' supply of oil at current demand rates. The report did not rule out new discoveries, but where the API said 57 percent of the United States' land had oil possibilities, the FOCB said 43 percent was "positively barren," and there was no reason to assume "the remaining 1,100,000,000 acres . . . or any large part of them, will be found oil bearing." The industry was guilty of wasting gas and leaving 75 percent oil in the ground. True, more crude might be gotten through secondary recovery methods, but there was "no positive assurance" these would be especially successful.[18]

Then where would the United States get future oil supplies? The FOCB turned to the familiar idea of using oil shale to produce petroleum liquids and recommended U.S. acquisition of reserves in foreign countries. Developing use priorities would be important to ensure future supply; here the FOCB asserted that use of fuel oil seemed to be giving way to use of coal! Apart from this bizarre perception, the only relatively new element in FOCB conclusions was the idea that future oil supply would be enhanced by better control of flush production in new fields. Without mentioning him, the FOCB took Henry Doherty very seriously.[19]

It was not, however, prepared to adopt Doherty's suggestion that, failing other action, the federal government should impose unitization. Instead, the FOCB urged oilmen to use voluntary cooperative agreements to limit development in flush new fields and head off overproduction. Oilmen should not take antitrust as "an actual or imagined or pretended barrier to cooperative action." Failing voluntary agreements, states ought to act to prevent waste of natural resources. Surely it was legitimate, in such matters, for states to protect property owners from others who would waste or destroy "common property."[20] This term, however, breezily presumed a reform on Dohertian lines. Indeed, in its confident recommendation of movement toward production control, the FOCB showed an optimism in striking contrast to its gloomy view of the industry's current state.

Though the FOCB much preferred voluntary industry or state action on conservation to federal measures, it did endorse the idea of petroleum reserves for defense. Federal authorities could and should exercise tighter control over oil discovered on public lands; they might, for example, hold up oil production from this source in times of overproduction. And there was imperative need for more guidance from federal bureaucrats, those men of science who gave so unstintingly of their time to compile the FOCB report. The USGS should continue and broaden research on oil accumulation, the Bureau of Mines should continue its work on production and refining, and the Bureau of Standards, a new entrant to the conservation discussion, should work on utilization of petroleum products. These agencies should have constant contact with industry members and state governments, helping them solve conservation problems.[21] Cooperation was the paramount need, and federal experts would be crucial to its success.

Overall, the first FOCB report reaffirmed two decades of conservationist criticism of the petroleum industry. If it was less strident than some critics on the themes of wasteful extravagance and improper use of oil, it clearly condemned the status quo. Still, it did not recommend immediate federal intervention in the operations of the industry; its emphasis on voluntary cooperation was acceptable by 1926 to many oilmen, though by no means to all. From an industry point of view, the report could have been much worse. Even so, the editors of the *Oil and Gas Journal* called it a "dangerous report," with "disturbing" suggestions.[22]

Whereas *American Petroleum Supply and Demand* got relatively little attention outside the industry, the FOCB's 1926 report encouraged industry critics to renew conservationist alarms. Among journals, the most ascerbic commentary came from the *Nation*, which reminded readers that the petroleum industry treated oil "as a drunken sailor treats his money"; it was disappointed that the FOCB could still believe private enterprise in oil was acceptable policy.[23] Other critics, such as Stuart Chase, accepted the FOCB's forecast of oil famine in six years as good news. Inclined to tie oil with a consumer culture they despised, they saw a "gasless America" as an appealing possibility: there would be "no taxis, traffic cops, Dixie highways, one-way streets, better Buicks, Standard Oil companies, filling stations, schoolgirl complexion billboards, airplanes, football stadia (for how could they be filled?), five-ton trucks, Klaxon horns, 25,000 new graves a year, *Saturday Evening Post* advertising, Fifth Avenue buses," and so on.[24] If oil was responsible for civilization as Americans knew it, then it had produced Charles Eliot Norton's world of "shoddy and petroleum," a crass, tawdry commercial world where the automobile encouraged idleness and extravagance. Running out of oil would be a social benefit!

The connection of oil and the evils of consumer society was developed at length by University of Kansas economics professor John Ise in *The United States Oil Policy*, a 1926 work much cited by subsequent writers on petroleum conservation. A Kansas farmer's son, Ise grew up in a state where populist orators vied for the headlines with independent oilmen fighting Standard Oil; of the two, the populists made the greater impression on him. Once Ise received his Ph.D., he spent his whole academic career teaching at the University of Kansas. Thus, as with academic economists such as Richard T. Ely and George Ward Stocking, several generations of students were exposed to his perspectives on conservation. They were profoundly negative on the petroleum industry.

Though Ise echoed many of the ideas of Progressive conservationists, at heart he was a preservationist, who did not want petroleum used at all. For him, oil taken from the ground was "merely consumed, that is to say, destroyed." To speak of oil field development was a misnomer, for it was really drained and wasted.[25] Like George Ward Stocking, Ise saw a great deal of waste in the oil industry, citing the drilling of too many wells and speculation, just as Stocking had done a year earlier. But Ise was much more concerned with the moral damage done by oil than was Stocking, and it was moral damage rather than either economics or public policy that was the central concern of his book.

As Ise saw it, the history of the petroleum industry amounted to one long moral disaster for America. Instead of Ida M. Tarbell's pre–Standard Oil Arcadia, Ise saw the industry's earliest years as a time of utter chaos. His pioneer oilmen, heedless and inefficient, wasted millions of barrels of oil, "recklessly and wantonly" ruining oil reservoirs in the process. They lured the innocent to invest in their destructive ventures and seduced the hardworking into abandoning productive labor for oil, while they themselves not uncommonly ended up overextended and ruined. No wonder they were helpless when an efficient competitor in the form of Standard Oil came along. Standard Oil's monopoly was preferable to "intolerable" competition. For that matter, the emergence of a monopoly such as Standard Oil was an inevitable economic development, "even if there had been no John D. Rockefeller." In short, Ise turned Ida Tarbell upside down and used the emergence of Standard Oil as evidence that the oil industry was "in many respects a natural monopoly." Any good Progressive could recognize that view as a brief for public utility regulation of oil.[26]

But it was not only oil production and producers that made the petroleum industry "a gigantic system of wrong." Oil inevitably led to speculation, luxury, and extravagance. People in booming oil regions became money-mad, out to get rich quickly. Those who did get rich were extravagant, buying "furs and silks and gowns"—here Ise used traditional gender-oriented imagery—and shaming "poor merchants and doctors and lawyers" who could not indulge in such finery. Everywhere cheap abundant oil encouraged waste by consumers, who used it for "unimportant purposes" such as driving automobiles. And because it was turned into fuel for autos, oil indirectly contributed to making life "cheap and shallow and superficial," at the expense of "thrift and economy and some of the modest virtues." For Ise, the automobile represented the prime example of mindless destructive extravagance, being used by "fat-bellied bankers and bourgeoisie . . . by gay boys and girls in questionable joy rides . . . by smart alecks who find here an exceptionally flashy and effective way of flaunting their wealth before those not so fortunate as themselves" — anything but self-sufficient manhood, this gendered language implied. Look into the moral decline of modern America and, if one believed Ise, there was cheap oil at the bottom of it.[27]

So, what should be done? Certainly, oil should be under heavy regulation, ideally as a federally managed public utility. Of course, use should be prioritized. In fact, Ise would have preferred to see oil no longer used as a fuel; if that meant the end of the automobile, that would be no loss. Failing so extreme a step, oil should at least not be used as fuel for ships, trains, or "unessential" automobiles. Cheap oil had to go. Whenever possible, American oil should be kept in the ground and necessary oil bought from foreign countries. There should be no oil exploration or production on public lands. In short, Ise condemned root and branch of the petroleum industry as he knew it; his ideal had much more in common with that of some of the more extreme recent environmentalists than with fellow economists such as Ely and Stocking. So extreme a position left little room for practical action.[28]

Confronted by conservationists at the FOCB and unfriendly preservationists such as John Ise, industry leaders of the late twenties tried to come to terms with conservation. One of the livelier arenas in which they did so was the Petroleum Division of the American Institute of Mining and Metallurgical Engineers (AIME), at whose meetings one can see industry members both challenging and adapting conservationist ideas and discourse. Thus, at the division's meeting in 1927, N. S. Reavis challenged conservationist focus on dwindling oil supply, but to do so, he borrowed from conservationist rhetoric and blamed the industry for speculation, "excesses and extravagances," "boom methods and bonanza ideas." He reasserted the industry's economic perspective in what was perhaps an unfortunate image from a public relations point of view: "We must take our cue from the old Standard Oil group and concentrate our energies on getting that penny or two a gallon profit on our manufactured product."[29]

Arthur Knapp of the Philadelphia United Gas Improvement Company realized that a tremendous amount of subjective understanding went into terms such as "conservation" and "waste," and he tried to redefine them in terms

compatible with industry economics. For Knapp, conservation was "the preservation of natural resources for economical use." Waste was "useless or unnecessary expenditure." But trying to define "unnecessary" or "economical" could,
as Knapp recognized, be a difficult proposition. In the end, Knapp fell back on
the bottom line: "Dollars and cents are the final measure of necessity." If one
tried to connect conservation with some aspect of the future, the one to choose
was future profit: "Sentimental considerations of the needs of future generations does not appear practical. It is impossible to conceive what specific needs
they will have or to measure these needs."[30] Unfortunately, Knapp overlooked
what was at the heart of most oil industry discourse of the later twenties: How
did one decide what course would yield future profit? Was future profit best
secured by rapid or slow development? By drilling more or fewer wells per
tract? By competitive drilling or unitization? Perhaps this route to conservation was as elusive as that involving the needs of future generations.

As other industry participants talked about waste, they usually had in mind
Knapp's view of it, albeit enlarged beyond narrow monetary return to include
return of produced barrels of oil — which, of course, would represent ultimate
monetary return. Thus, oil left unrecovered in a reservoir whose gas energy
was exhausted represented waste. They were inclined to dismiss critics' charges
of physical waste of petroleum, simple loss of fluid petroleum, or, apparently,
anything else; as former USGS employee J. B. Umpleby, long a production
consultant to Pennsylvania operators, put it, "Incident to physical operation
there is remarkably little waste." What Umpleby and many others frequently
noted in a time of falling crude oil prices was unnecessary costs. This led
Umpleby and his colleagues to begin to sound somewhat like George Ward
Stocking as they complained that too much competition was the source of
unwelcome costs. The physical waste of oil was negligible in comparison with
the waste of money in producing it.[31]

Even as they lamented the cost of exploration and development, as well as of
filling stations on every corner, oilmen realized that any moves to diminish
competition in their industry, for whatever purpose, would be politically unacceptable. At an AIME session in 1927, Walter van de Gracht of Shell reminded
everyone of the dreaded "M" word, monopoly. "If a thing like that is started
our attorneys have to tell us that if we want to keep out of jail we had better
stop." He added, "Of course, monopoly in itself is a terrible bugaboo in politics, but really, monopoly in itself is not such a bad thing." Getting politicians
to agree that monopoly was not such a bad thing was, as van der Gracht readily
admitted, probably impossible.[32]

However, politicians were not as familiar with the term "unitization," and
even though many oilmen had initially disagreed with Henry Doherty, more
and more of them could see the usefulness of a concept that set aside antitrust

menace with the blessing of conservation. In October 1929 the AIME decided to devote a whole section of its Petroleum Division meeting to advantages of unit operation, leaving no doubt of its general opinion of such practice, and discussion of unitization was an important part of the group's proceedings in the early thirties. Much technically oriented discussion focused on emerging understanding of reservoir pressure in field operations, of the forces in oil-bearing formations pushing petroleum to the well bore; engineers tried to describe the propellant roles of gas in reservoirs, gas dissolved in oil, and water adjacent to or underlying oil formations.[33] Quite apart from the practical application of such findings to the recovery of oil, as a concept reservoir pressure was profoundly useful to those who argued against unrestrained competitive drilling of oil fields, for it was something such drilling diminished. Its diminution not only left oil unrecovered in the ground, an "economic waste," but could also represent real property loss to participants in oil field development, in effect, physical waste. If the law said oil belonged to the operator who produced it, reservoir pressure, by contrast, could be seen as something belonging to all, to be shared by all in order to produce oil. Following this logic, unrestrained competitive drilling on the part of some operators could be taken as appropriation of what was not justly theirs — the means of pushing oil to the wellhead. Thus discussions beginning with charts and equations readily broadened into nontechnical discourse in support of unitization, government intervention, and grander objectives.[34]

Considering industry problems, the scientists were critics of competition. The villain of the piece was, of course, legal convention in the form of the law of capture, to which, as J. B. Umpleby put it, "more ills of the petroleum industry can be traced . . . than to any other cause." Similarly, L. C. Snider, who worked for Henry Doherty, argued that the law of capture made petroleum production into a situation in which "haste makes waste" and operators sought quick returns "with a complete disregard for small economies."[35] Phrases of this sort show that however up to date the science of AIME participants, they were as prone as nonscientists to fall back on traditional formulas drawn from public discourse, in this instance the condemnation of extravagance. Snider also condemned the speculation he saw endemic in the petroleum industry, "the gambling attitude" characteristic of petroleum production, yet another result of the law of capture. Unit operation would mean predictable return over long periods; without early, spectacular returns from competitively drilled wells, there would be less speculation in the oil business. For that matter, with unitized production, "the economic, social, and moral wastes due to the boom oil town can be eliminated." Here was an agenda going well beyond maintenance of reservoir pressure.[36]

Among scientific enthusiasts for unitization in the early thirties, Earl Oliver was a zealot. The son of a Pennsylvania pumper, he headed the AIME's study of unitization in 1929–30. In the early thirties, the *Oil and Gas Journal* also offered him a forum, and he wrote a series of articles promoting unitization for the periodical, with editorial endorsement.[37]

Oliver told anyone who would listen that the American petroleum industry was unstable, wasteful, and, in a word, "sick" because of the law of capture. Its wastefulness, rather than efficiency, resulted in "ruthless destruction of the nation's irreplaceable natural resources and economic chaos in the industry."[38] Such waste was incomprehensible from a civilized point of view: "To the man from Mars there is perhaps little difference between mankind and the jackal family in some of their methods of appropriation." In fact, what John Ise blamed on the petroleum industry in general Oliver blamed on the law of capture in particular—disorderly oil boom towns, oil field crime, physical waste of petroleum, speculation, stock fraud, debased behavior were all part of "the price society is paying for perpetuating a poorly devised method of determining each owner's share of oil and gas in the pool underlying the land owned by him."[39] The law of capture could thus be charged with subverting morality and civilization. It amounted to "the law of the jungle."[40]

Oliver's solution to industry problems was to abandon the law of capture for shared property rights in oil pools and unit operation of oil pools, something he asserted all engineers supported. If engineers could run oil fields, they would eliminate waste and "fluctuations that characterize the oil industry." He admitted, however, that other measures, such as proration of oil production and an oil tariff, might be needed to supplement unitization on a temporary basis if the problem of too much oil on the market was to be resolved. And no progress could be expected on the part of industry unless it worked with government; solving industry problems would require "most sympathetic cooperation between government and industry, backed by an equally sympathetic public understanding." That, of course, had never existed. Oliver wistfully observed, "If we were functioning under a Mussolini or a Stalin, stabilizing the petroleum industry would be a much simpler task."[41]

Engineers could tell the public what had to be done, but it would be up to lawyers, Oliver admitted, to figure out how to make engineers' solutions legal—something he and other scientific proponents of unitization were confident lawyers could do. Lawyers had, in fact, been working on the problem. In 1927 the American Bar Association's Section of Mineral Law set up a committee on conservation, and lawyers such as Skelly Oil's W. P. Z. German, Carter Oil's James A. Veasey, and Robert Hardwicke began to discuss regulatory possibilities in general and unitization in particular. When they did so, they picked up conservationist discourse and tried to combine it with ideas advanced by engineers.[42]

Jurists commonly accepted the conservationist image of the wasteful American oil industry and the baneful effects of competition in production. Writing for the *Yale Law Journal* in 1931, for example, J. Howard Marshall and Norman L. Meyers, both of whom would serve on Harold Ickes's Petroleum Administrative Board, agreed that competitive drilling was a "vicious system," competition "heedless," and economic losses "appalling"; not surprisingly, these authors cited the work of Henry Doherty, George Ward Stocking, and Earl Oliver. Similarly, Donald H. Ford, writing in the *Michigan Law Review* the following year, argued that unitization in production was necessary because it would "bring sanity to an industry that has been well-nigh wrecked by a mad adherence to competition" and "dispense with the present wasteful methods of drilling."[43]

Lawyers, however, had to figure out how limiting production would not amount to price fixing. Some combined conservationist discourse with engineers' ideas about reservoir pressure to offer an alternative to the law of capture and to develop a notion of correlative rights in oil pools. Thus, speaking before the AIME in 1931, W. P. Z. German argued that the state had an interest in conservation of "irreplaceable oil and gas deposits" but it also had to protect the interests of property owners in oil pools. Rather than see petroleum as the property of whoever produced it, German argued that petroleum should be seen either as belonging to whoever owned land over it or as common property to be shared equitably by all mineral owners in an oil field. Given either of these alternatives to the law of capture, no one operator could legitimately act so as to deprive others of their rightful shares of production by greatly diminishing reservoir pressure. The state should exercise police power to uphold property rights and make certain that operators did not work against others' interests: "It is as much a part of the duty of the state thus to protect the several proprietors in a pool as it is to protect the general public in the conservation of these natural resources."[44]

In their understanding of reservoir pressure, engineers had certainly come up with something lawyers such as German could use in a general way. But as soon as engineers and lawyers came to discuss the practical operation of unitization, the limitations of engineering became evident. How, for example, would one determine the rightful or equitable share each operator would have in an oil field's production? German thought engineers could decide this on the basis of drilling test wells or taking core samples to determine, proportionately, how much petroleum lay under each property. But he thought this would be possible prior to development or in the early stages of field development, implying that engineers could plan reservoir operation almost as soon as oil was discovered. Even avid engineering proponents of unitization recognized that to understand a reservoir and determine how to apportion production engineers needed data from a number of wells, the more the better. And how would production from test wells be allocated in the absence of a working allocation formula? Would it be stored (and hence subject to evaporative or economic loss) until engineers developed the formula? Who would pay for test wells and storage? And until engineers worked out their understanding of a reservoir's characteristics, how would they manage it? In effect, German assumed engineers could do things on a reservoir's discovery that actually were possible to do only after considerable drilling and production. Questioned on this point, German could only admit that the practical implementation of unit operation was "in an embryonic state." He had hoped engineers would come up with answers.[45]

Thus, engineers trusted lawyers to make their schemes of unitized operation legal, and lawyers expected engineers to make legal concepts practicable: small wonder that the discussion of unitization had little immediate practical result. Engineers and lawyers alike saw a need to make unitization compulsory and envisioned action by state or federal governments. Both sides, however, minimized the difficulties in implementing unitization and overestimated its benefits to the petroleum industry. Not only was there the problem of allocating rightful shares of production, but, by eliminating the competitive advantage of early entrants in a field, unitization schemes did not reward prospectors' risks in opening new territory. With its methodical, planned drilling and control of production rate, unitization implied slower return on investment, a major handicap to smaller operators. For unitization to control national production and shore up prices, moreover, it would have to be mandatory in all fields, in all states. Someone would have to find a way to apportion production among states and divide up access to markets. That implied federal or interstate authority. Without it, unitization could not ensure that too much oil would not flood markets. Nor, without prices high enough to encourage discovery, could it ensure oil for the future. But to many observers who saw an industry in disarray, the answer of unitization seemed the soundest of any.

For that reason, by 1930 an increasing number of industry leaders supported compulsory limitation of production, whether by state-ordered proration plans or, preferably, state or federally mandated unitization. The API endorsed unitization in principle in 1929, and even such former Doherty foes as W. S. Farish and J. Edgar Pew of Sun Oil argued for it. Taking up unitization meant joining the conservationists on the undesirability of free competition, but it also, in effect, absolved the industry: until laws permitted agreements to limit production, antitrust laws made conservation impossible for the industry to attain. A substantial segment of the industry, however, did not accept such accommodation of conservationist discourse, and from mid-1929 they offered noisy dissent. Their objections were prompted by a bumbling federal attempt to act on some of the FOCB's 1926 recommendations.

From 1926 onward, the tide of crude rose higher and higher on the American market as giant field discoveries continued. In 1926 alone, oilmen brought in enormous production in Seminole, Oklahoma, and in the Permian Basin of West Texas. As prices of crude fell in response to mounting supply, oilmen began to experiment with voluntary production curtailment schemes, in which producers agreed to set a level of field production and then divide up, or prorate, total production among leaseholders. These schemes were fraught with problems such as the basis on which production would be prorated or the enforcement of production limits, and in fact they did not work well, with the possible exception of the Permian Basin Yates field. Yates was so isolated that those who did not want to accept proration found that they lacked pipeline connections to markets. Still, the API had been sufficiently encouraged by these experiments in voluntary curtailment to sound out the Justice Department as to whether it could endorse the FOCB's recommendation of industry-government cooperation for conservation to the extent of letting voluntary agreements escape antitrust scrutiny; if so, the API had a plan for cooperative action.[46]

Notwithstanding the FOCB recommendations, or, for that matter, President Herbert Hoover's own rhetoric about cooperation, the Justice Department declined to approve voluntary agreements. But Interior Secretary Ray Lyman Wilbur decided to move toward conservation of oil on public lands. Wilbur announced that his department would not issue any new permits to prospect or drill on public land and would cancel existing permits whose holders had no work actually under way. Having outraged Rocky Mountain oil operators, largely dependent on prospecting on federal lands, Wilbur proceeded to try to encourage industry-government cooperation with a conference of oil state governors in Denver, a site admirably located for demonstration on the part of those operators, and named Mark L. Requa conference chair. Wilbur hoped the conference would pave the way for some sort of interstate agreement to coordinate state petroleum-related laws and state efforts to limit production; Requa and George Otis Smith would address conference participants, thus putting the conference in a firmly conservationist context. The conference convened on June 10, 1929.[47]

Had the conference been composed of state governors alone, its outcome might have been inconsequential. But Rocky Mountain operators mobilized support among independent oilmen, and the governors' delegations included numerous oilmen unhappy with federal policy. They were not inclined to agree with Wilbur's and Smith's predictions that domestic oil reserves would be exhausted in a decade, but they were even less disposed to hear Requa admonish them that if the industry could not cut back production voluntarily, government coercion might be needed. Requa's provocation not only guaranteed there would be no interstate agreement forthcoming but also led directly to the emergence of a new voice in petroleum-related discourse; on the second day of the conference, independent oilmen led by Oklahoman Wirt Franklin left it to form the Independent Petroleum Association of America (IPAA).[48]

Claiming to speak for independent oilmen, the IPAA challenged the conservationist point of view by falling back on older, essentially monopoly-oriented discourse. Franklin, for example, saw talk of conservation as merely the device of big oil companies. For him, the Colorado conference revealed a scheme "to turn the markets for petroleum in the United States . . . over to a few large companies engaged in exploiting the petroleum reserves of South America and in importing the production thereof into the United States."[49] Hoover's conservationists were conspirators in the service of monopoly, something independents at odds with large companies had been looking for since Roger Sherman. Rather than being the Progressive bulwark against monopoly, conservation as Franklin described it promoted monopoly — a neat inversion of older discourse.

As he attacked "so-called conservation," Franklin dismissed the ideas that the United States was running out of oil and that Americans should cut back on petroleum consumption. "This is the oil age," he argued; "let us use our oil reserves while they are yet valuable, while we need them and before some new form of power is discovered." Far from keeping oil in the ground, the United States should be using it. He allowed that some conservationists might once have been well intentioned, but the conservation program had outlived its usefulness and "become destructive to a superlative degree." Franklin also doubted that overproduction was an industry problem. Too much oil on the market was really a matter of too much foreign oil brought to American shores. It was this oil that had built up in storage, pushed prices of crude below producers' costs, and made futile the efforts to stabilize industry economics through domestic proration. What the industry faced was not overproduction but oversupply, for which imports were responsible.[50]

Even as he took issue with conservationist ideas, however, Franklin used some of them to advance the interests of small producers, most particularly in developing a case for preferential treatment of stripper well operators. According to Franklin, there were at least three hundred thousand small oil wells producing some half million barrels of oil a day. Falling crude prices were forcing operators of such wells to plug and abandon increasing numbers of them, yet these small wells were "the backbone of the oil industry, its very lifeblood." Once abandoned, these wells would never again be opened, nor would it pay to drill new wells in their areas; in short, one could expect that, if low prices continued, production of that half million barrels a day would be lost forever, "the most serious blow to conservation of oil in the United States which could be imagined." To save these wells from abandonment would be "true conservation."[51]

Because the case for stripper production developed by Franklin and many others thereafter would become not only an enduring part of the petroleum-related discourse but also a justification for special treatment of a segment of the petroleum industry, it is worth further examination. Up to the early thirties, conservationist attention to oil fields tended to focus on what happened to fields in flush production, in the early stages of development. Singling out older, settled production for concern marked a shift of emphasis in conservation discourse. It also represented yet another example of seeing small businesses in a positive light, a familiar part of industry discourse about oilmen, now applied to oil wells. Thanks to Franklin and the IPAA, small producing wells came to be seen as the special province of small independent operators, and, indeed, many small operators did run stripper wells, and many belonged to the IPAA. But many major companies also operated strippers; then as now, settled production was not the exclusive preserve of small independents. Compared with flush wells of the sort that would be common in East Texas in the early thirties, small production of stripper wells made them high-cost operations, but stripper advocates avoided mention of that. Instead, they stressed how great a volume of oil stripper wells produced and tied it to a prediction of doom: such production could be irretrievably lost.

Why irretrievably? In the industry's early decades, operators commonly said that a producing well, once shut in for a period, would not produce profitably again; this observation may have originated in the experience of Pennsylvania producers of the late 1860s who tried shutting in production in response to low prices, but it was still in circulation in the mid-twenties, and the implication was that it would be physically impossible to restore the production of a shut-in well.[52] In the absence of well work-over technology or production engineering, such expectations may well have been realistic. In any event, the idea that stripper production might be irretrievably lost put it in a special category, one that made it essential rather than inefficient.[53]

While Wirt Franklin and the IPAA developed the case for stripper preference, they took fullest advantage of their claim to speak for independents by harking back to the old image of the evil Standard Oil Company: after all, this had worked for independents before, and it would be used continually by various independents arguing their positions before Congress during the 1930s — even when those independents took different sides on issues. For example, when prominent attorney and Wichita Falls, Texas, oilman Orville Bullington, Republican candidate for governor of Texas in 1932, testified before the Senate Committee on Commerce in January 1931, he began his testimony by identifying himself as "an independent producer of oil," "not connected with any of these other standard [*sic*] units." This introduction allowed him to launch into what he called "a brief history of the independent oil business," in which, even though he allowed that the history of Standard Oil was "too well known to need repetition," he repeated the old laments of monopoly, going so far as to maintain that before dissolution, Standard Oil had "eliminated practically all competition." Bullington used this as a prelude to argue that existing Standard Oil companies were using imported oil to drive American independents out of business. Then the consumer would "be left to the tender mercies of the oil trust, and robbed without stint or limit." Long before that, thousands of workers would be unemployed and the half-million-barrel-a-day production from stripper wells would be lost forever.[54]

Taking the time-honored step of blaming big companies, especially Standard Oil companies, for independents' problems, the IPAA's first step toward helping oilmen to solution was protection, to choke back the volume of imports by a dollar-a-barrel duty on imported crude, an ad valorem tax on refined products, and an embargo on imported crude and products, or a combination of all three; it pushed hardest for the dollar-a-barrel tariff. But when given a forum, the IPAA would argue the need for pipeline divestiture and ending major company ownership of retail outlets; it linked imports, pipelines, and retail sales in a pattern of price-fixing by majors, in effect, the forces of monopoly at work.[55] The IPAA could rally support on these issues from many independents. When it came to addressing the question of what course to take on state conservation measures and, specifically, proration programs, the IPAA found it difficult to reach anything like consensus.

Although the early history of the IPAA and conventional wisdom support the notion that as a group independent oilmen fought production limitations or proration schemes while major companies supported them, oilmen did not line up as neatly as this on the issue. Considering the desirability of proration, what the oilman, large or small, had to reckon with was how it would fit with individual business strategies. If a firm could offset low crude oil prices by producing and selling a tremendous amount of crude, or by purchasing large amounts of crude at rock-bottom prices, refining it at low cost, and selling products at competitive prices, that individual firm, independent or major, might do quite well. When the East Texas field developed, it offered oilmen those kinds of opportunities. Oilmen who enjoyed them were understandably unenthusiastic about limiting production through proration or any other means. But oilmen, large or small, whose production was modest and not likely to increase, whose refineries were far from abundant crude, or who had much stored oil bought earlier at higher prices faced increased disadvantage as production from Texas and Oklahoma flooded markets. Their business strategies could not meet competition from those who could take advantage of low prices and low costs. Large or small, independent or major, those losing from the flooded crude markets were likely to see proration — and perhaps even more vigorous measures to control production — as absolutely essential.

Because one's position on proration depended on individual situations and strategies, scholars should recognize what the IPAA soon discovered: independents (and majors) did not share identical positions on proration or regulation. Initially the IPAA waffled on the issue. It argued that proration was useless in the face of sizable imports, but it said it was not opposed to conservation laws in general. Similarly, when some IPAA members formed the Independent Petroleum Association of Texas, they claimed they did not necessarily reject proration but opposed the kind of plan the Texas Railroad Commission wanted in East Texas. IPAT's independent members emerged as determined opponents of proration programs, while the independents in the Texas Oil and Gas Conservation Association fought as determinedly for it. The Independent Oil Operators of Oklahoma, formed in January 1931, immediately announced that proration was to blame for all the ills of the oil industry; at the same time, the Mid-Continent Oil and Gas Association, whose membership included majors and many independents, endorsed it. Indeed, the last group held that fairly administered proration was the only means the industry had "to protect the non-integrated producer against discrimination and drainage and thus to prevent waste."[56] No wonder the IPAA initially waffled on proration.

The IPAA's aggressive attack on conservationist discourse, both inside and outside the petroleum industry, showed how older, antimonopolist discourse could be used to oppose conservationist objectives, in this instance, to cut back oil production. As in the contest of Pennsylvania independents against Standard Oil, it could be used to defend the interests of high-cost producers, such as stripper well owners, against the advantage of lower-cost producers, owners of flush production, for example, but the IPAA's attack also pitted a developing area of policy, oil conservation, against one older and better established, antitrust; one mandated production limitation, whereas the other effectively forbade it. In paradoxical contrast to the events of only six years earlier, using the antimonopoly cry now justified avoiding government interference with the oil industry rather than mandating it.

In looking at the discourse on the petroleum industry in the later twenties, two general characteristics are striking. First, people at least ostensibly addressing the same subject, conservation of petroleum, shared remarkably little common ground in terms of what they meant and how to achieve it. Second, participants in discourse addressed topics with a remarkable lack of practicality.

These two characteristics of discourse helped people keep talking past one another rather than arrive at common understandings on real problems. Chief among real problems the industry faced was a mounting supply of oil that encouraged go-for-broke production in the field and, at the other end of the pipeline, gasoline price wars. But while real problems got worse, the FOCB and API were squabbling over whether the United States' oil reserves were half full or half empty. John Ise and Stuart Chase were telling Americans the nation would be better off if they did not use oil. Henry Doherty's panacea of unitization was enthusiastically embraced by engineers and lawyers, each group thinking the other could solve problems blocking achievement of it. Some oilmen began to warm up to unitization, but when they sounded out Hoover's Justice Department, they came up against the old antitrust barriers to industry agreements. Last but not least, Wirt Franklin's followers saw industry problems in traditional terms of larger companies trying old monopolistic tactics — chiefly importing oil — and dismissed conservation as a camouflage for monopoly.

Thus, all parties kept talking while a relentless tide of crude oil rose higher. The tide of crude depressed prices and rates of return on investments. It cut the return of royalty owners. It diminished the value of a steadily growing volume of oil in storage and lowered the value of reserves. It led to pipeline company curtailment of purchases. It brought fierce competition among refiners and retailers as it permitted deep cuts in product prices. Across the board, it became harder for industry members to break even. And this was long before the industry entered the most severe crisis in its history.

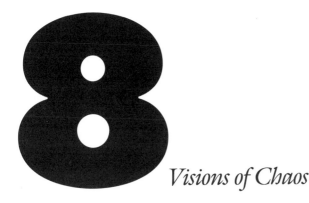

*Visions of Chaos*

Between 1930 and 1935 a great many oilmen, from the ranks of both majors and independents, did what would have seemed entirely incredible a decade earlier: they decided the industry needed some form of government regulation. And in arguing for regulation, they attacked unregulated competition — "chaos" — and refined their own definition of conservation, a definition tied to industry economics. Their understanding of conservation ultimately became the basis for policy of the Texas Railroad Commission, which played a key role in the controversy over industry regulation. But, as we shall see, federal officials such as Harold Ickes also advanced a case for regulation on the basis of conservation, and their understanding of conservation was, as Ickes demonstrated, grounded in the Progressive faith in government regulation by experts — laced with a liberal dose of oil according to Henry L. Doherty. The result was a struggle in discourse between two camps calling themselves conservationist and a third group continuing to wave the antimonopoly flag. As if that were not enough, some contenders, such as the members of the Texas Railroad Commission, raised the cry of states' rights and state sovereignty. In the course of contention, several people went through a sea change in identity: Henry Doherty went from industry pariah to industry visionary; Wirt Franklin went from being the foe of monopoly to the champion of federal regulation; and Harold Ickes, from his appointment as secretary of the interior, came to aspire to be an oil czar. In industry annals, the time was indeed a strange and volatile one, far too volatile for many oilmen's taste. No wonder, as they looked at their times and their industry, the term oilmen most often used to sum up what they wished for was "stabilization." For many oilmen conservation was to be part of that broader objective. For conservationists, conservation was ostensibly the objective. And for many antimonopolists and states' rights advocates, either objective was suspect.

As part of their adaptation of conservationist discourse, the kind of connection oilmen would make between conservation and stabilization was most clearly developed by Leonard M. Logan Jr. in his *Stabilization of the Petroleum Industry*, which appeared in September 1930. Long a columnist for the *Oil and Gas Journal*, in 1930 Logan was an associate professor of economics at the University of Oklahoma. Thus his experience bridged the gap between industry and academe. Unlike most contemporary commentators, he was able to perceive an essential difference between the discourses of the two groups, that of industry based in economics, "exchange value," and that of many of its academic and journalistic critics based in morality, "utility value." To Logan, the latter perspective was flawed by pervasive subjectivity, and when it tied recommendations for action to projections of the needs of future generations, it lost credibility: "Wants of today are not by any means the wants of tomorrow." Yet as Logan argued that diminishing petroleum supply would lead to the reservation of oil "for higher uses," he, too, accepted the normative prioritizing common to conservationist critics of the industry, simply giving it an economic camouflage.[1] Like Logan, industry leaders would increasingly use normative discourse in advancing their points of view.

Like so many commentators before and after, Logan began his study of the petroleum industry with a recital of the sins of Standard Oil, only with the John Iseian spin that Standard Oil did what many of its competitors were doing but did it better. And because the old Standard Oil came to have considerable control over crude oil prices, it brought a measure of stability to the industry. This was lost with dissolution, and stability gave way to "nearly ruinous" competition.[2]

By stabilization, Logan meant a condition in which production was roughly in line with consumption, where supply and demand were in the kind of equilibrium that meant "reasonable profit," and there were no wild surges of boom and bust. Stability was about profits and predictability, the one intertwined with the other, and the way to stabilization was production control through voluntary agreements of industry participants. Unbridled competition as sanctioned by the law of capture and laissez-faire produced a chaotic industry, and if the industry was ever to see better days, competition would give way to cooperation. In practical down-in-the-field terms, that cooperation ought to take the form of unit operation of oil pools.[3]

Logan's criticism of competition and his advocacy of unit operation clearly showed his debt to the leading conservationists of the twenties, and he quoted industry critics such as Henry Doherty and Mark L. Requa at length. He did not fall back on doomsday forecasts of oil famine, but he did stress that oil was "an exhaustible resource" and the industry was wasteful of it. But for Logan, waste was "economic waste," a term much used in the following years and meaning the situation in which the economic value of the goods produced was less than the cost of producing it: a situation that would unhappily become all too common in the oil industry's next few years. Thus, Logan offered an understanding of waste with which industry participants could agree; waste was about the bottom line and return on investment rather than consumption habits or rowdy oil towns or what future generations might want. More sensitive to different channels of discourse than other commentators, Logan sought to shift the meaning of conservation; for Logan it was part of stabilization, and if one achieved that, there would be conservation because there would no longer be waste.[4]

Leonard Logan came closer than other observers to synthesizing what was by 1930 a well-defined channel of conservationist discourse with the evolving industry understanding of conservation, an understanding no longer entirely hostile but wary of threats of government control. He understood that both inside and outside the industry the idea of cooperation to limit production would meet resistance. But, as he saw it, without cooperation to solve problems the industry would face coercion and government control. Such a lesser-of-evils argument would often be repeated in the thirties.[5]

Writing when and where he did, Logan had good reason to wonder if the oil industry would have to be coerced to limit all-out production. By 1930, not only had oilmen talked about conservation and stabilization, but in more and more oil fields they had tried proration schemes to slow the flow of crude oil to the market. They learned, however, that in any field proration was adopted, some holdouts would cheat or go to court to resist limitation. They also learned that even when operators cooperated to limit production, fresh discoveries challenged the effect of existing limitations. Thus, having tried to prorate production in the greater Seminole area in Oklahoma and in the West Texas Yates and Hendrick fields, in 1929 oilmen had to start over again at Oklahoma City. By 1930, when average daily national oil production was 300,000 barrels over what markets could absorb, the field brought in discoveries that included 50,000-barrels-a-day "elephant" wells. No wonder many oilmen were skeptical about whether field proration could stabilize markets.[6]

Since it was easy to imagine what would happen if proration was ended, however, oilmen kept on trying to make field proration work. On September 12, 1929, Oklahoma operators cooperated with a field shutdown of a month ordered by the State Corporation Commission. When production resumed, it was cut back to 60, 50, and 25 percent of capacity, until, by the middle of 1930, its production was only 8⅓ percent of capacity, or only 86,500 barrels per day of the estimated 600,000 it was capable of producing. The cutback let one journalist call proration an "outstanding" success.[7] Such efforts by the industry to set its own house in order would amount to nothing if comparable discoveries continued. As bad luck would have it, several months later, on October 3, Columbus Marion "Dad" Joiner brought in his Daisy Bradford Number 3: the discovery well of the East Texas field.

What Joiner discovered was a truly giant oil field, the largest discovered in the United States to that time, a field ultimately some 134,000 acres in extent and stretching over five Texas counties. Because major and large independent companies had not taken the area's potential very seriously, acreage was readily available to small independents, to be had in small tracts at little cost. Moreover, because most production was relatively shallow, around 3,600 feet, and the field offered no unusual drilling problems, wells were cheap and quick to complete. The location of the field near urban markets and the physical characteristics of the sweet, high-gravity crude that had a high yield of gasoline created niche opportunities for small refiners.[8] If ever there was a small operator's paradise, East Texas was it. When falling oil prices made it harder to pay bills, a field offering a great deal of oil at exceptionally low cost was the answer to a small operator's prayer.

Within a matter of months, however, the oil promoter's dream became the oil industry's nightmare. By the end of May 1931, as hordes of oilmen descended on East Texas to try their luck, the new field's daily production reached 350,000 barrels, from about 700 wells. By June 20 more than 1,000 wells flooded markets with crude oil and drove prices down to twenty cents a barrel. In response to the glut, many buyers gave up posting prices, and much oil sold for a dime a barrel or less. Crude prices in other regions could not hold up under such pressure; thus, in July, the posted price of West Texas–New Mexico crude, for example, dropped to ten cents a barrel—one-tenth of the price of a barrel of water in that arid region.[9] All over the United States oil producers whose production was costlier and less prolific than that of East Texas faced ruin, at least if nothing was done quickly to dam the deluge of oil.

In the face of looming catastrophe, both the oil industry and the state of Texas seemed impotent, for there was no easy way to restrict production. Voluntary proration worked in places such as the West Texas Yates field, where limited market outlet and relatively few field participants made it practical to reach voluntary agreements. East Texas was not that kind of place, and from the beginning, industry participants doubted that any effective production limitation would come about through voluntary agreement. Nor were the elected Texas Railroad Commission members eager to coerce operators: political support for cutbacks was limited, and legal basis for such action was questionable. Under the Conservation Act of 1929, the commission could intervene to halt the physical waste of petroleum, but it was specifically forbidden to act on the grounds of economic waste — production in excess of market demand. After the legislature passed the Common Purchaser Act of 1930, the commission attempted to prorate production, only to see its orders both enjoined and ignored. Nevertheless, in April 1931, the commission ordered the limitation of East Texas production to ninety thousand barrels a day. The result was predictable: the order had virtually no effect on mounting production.[10]

In July 1931, with no realistic prospect of either industry or Railroad Commission reining in East Texas, Texas governor Ross Sterling, former president of Humble Oil, called a special session of the legislature to address the conservation problems posed by East Texas. As he put it, "A grave crisis confronts the state in the conservation of its natural resources. The earth's reservoirs of oil and gas are being drained and virtually thrown away, and enormous underground waste is resulting from the orgy of disorderly production." Having thus described a moral crisis, Sterling allowed that the legislature might discuss resources such as soil and water, but he made it clear that oil was top priority: "The oil industry . . . is demoralized and tottering on its foundations; thousands of people, directly dependent upon the industry, are going bankrupt . . . all due to the wanton waste of oil and gas." That could not continue: "Texas of today owes a solemn moral obligation to Texas of tomorrow not to exhaust and dissipate its resources needlessly so as to deprive oncoming generations of their benefit."[11] Here was normative conservation rhetoric worthy of George Ward Stocking and John Ise put to use to aid the oil industry, long a conservationist target.

If Sterling aspired to save the industry by controlling East Texas, his design was soon frustrated by a feisty federal judge, Joseph C. Hutcheson. A Progressive Democrat and mayor of Houston before his appointment to the federal bench by Woodrow Wilson, Hutcheson was obdurately unwilling to accept adaptations of conservationist rhetoric, inclining instead to see it as camouflage for self-interest. To the judge, the paramount legal issue was still competition. He would not be party to the operation of a government-sponsored cartel, an association of producers to sustain prices. Thus, when he set aside the Railroad Commission's East Texas orders in *Alfred MacMillan et al. v. Railroad Commission of Texas*, Hutcheson fulminated against "the artificial forcing of prices by governmental action in cooperation with those engaged in the oil industry, interested in raising prices, either by stimulating demand or by keeping the supply in bounds." Not having any of that, the judge saw no relation between the Railroad Commission's orders and lawful conservation; the commission had gone beyond physical to economic waste, a matter of prices and profits. Adhering to the traditional antimonopoly perspective of his party, Hutcheson forbade the commission to act on anything but the demonstrably physical waste of oil. In response, Sterling and the legislature tried to carve out a sphere of operation for the commission by imagining every conceivable type of physical waste and prohibiting it, in the Anti-Market Demand Act of 1931. But until the commission could devise field orders to implement the new Act, East Texas remained unregulated. The standstill led the frustrated governor to take matters into his own hands in mid-August 1931, declaring East Texas on the brink of civil rebellion, placing the field under martial law, and sending in the National Guard to shut down its wells.[12]

Across the state line, Sterling's Oklahoma counterpart, William "Alfalfa Bill" Murray, took similar action. It was one thing for Oklahoma production to depress crude prices, another to see prices fall to below fifty cents a barrel because of a flood of East Texas oil. Pressured by Oklahoma producers, on July 28, 1931, Murray said that unless the price of oil rose to one dollar a barrel, he would close down all prorated fields. To this, Harry Sinclair, a large regional purchaser and Murray's special bête noire, opined, "All the proclamations and troops in the world will not add one cent to the price of oil." Murray sent out the militia on August 4, and 3,106 wells closed; stripper wells, of course, were exempt. As justification for his action Murray said that low oil prices were keeping money from Oklahoma schools and that he would not only work for schools but fight monopoly by the shutdown.[13] Murray was as old-fashioned as Sterling was up to date in his oil-directed rhetoric, but both governors kept oil in the ground and off the market. They did so, however, by armed force, clearly an unacceptable policy for the long run.

In the shorter run, using troops to control oil production was also unacceptable in court. Governor Ross Sterling allowed the East Texas field to resume production on September 5, 1931, but the Texas militia remained in the field to help the Railroad Commission enforce its proration program. The federal bench soon showed that it had not changed its position on proration; little more than a month later, it enjoined the commission from limiting production of five wells operated by the Brock-Lee Oil Company. Sterling responded by removing the commission from proration management in the field and ordering the militia to enforce existing field rules. This move brought federal judges Randolph J. Bryant and Hutcheson to declare proration by martial law illegal in February 1932. Once again burdened with the task of regulation, the Railroad Commission issued new rules, which were immediately challenged, and in October 1932, federal judges threw out the commission's entire East Texas proration program.[14]

At the heart of the difference between Texas officials and the federal judges was the notion of underground waste. Taking the direction that industry conservationist discourse had taken since 1929, the commission treated practices reducing ultimate field recovery, such as sudden reduction of reservoir pressure, as physical waste, even though one could not see that waste taking place. For the federal judges, working with the older idea of economic waste the industry used in the mid- to late twenties, barrels of oil left in the ground were economic waste, and for them physical waste had to be visible, or least empirically demonstrable. There had to be engineering data showing how many barrels of oil would be left unrecoverable in a reservoir, a calculation based on precise measurements of reservoir pressure. Such a calculation ultimately would rely on technology the industry did not perfect until late in 1932; to that point it could not demonstrate physical waste to the satisfaction of Judge Hutcheson and his colleagues.[15] Commission and industry references to physical waste thus left the judges unmoved: they saw the commission and oilmen in cahoots to raise prices and hence respond to economic waste — which was prohibited by Texas statute. In short, until technology could demonstrate that old understandings about waste should be replaced by new ones, Railroad Commission limitation of East Texas production would not stand up to court challenge.

While lawyers wrangled over East Texas in court, out in the field itself there was considerable inducement to operators to ignore authority and produce wells wide open. Maximal production was one response to low prices; if prices rose, as they did in late 1931 and 1932, one was that much further ahead. Carved into a myriad of leases run by hundreds of operators, East Texas's pine woods would have offered a challenging regulatory environment on any terms. Once it was clear the courts would block Railroad Commission orders, any operator producing more than his allowable had an excellent chance of getting away with it. The oil produced could be sold to any of dozens of small refiners ready to buy crude at rock-bottom prices. Thus, whatever the commission said about field output, since no one strictly monitored the production of every well or the terms on which the production came to refiners, from mid-1931 through 1934 it was practically impossible to say exactly what the East Texas field's daily output really was. Throughout the period many operators produced "hot oil," oil produced in violation of Railroad Commission rules.[16]

If the regulatory environment the commission faced was, to say the least, challenging, its inability to come up with a viable proration plan and stick with it made a difficult situation worse and undermined its credibility as a regulatory agency. Because it could find no consensus on which to base its field rules — indeed, there was no consensus among industry members, let alone between the industry and area land and royalty owners — the commission floundered from one proration scheme to the next, always granting so many exceptions to its rules as to make them inoperable. Between mid-1931 and mid-1933, the commission tried prorating production among operators on the basis of a combination of acreage held and well potential; on acreage alone; on per well allowable; on per well allowable taken with bottom-hole pressure; and on per well allowable taken without bottom-hole pressure: none of which checked drilling. As the number of producing wells in the field mounted, the commission could only keep to a given level of field production by cutting back operators' allowables, however calculated. By August 1932, as it tried to hold total daily field production to 325,000 barrels, the commission had cut per well allowable to 43 barrels a day. That was a small fraction of what most wells could produce and yet another inducement to disregard the rules. But to raise total field output would threaten prices; when the commission caved in to popular pressure and set daily field allowable at 750,000 barrels in April 1933, prices plunged back down to ten cents a barrel. Few operators could make profits at that price.[17]

To some extent, moreover, the continued failure of the Railroad Commission to regulate East Texas production contributed to an atmosphere of open lawlessness in the field. Even good citizens who were ready to abide by regulation found it hard to stick with it when neighbors who violated the rules could drain oil from their leases. Knowing a "hot oil" producer was doing this with impunity led to vigilantism: suspect gathering lines were dynamited; one major company, discovering that a small refiner was siphoning crude from its pipeline directly into his plant, arranged for a delivery of cement in the dead of night and pumped it into the works.[18] Taken with the normal rowdiness of life in a booming oil field, with round-the-clock drinking, gambling, and carousing, disregard of regulation and subsequent vigilantism made what was happening in East Texas seem frighteningly out of control.

That, in turn, made East Texas a tempting subject for popular journalists. They used it not only to depict the shocking effects of oil development but to indict the industry on charges of gross mismanagement of petroleum resources. Perhaps most colorful of the popular journalists was *Collier's* writer Owen P. White, a Texan who had written exposés on state politics in the late 1920s.[19] As White put it in "Drilling for Trouble," "When you strike oil, you let loose Hades." Before oil, East Texas was "a place to delight the soul," "calm and peaceful, the sort of place where every one went to one or another of its small churches." Then came oil. East Texas was "inundated with sinners," and those small churches were used as flophouses for "bums and floaters." Two years later White decided that the East Texas boom represented a "haphazard drilling orgy." As for oilmen, "They are taking one of the greatest *irreplaceable* natural resources that this country possesses and, in the cause of irrational cupidity and by unscientific production processes, are literally throwing it away . . . in short, robbing the public!" A matter of such importance could not be left to states such as Texas; as White put it in 1935, "Texas doesn't give a whoop as to what happens to the oil resources of the rest of the United States."[20]

White's progress toward advocacy of federal control reflected the general tenor of journalists' coverage of the oil industry's plight in the early thirties. Journalists usually pointed out that the United States had a limited supply of oil and that waste had been profligate, the common themes of the twenties, but they also saw the industry as incompetent, "the Balkans of American business," as a feature writer for the *World's Work* put it. For this journalist, the United States had oil "awaiting the turn of a valve. . . . To take only what is needed for today and leave the remainder for the exigent tomorrows should be a matter as simple as drawing the water for a morning tub."[21] The image of solving problems of supply and demand as simply as turning on a tap or opening a valve, like the image of an underground oil hoard for Uncle Sam of twenty years earlier, reduced industry complexity to apparently elementary common sense. It was an image so alluring that it ought to translate into operative terms, so "obvious" that only incompetence or connivance kept oilmen from acting on it.

To journalist John T. Flynn, the vast majority of oilmen were both incompetent and conniving, incapable on their own of managing their industry in an orderly fashion. He saw them as motivated by selfish interest and their talk of conservation directed simply at keeping gasoline prices high; as he told *Collier's* readers, "Most oil men don't care a hoot about saving oil."[22] Nor had they ever shown any regard for the public interest, and that led Flynn to an indictment not only of oilmen and their industry but also of the free market, American capitalism, and Judaeo-Christian ethics, all as part of his 1932 biography of John D. Rockefeller, *God's Gold*.

Flynn read the work of all Standard Oil's most notable critics — Roger Sherman, Simon Sterne, Henry Demarest Lloyd, George Rice, and, especially, Ida M. Tarbell, whose *History* he thought "the ablest document of its kind ever produced by an American writer." Indeed, Flynn relied heavily on Tarbell's presentation of Rockefeller for his own portrait of the man — as a loner, virtually humorless, and intensely avaricious. But East Texas and twenty years of conservationist rhetoric stood between Flynn and Tarbell, and to Flynn Standard Oil's competitors, the stalwart independents Tarbell cast in the roles of republican yeomen, were stupid and narrow minded, "inefficient and wasteful, ignorant and utterly oblivious of the problems in an industry produced by a wholly new set of conditions." Early Pennsylvania oilmen, like their successors, cared only for "indefensible profit"; they never saw oil as "a gift from nature to the nation" or reckoned that "the public had rights superior to their own."

Reminiscent of current events, early oilmen overproduced oil and then "clamored for all sorts of help" to cope with a situation in which "the forces of trade and industry had become wild." They were far from manly, self-sufficient masters of their affairs.[23]

In the maelstrom of oil industry "disorder, chaos, waste, incompetence" appeared John D. Rockefeller, a man with a plan for a way out of competitive chaos and, as Flynn applied gendered imagery of heroism, "the courage of the great commander who does not shrink back from the sometimes cruel need incident to carrying a great plan forward." Borrowing Tarbell's image, Flynn saw Rockefeller crushing competitors "with the swiftness of a Napoleon," but he was also building the monopoly that would set aside the kind of unrestrained oil industry competition that was "a crime against order, efficiency, economy." Rockefeller's success with Standard Oil was not the result of cheating competitors such as the widow Backus, whose tale Flynn retold, or taking rebates that, though Flynn thought Rockefeller introduced them to the industry, everyone else got anyway. Far from breaking the rules of the game, Rockefeller played fair and square: he compensated competitors he bought out fairly; he dealt honestly with his customers; he never abused investors through overcapitalization or watered stock; and he paid his workers well. Overall, of the late-nineteenth-century fortunes, his was "the most honestly acquired." And the success of his company was a measure, not of skulduggery, but of "immense efficiency," ruthless energy, and a level of business acumen his ignorant and inefficient competitors lacked.[24] Here was a picture of a powerful, resourceful player, one who adhered to at least most of the code of nineteenth-century manhood by being a square dealer — a great contrast to Tarbell's devious conniver.

Then why was Rockefeller condemned when business titans such as Andrew Carnegie and J. Pierpont Morgan, guilty of far more reprehensible practices, were tolerated, if not lauded, by the public? Flynn stressed that Standard Oil's competitors had been extremely vocal and had gotten a great amount of public attention. But Rockefeller's real piety had also worked against him, for he seemed the perfect hypocrite, the devout Christian who stopped at nothing to get the upper hand in business. For Flynn, however, Rockefeller's behavior was not hypocritical but consistent with the Old Testament values to which he subscribed; those very values promoted avarice, fraud, and ruthless destruction of adversaries. Rockefeller's God was a "selfish, jealous, pitiless Deity" whose worship encouraged "a low order of ethics."[25] It was not that Rockefeller did not take his religion seriously but that he took it far too seriously for anyone's good. In his attempt to end disorderly competition, Rockefeller simply went about things the wrong way. The right course consisted of substituting government regulation and planning for the kind of flawed control a person such as Rockefeller could bring to an industry.

With the inauguration of Franklin D. Roosevelt in 1933, there was reason to believe that the American oil industry would see some sort of new order. Oilmen and their critics alike clamored for change. When Roosevelt appointed Harold Ickes secretary of the interior, conservationist critics of the industry, as well as antimonopolists, rejoiced. Ickes had been a Bull Mooser, then a supporter of the late Senator Robert M. La Follette, the kind of old Progressive FDR expected to broaden his base of support. Ickes had fought monopoly in the Chicago People's Traction League, directed against Samuel Insull. He was also an admirer of Gifford Pinchot, and once in office, he appointed two Pinchot associates, Harry Slattery (Pinchot's secretary at the Forest Service and lobbyist for the National Conservation League) and Louis R. Glavis, to executive positions at Interior. Anyone looking at Ickes's background and his appointments could readily surmise his position on oil, though what he might be able to do was less clear.[26] Notwithstanding his lack of firsthand knowledge of the petroleum industry — he came to Interior knowing virtually nothing about it — Ickes decided that he could be the industry's savior if the power of decision making for the entire industry could, in effect, be put in the secretary's hands. He wished to be, to use the term that emerged before he actually took office, "oil czar."[27]

Though Ickes proved adept at exploiting familiar themes in oil-related discourse, neither he nor his assistants had enough acquaintance with industry operations to define the powers of a would-be czar. Asked what a stripper well was, for example, Ickes said he did not know; all he knew about them was that they had to be pumped, so he assumed a stripper well was any well on a pump. Similarly, department solicitor Nathan Margold, protégé of Justice Felix Frankfurter, knew that natural gas had "some connection with the production of crude petroleum," but he could not be more specific. More to the point, having told the House Ways and Means Committee that Ickes would need power to allocate production among states, Margold admitted that no one had figured how to do it. The secretary would set crude oil prices, but Margold did not know how this would be achieved, either. Without knowledge sufficient to provide details of a plan, in fact, what Ickes and Margold hoped for was a legislative blank check for action. Margold admitted, "It has been the policy of the President to have the broad power delegated and have the thing worked out by regulators."[28]

For all Ickes liked the prospect of carte blanche, what he and his aides desperately needed in 1933 was clear guidance from the oil industry, a unified position on what could be done, which they could then execute. But that, of course, was the problem: had the industry been able to agree, there might never have been pressure for federal intervention or talk of an oil czar. Unable to draw on either industry consensus or suitable organizational capacity at Interior, Ickes moved into a more familiar political domain. He called a Washington conference in March, of governors of oil-producing states, to discuss "re-establishment of normal business" in the industry. Presumably, Ickes could sit with a dozen governors, iron out disagreement, and arrive at consensus as to federal policy changes. As in 1929, however, the guest list rapidly inflated to include a large number of oilmen representing both majors and independents, and the result was a babel of familiar themes: there was too much domestic oil; there was too much imported oil; there was too much illegally produced oil; there was too little oil reaching the market because of pipeline monopolies; action should be taken by states; action was required from the federal government. In Ickes's favor, many oilmen, representing both major and independent sectors of the industry, supported some measure of federal control of oil by March 1933. But the conference was more inclined to endorse state than federal remedies. There was remarkably little support for Ickes's elevation to "oil czar."[29]

Turning to Congress, Ickes had Margold draft legislation that took form as the Marland-Capper bill, giving Interior power to allocate production of petroleum and products among states, to control imports, to prohibit interstate transport of petroleum produced in violation of state and federal law, and to fix petroleum prices. The bill created the oil czardom such as it was. It also regulated oil additionally and separately from the regulation provided in the National Industrial Recovery Act (NIRA).[30]

During hearings on Marland-Capper, there was little originality about what Ickes told Congress; he put familiar themes as dramatically as possible, much in the manner and perspective of Mark Requa years earlier. Oil was essential to civilization, but the industry lacked "that reasonable control" necessary for its own well-being and that of the nation; it was running "amuck [sic]." Not that he would admit to concern about oilmen; were oil not so vital, "we might with complacence sit by and watch the producers kill themselves off." Oil, however, was an irreplaceable resource: "We cannot permit men, even if they do invoke the sanctity of private property, to waste, yes, even to permit the flow into the gutter of what may in time prove to be the very life-blood of the Nation." And this rhetoric was restrained compared with what he said about oil after Congress rejected Marland-Capper. In a special feature he wrote for the *New York Times*, Ickes accused the industry of profligacy and squandering resources; echoing Pennsylvania geologists of a half century earlier, he warned, "Unless we put a stop to this wanton waste, this profligate dissipation of an indispensable natural resource, our children will feel for us the pitying contempt that we will so richly have earned." Here was the old gendered imagery of luxury and extravagance turned against a whole industry. The industry was guilty of unscientific exploitation, overproduction, reckless and improvident methods of capture, ruthless dissipation of natural gas, and last — but not least — greed. To use the ultimate moral summary, the industry was in chaos. And yet, were the industry regulated, all this could be corrected; demand for petroleum could be met "almost as simply and economically as turning on and off a water tap in a city water system." What was needed was the legal empowerment of Ickes with discretionary control of the industry. Closing with a threat, Ickes added that the longer such legislation was postponed, the more drastic it would have to be, and hence: "The oil industry would be serving its own good by cooperating to obtain such legislation."[31]

Despite his best effort, Ickes did not become oil czar, but when the NIRA-mandated oil code was withdrawn from National Recovery Administration (NRA) jurisdiction, he became oil administrator, with two advisory bodies, the Petroleum Administrative Board and the Planning and Coordination Committee, to help him with his new duties. The Petroleum Administrative Board was headed by Interior Solicitor Margold and included Yale jurists Norman L. Meyers and J. Howard Marshall, whose *Yale Law Review* articles indeed took them a long way; J. Elmer Thomas of Fort Worth; and John W. Frey of the Department of Commerce, who would later coauthor the official history of the wartime Petroleum Administration for War. A more unwieldy body, the Planning and Coordination Committee included James A. Moffett of Standard Oil of New Jersey (SONJ), independent M. L. Benedum, and Donald Richberg as "representing" the government, and many of those industry participants already prominent in discourse. Many of these persons were already on record as favoring regulation, so their perspectives were, in at least one respect, in accord with Ickes's objectives.[32]

Looking at what the New Deal did about oil in 1933, one sees how many venerable themes in public discourse on oil found form in legislation and public policy. Sections 9(a) and (b) of the NIRA focused on the time-worn bogey of pipelines and divorcement. The most enduring provision, 9(c), permitted the federal government to stop interstate transportation of oil produced or withdrawn from storage in violation of state conservation regulations. On the face of it, this addressed the problem of hot oil, but it also embodied the older conservationist impulse to keep more oil underground as well as the yet older suspicion that stored oil was hoarded to force prices down.[33]

The codes of "Fair Competition" required by the act for oil and other industries reflected old fears of monopoly. With respect to oil, the code's preamble incorporated thirty years of conservationist discussion, calling for the prevention of both physical and economic waste. The body of the code let the oil administrator allocate production of petroleum and products among states — essential to the enforcement of state regulation and addressing the old problem of geographical rivalries within the industry. He could control levels of imports, an obvious concession to the Independent Petroleum Association of America (IPAA) and to traditional protectionist thinking, though at odds with the conservationist's strategy of pumping foreign nations dry first. Targeting imports, like regulating withdrawals from storage and threatening pipeline divorcement, reflected old impulses aimed at offsetting the economic advantages enjoyed by large corporations. By contrast, that part of the code requiring the oil administrator's approval of plans for development of new oil pools, with applicants to submit maps and geological data, worked against smaller producers; major companies could afford to work up this material more easily than small independents because they ordinarily developed larger tracts, over which they could spread the related reporting costs, and they regularly employed the technical staff necessary to comply. The code's emphasis that plans would protect correlative rights showed a clear acceptance of the position that Henry Doherty, the Federal Oil Conservation Board, and the American Institute of Mining and Metallurgical Engineers had been urging for the past decade. The code also included conservationists' normative prioritization: having plans approved by the administrator would serve not only to end waste, especially of gas, but also to put production to "beneficial use."[34]

Seen in the historical context of discourse, what was new about the New Deal for oil was the role of the federal government and, especially, the oil administrator in industry operations. Otherwise, a variety of nostrums, many decades old, were revived to address a problem unprecedented in scale rather than nature. Some of them, such as making pipelines independent of major company control, had little to do with the oil glut and made no operational sense. Their value was rhetorical and political. Others, such as controlling interstate shipment of oil, could be translated into operational terms that might and, to some extent, did affect what was happening in East Texas. Somewhere in between was the scheme to have Washington approve all new field development, which sounded more practical than it was; Washington was in no position to implement that. In specific fields such development plans, safeguarding correlative rights and preserving reservoir pressure, would be based on drilling no more than ten wells, whose data would offer all information necessary to determine how wells should be spaced, at what rate they should be drilled, and how fast oil would be produced. If this was fanciful in an industry just beginning to understand how to go about reservoir engineering, the idea that doing all this would harmonize reservoir energy with market demand was wishful thinking. Ickes and his friends, in effect, took what engineers said they wanted to do as evidence of what they could do and made it policy.[35]

In practical terms, however, the oil administrator needed legislative reinforcement. The NIRA's duration was to be two years, and should it expire without some additional legislation, Ickes's reign as oil czar would be over. He needed specific legislation, modeled on Marland-Capper, to consolidate his position. Ickes and Congress also needed to figure out what precisely should be done for or to the oil industry on the federal level. To that end, in 1934, Oklahoma congressman Wesley Disney and Senator Elmer Thomas introduced bills similar to Marland-Capper. By this time, yet another channel of discourse, venerable in public affairs but relatively novel as applied to oil, had surfaced, the discourse of states' rights.

A wide range of questions could be raised about the constitutional propriety of many New Deal measures, among them the Thomas-Disney proposal. How, especially in peacetime, could one justify letting the federal government intervene in production of oil within the confines of a state? Wasn't this a clear invasion of state sovereignty? Working with both a measure and arguments developed by Nathan Margold, Congressman Wesley Disney argued for a national interest in petroleum. Certainly this was orthodox conservationist discourse, but it carried no weight with states' rights advocates in and out of Congress. A Dallas attorney, for example, insisted that the Constitution gave Congress no power to regulate "oil, haircuts, woodchoppers, or anything or anybody." Speaking for Texans, "descendants of the patriots of Goliad, the Alamo, and San Jacinto, who fought and died to establish a great commonwealth," he said Texans did not want to give up rugged individualism for Nazism.[36]

The grand master of states' rights rhetoric to emerge from the controversy was unquestionably Ernest O. Thompson, of the Texas Railroad Commission. Former mayor of Amarillo, Thompson had no direct experience of the oil industry before he was appointed and then elected to the commission in 1932, but as a graduate of the University of Texas and onetime protégé of state supreme court justice Reuben R. Gaines, he knew many of the right people, including federal district judge Randolph Bryant. Thompson was one of the few who successfully straddled the discursive division between conservation and monopoly; he supported conservation in the understanding of it the industry would take, but he argued that federal management of conservation would open the door to monopoly.[37]

When Thompson testified before Congress on Texas, he referred to "the sovereign state of Texas," a place with more oil reserves than any other state in the United States or "any other nation in the world." Indeed, he gave Texas a feminine personification, like Britannia or La France; thus, of the state, "she has enacted one statute after another." Having put Texas on a par with independent nations, Thompson argued that in "carrying out her own policies essential to the protection of her interests," Texas was also serving national policy by preserving an irreplaceable resource indispensable for defense. While insisting on Texas the "sovereign state," Thompson made it clear that he had no intention of yielding an inch of Railroad Commission regulatory authority to any Washington agency, refusing to accept, as he put it, "the noose of federal control around our necks."[38]

Nor was Thompson about to be told what to do by Harold Ickes; he seems to have delighted in challenging the secretary of the interior. Characteristic was his sparring with Ickes over a wild gas well in the Conroe field. Ickes sent a preemptory telegram ordering the well be shut in. Thompson called Ickes long distance to say, "Mr. Secretary, I have read your telegram to the wild well in Conroe and it is still blowing. Do you have any other further suggestions?"[39] Thompson appears to have viewed Ickes with a mixture of contempt and deep animosity. When Ickes threatened a meeting of the American Petroleum Institute (API) with nationalization in November 1934, Thompson exulted, "Old Ickes and federal control are both through. That speech killed them."[40]

With a view to trying to find some consensus on industry questions, Maryland congressman William P. Cole launched an investigation of the petroleum industry. Among other objectives, Cole's investigation aimed at finding out if there was too much oil on the market, if oversupply caused abandonment of stripper wells, and if oil was being wasted or put to inferior uses — all questions from conservationist discourse.[41] The result of these congressional projects was the most intensive legislative airing of the oil industry's various problems in history, hearings generating thousands of pages of testimony, a veritable time capsule of public discourse on oil to which the usual federal bureaucrats and lobbyists contributed their opinions and in which industry members and state officials also took extensive part. Those who participated in this forum usually gave their assessment of the industry's current condition and suggested what Congress might do.

From the onset, Cole's committee faced the problem Ickes's governors' conference faced, the absence of an industry consensus of any sort. Amos L. Beaty, now on Ickes's Planning and Coordination Committee, supported federal control, "either control or chaos," as he put it: "We need central control, and nothing else will really be effective." Phillips Petroleum president Frank Phillips did not want an oil czar, but he did want a federal board to run oil and horrified most industry members by suggesting that "the resources of the industry should be nationalized." By contrast, W. S. Farish of SONJ preferred state regulation to federal control. His colleague, Wallace E. Pratt, a vice president of SONJ, also preferred state regulation but thought federal intervention was necessary to make state regulation workable. The majors did not agree, and executives of the leading firms did not, either. Opinions differed within a single oil family: though both J. Edgar Pew and J. Howard Pew of Sun Oil opposed federal control, the former thought state-mandated proration would serve stabilization and conservation, but the latter opposed any limitation of production as leading to higher consumer prices.[42]

Independents were also divided. Many spoke out strongly for federal control of the sort Ickes hoped to obtain, arguing that nothing on the state level had worked and that the NRA was inadequate. Most prominent among this group were IPAA leader Wirt Franklin, Charles Roeser, and H. B. Fell. Pure Oil president Henry M. Dawes wanted federal legislation to limit production but shied away from plans for an oil czar. Oilman-politician E. W. Marland, however, now rejected all federal intervention and called overproduction the only protection the public had against inflated prices. His fellow Oklahoman Jake Hamon thought federal control of oil would put small independents out of business; as Hamon saw it, "The oil business is in good shape. I think we have all the laws we need."[43]

A strident minority of independents spoke out even more vehemently against federal regulation, seeing it simply as a way to put the oil industry in the hands of major companies and thus achieve monopoly. As Elwood Fouts, speaking for H. R. Cullen, J. S. Abercrombie, John Henry Kirby, and other Houston oilmen, put it, "What is sought [by the majors] is to better the price, and tighten the monopoly, and deliver the whole into the possession of the petroleum barons of this country." Royalty broker J. Edward Jones of New York agreed, alleging that there was no overproduction; harking back to the dark days of John D. Rockefeller and his rebates, he claimed that major oil companies supported giving Ickes power because he would turn the industry over to them. Speaking for the Independent Petroleum Association of Texas, Jack Blalock said Ickes "did not know an oil well from a Sears-Roebuck catalog"; there was no overproduction, and oilmen were prosperous. The only reason the federal government had gotten into oil's problems was that there was a "conspiracy on the part of a few major oil producers that are trying to protect their investment in the stripper well states, asking you to hold Texas down." Possibly none of these opponents of monopoly noticed that so many representatives of major companies opposed giving power to the very man who was supposed to hand the industry over to them. It is more likely that these independents, like most witnesses at federal hearings, simply read "canned" position statements, prepared long in advance of the hearings. Because the various independents and majors made no attempt to listen to one another, and because congressmen were also aligned with prehearing positions and factions, Cole's hearing was largely liturgical and symbolic.[44]

Supporters of regulation, whether on the federal or state level, really needed a way to argue against a free marketplace for petroleum without appearing to support what so often was opposed to it, monopoly. For this object, the nebulous concepts generated by three decades of conservation-directed discourse served admirably. So did the idea of chaos; to say that the petroleum industry was in chaos rather than hard times implied the kind of moral breakdown government was supposed to prevent. In brief, regulation advocates painted the oil industry's situation as chaotic; they charged the industry with waste; and they said regulation would serve conservation. Lest anyone see regulation as the harbinger of control of all industry, they stressed the oil industry's anomaly; oil needed regulation because it was so different from everything else. All these ideas — chaos, waste, conservation, anomaly — took discussion into a normative arena in which one set of values, antiwaste, could be pitted successfully against another, antimonopoly. But ideas such as waste and conservation were also capable of being defined variously.

By contrast to the API's stand on waste a decade earlier, in 1934 many oilmen were willing to own up to industry wastefulness of virtually any variety conservationists had ever mentioned, in order to make their case. They referred to oil lost in storage or spilled, gas dissipated without use, inefficient use of reservoir energy, drilling unnecessary wells, building too many gas stations; physical or economic, as waste they would admit it all. Nor did they overlook the threat of yet more waste in the form of abandoned stripper wells, which, they assured congressmen, caused production to be forever lost. They differed, however, on the question of whether waste was also putting oil to "inferior uses." C. B. Ames of the Texas Company and Wirt Franklin were willing to support the usual condemnation of oil used as boiler fuel. But J. Howard Pew flatly resisted normative prioritization by experts, denying that any use was inferior to any other, and W. S. Farish agreed that no uses of oil ought to be prohibited. Differ as industry spokesmen did, once they admitted that waste existed, it was a short step to argue that regulation was necessary in order to end it.[45]

The positive construction of ending waste was promoting conservation, and here what the political and industrial proponents of regulation meant in using the term was directly tied to their objectives in regulation. Harold Ickes, for example, did not define the term but used it as opposed to "wasteful depletion."[46] Clearly his understanding of the term was grounded on the old Progressive fear of oil shortage and subsequent high prices to consumers, a menace to be avoided by keeping oil in the ground. Many oilmen took pains to criticize such a view of conservation as "hoarding," a nicely chosen term implying such negative features as stinginess and miserliness. It was pointless to keep oil in the ground against future need when one could not know what future need would be. Rather, they argued, "true conservation" implied use; as

J. Edgar Pew put it, "Conservation assumes wise economic, judicious utilization." W. S. Farish seemed to agree: "Conservation is efficient use, and that is all it means."[47]

At the same time Farish and others stressed use, they tied economics to conservation. Conservation served stabilization. They hoped that balancing production and demand would restore prices to higher levels. In short, conservation was another way of looking at industry economics. This was the essence of how oilmen adapted conservationist discourse, quite at odds with conservation as viewed by people such as Gifford Pinchot. Farish talked of "conservation in the sense of getting efficient use and full value, fair value, for a natural resource. . . . I contend that stabilization of industry as it can be attained legally under our system of government by state action, is true conservation in that it brings into efficient use and it prevents physical waste and economic waste." As for stabilization, it meant "the maximum yield out of oil pools at the lowest cost consistent with the interest of the public and the industry, so as to avoid waste." For Farish, conservation and price stabilization were "synonymous terms. . . . They mean practically the same thing." Such justification of stabilization was echoed by Wirt Franklin: "Without stabilization of the industry there can be no conservation." Even more directly, Charles Roeser affirmed, "I think conservation and price are wedded for life." These oilmen thus used the understanding of conservation asserted by Leonard Logan to advance what they saw as economically imperative for their industry.[48]

Needless to say, not all listeners bought the idea that conservation should be identified with the economic well-being of the oil industry. Congressman Charles A. Wolverton wondered whether oilmen were using "a smoke screen of conservation" to achieve stabilization. He told Wirt Franklin, "Your theory of conservation is entirely different from that entertained by those who term themselves 'conservationists.'" Questioned by Wolverton, Kansas State Corporation Commission assistant Marvin Lee agreed: "If you could pass a law putting the price of crude to about twice its present price . . . you would not hear anymore out of [oilmen]. They would not worry about conservation." To these observers, the oilmen's understanding of conservation was unacceptable, and they were duplicitous for advancing it. Indeed, in the view of many congressmen, conservation and stabilization, far from the same, were at odds. The former served the common interest whereas the latter advanced special interest.[49]

Even as they embraced it, advocates of regulation recognized that they asked for a degree of government regulation of business, whether at the state or federal level, that, apart from railroads, was unprecedented in peacetime. It was a measure of intervention that, ten years earlier, only a maverick such as Henry Doherty had supported. To counter argument that such regulation would set a dangerous precedent, oilmen in favor of regulation stressed industry anomaly. Using the well-worn argument that the law of capture made the oil industry unique, they argued not only that the industry needed political help to solve its problems but also that no other industry could be seen in a comparable quandary. Oilmen in the future would have reason to regret stressing the anomaly of oil, for as the perspectives of economists such as Richard T. Ely demonstrated earlier, that could justify singling out the industry for punitive action and burdensome control. The argument from anomaly, however, was a familiar one — to oilmen and their critics alike.[50]

Opponents of regulation could not argue for things such as waste, but they could argue that federal regulation was unnecessary, that it would be an aid to monopoly, or that state regulation was adequate — or some mixture of all these. There were nonconformists such as J. Howard Pew, who, in a splendid burst of gendered discourse, told the Cole committee he didn't want a nurse for his business, nor did he want anyone else to have one.[51] More conventionally, Joseph Danciger, that obdurate foe of the Texas Railroad Commission, saw all production limitation, federal or state, as a step toward putting independent refiners and producers out of business.[52] But the master of condemning federal regulation was unquestionably Ernest Thompson, with his battle cry of states' rights.

Raising the cry served Thompson admirably, in terms of both heading off Ickes and appealing to Texas electors. Federal regulation of oil would "wrest from the states their proper sovereign power over a purely internal affair and . . . give it to some agency in Washington." And to what end? There was nothing the federal government could do with respect to conservation that the Railroad Commission could not do or was not already doing. What it had done in East Texas was "a splendid achievement in conservation." To the commission's critics, who argued it had not enforced its proration arrangements, Thompson responded acerbically, "I think we are enforcing proration of oil at least as well as the Federal government enforced prohibition." Moreover, he argued, oil's problems had been greatly exaggerated, as oilmen cried "chaos" rather than "wolf": "Our experience has taught us that the oil business thinks that it has to be saved from its own 'chaos' about every 60 days. . . . I do not know what this chaos is, but it is always present in the oil business." Indeed, oilmen were "the richest people in our state. I know of no others who are so prosperous as in East Texas." It was not genuine distress or concern for conservation but the desire for ever greater profits that had prompted the industry supporters of regulation by Washington.[53]

Regulation from Washington would, Thompson maintained, simply mean more control of the industry by the big companies. Take, for example, Ickes's support of unitization, to be directed from Washington: "It would be the major companies' dream realized because there would be no need to consider the little independent who always leads in the drilling of oil wells in the new field. Unitize him and you put him out of business." Ultimately, big companies would "control production, and control the pipelines, and the refineries, and the outlets to the consumers to the ultimate money benefit of themselves." That would bring gasoline prices so high that people would not be able to buy it. Following this line of argument, Thompson effectively styled the role of the Railroad Commission—which only one year earlier faced a state legislative challenge to take away its authority over oil and gas on the grounds of its ineffectiveness—as the protector of state sovereignty, natural resources, consumers, and independent oilmen. To develop this image for the commission, in the face of what had been written and said in response to events in East Texas over the past three years, was a rhetorical accomplishment of the first magnitude.[54]

While Thompson worked to rehabilitate the Railroad Commission's public image, the commission slowly recovered from the nadir in its standing reached in 1933, when the Texas legislature very nearly took its oil regulatory functions from it to give them to a state natural resource commission. In November 1932, the Market Demand Act had effectively reversed legislation of the previous year and allowed production limitation to prevent production in excess of market demand. Late that year, oil field technologists finally perfected devices to measure bottom-hole pressure, and that made it possible to work up data supporting the idea of waste resulting from diminished reservoir pressure. It also made possible the engineering demonstration of the water drive at work in the East Texas field. In short, engineers could present federal judges with abundant data in support of production limitation in East Texas, and that data was extensive enough to change the jurists' minds. As Judge Hutcheson put it in the 1934 Amazon case, "All this vast amount of evidence submitted in favor of the Commission's findings, is too ponderable to be brushed aside as no evidence at all," a graceful reversal of his former position. The courts thus came around to commission regulation of production.[55]

Court acceptance of Railroad Commission regulation was only part of the progress to recognition of that body's authority. Notwithstanding the 1933 passage of a state law making it a felony to produce oil in violation of state regulation and the provisions of the NIRA and Petroleum Code prohibiting transportation of illegally produced oil over state lines, enormous amounts of hot oil continued to be produced in East Texas. Establishment of the Federal Tender Board in the field during the latter part of 1934 helped stem the flow to some extent, but dramatic reduction did not take place until 1935, when interest shifted to other regional discoveries. By that time, the Texas legislature had also given the Railroad Commission authority to stop shipment of products of illegally produced crude, as well as the illegal crude itself; declared illegally produced petroleum contraband subject to seizure by the state; and gave the commission power to prorate natural gas production to market demand. Overall, the commission emerged from the East Texas crisis with substantial statutory enhancement of its powers.

None of this would have been effective, however, had not most Texas oilmen decided that they could overcome their doubts about the commission and work with it. Not only did they find that commissioners could be generous about interpreting regulation to harmonize with what they wanted—waivers on well spacing regulations were a good example—but when the Railroad Commission talked conservation, it shared industry understandings about what that was: the maximum yield out of oil pools, at the lowest cost consistent with public and industry interest, and avoidance of waste.[56] It took the industry's adaptation of conservationist discourse, as represented by people such as Leonard Logan and W. S. Farish, and made that adaptation operational. When the commission regulated how many wells could be drilled on a forty-acre tract, for example, it could cut back on the "unnecessary drilling" the industry condemned. When it set well and field allowables, it ended uninhibited competition. Most of all, when it arranged statewide proration of production with reference to what petroleum purchasers told it they were willing to take, it balanced production and market demand. Here was the stabilization the industry talked about, whether or not one thought of it as "true conservation." And because it was squarely tied to operations and economics, it worked.

With respect to keeping prices at profitable levels, the system was not infallible. Market forces elsewhere could upset prices. Thus 1939 boom production in Illinois, which did not limit production, led Humble Oil to try to respond to the market by lowering prices. The commission's response showed its sense of politics and priorities; it ordered field shutdown until the purchaser came to heel and agreed to rescind price cuts.[57] This was not exactly fixing prices: but at a time when Texas was the largest petroleum producer in the nation, what the commission did definitely had a paramount influence on them. If one subscribed to stabilization, that was how things should be.

As much as it might defend the sovereignty of the state of Texas, however, all on its own the Texas Railroad Commission could not guarantee stabilization, since Texas was not the only producing state in the Union. In lieu of federal government direction, what states' rights advocates opted for in the thirties was interstate cooperation. In February 1931, representatives from Texas, Oklahoma, Kansas, and New Mexico formed the Oil States Advisory Committee with a view to some cooperative effort to limit production.[58] State legislatures were initially unenthusiastic about OSAC's model bill to formalize cooperation, but by 1934 it became clear that some form of interstate alliance or body might be a useful counterpoise to direction from Washington. It was equally clear that states' rights diehards would not give up control of oil production to such a body any more than to Washington — in particular, Texas governor James Allred and Ernest Thompson were adamant. So, when the Interstate Oil Compact Commission was established in 1935, it had no coercive powers. Texas, Oklahoma, New Mexico, Kansas, Colorado, and Illinois joined the IOCC, whose purposes included getting states to pass conservation statutes to head off physical waste, and to agree on a total production of crude then allocated among members. This kind of body, one of the recommendations of the Cole committee, won congressional approval in August 1935. As William Childs has described, the IOCC became a bastion of states' rights opposition to oil schemes from Washington.[59]

Apart from the creation of the IOCC, the enduring federal legacy of months of sound and fury about oil in and outside Washington, the fruit of hundreds of opinions, thousands of pages of testimony, and many federal dollars was the Connally Hot Oil Act. On January 7, 1935, the United States Supreme Court struck out Section 9c of the National Industrial Recovery Act; it declared the entire NIRA unconstitutional in June, thus sweeping away the NRA and Ickes's aspirations to oil czardom. But the loss of Section 9c meant that there was no longer federal power to control interstate shipments of oil. Senator Tom Connally, of Texas, thus introduced a bill to prohibit interstate shipment of oil produced in violation of state conservation regulations. The bill provided for a two-year, renewable, federal control; after several renewals, the Connally Act was extended indefinitely in 1942. And that was all.[60]

So modest an outcome is tantamount to saying that the conservationist elephant labored and brought forth a gnat. How could so much discussion of policy and regulation have come to so little? Scholars have advanced a number of opinions. Gerald Nash, for example, suggests that though few measures emerged, the New Deal had a profound ultimate impact on oil policy, in part because it helped "to crystallize a consensus among the industry and in Congress."[61] But on this tantalizing note Nash leaves us. What consensus? That most people didn't want Harold Ickes running oil? And what ultimate impact? The Connally Act? In short, such a position stops short of explanation. Richard Lowitt explains that little was done because reform ran aground on "massive pressures from oil-producing states" and approaches to problems that were "favorable to the major integrated petroleum companies and the large independent producers." This perspective is understandable given Lowitt's near total reliance, as he says, on what Harold Ickes wrote.[62] If massive pressure from oil states means resistance from bodies such as the Texas Railroad Commission, certainly this resistance did not help New Dealers get things done. As for the big oil/special interest argument, what can we then make of the variety of oil opinions aired in 1934 before the Cole committee? The whole problem facing that committee was that oilmen large and small could not agree on what they wanted. They couldn't even agree on whether they wanted to work with Ickes or not; Frank Phillips, no small producer, was willing to let him manage all national resources, including oil. Before we say policy ran aground on special interests, we need to identify who and what those interests were.

John Clark goes along with the idea of interest group pressure and remarks that interest groups "contested the formulas adopted in 1933, and in key areas sabotaged the system." But he also says that special interests couldn't form effective alliances with a broader public, leaving questions unanswered about how they got things done. More to the point, however, he also argues that the oil code suffered "from the absence of a clear idea of what was intended." Roosevelt and his planners, like Ickes, simply didn't have a vision of "future structure," of where they wanted to go. The NRA amounted to "the ephemeral supremacy of planners in search of a plan."[63]

For this position the evidence is, as we have seen, abundant. It was not only Ickes and his planners but also oilmen of all varieties who approached pressing problems with a variety of ideas, old and new, that did not add up to workable policy. If one believed traditional conservation thinking, for example, one wanted to change competitive ways of producing oil. If one was afraid of monopoly, any challenge to competition was unacceptable. If one believed oceans of oil produced below cost was wasteful, one could argue that proration to stop it was in line with conservation. If one believed conservation was only served by keeping oil in the ground rather than producing it according to demand, proration was not conservation, but it might be price-fixing in disguise. If one believed conservation was served by importing foreign oil, one had to answer to those who said it was a device of monopoly to ruin domestic producers. If one believed in cheap oil for American consumers, as Harold Ickes said he did, one could hardly defend cutting back production — as Harold Ickes did. And so it went. Out of the tangled discourse about oil in the early thirties it is hard to see how federal regulators could have come up with something that worked, let alone that satisfied everyone.[64]

No wonder that, after 1935, New Dealers interested in reform began to look back to old-fashioned antimonopoly ideology. Antitrust action had fewer ideological pitfalls in its train. Unlike conservation, everyone could agree on what antitrust was, if not on what justified it.

## *Monopoly Revisited*

In the later thirties the focus of oil-related discourse shifted, from conservation, which had dominated it for almost a quarter of a century, back to the perils of monopoly. Conservation was not forgotten; Franklin D. Roosevelt's Natural Resource Commission contained a section that reported on petroleum in 1939, repeating many familiar themes. But when industry critics of the later 1930s looked at conservation in the petroleum industry, they inclined to view its achievement as warped by cynical profit seeking on the part of industry giants; the theme of profit taking led naturally to the old refrain of monopoly.

Just as in the late-nineteenth-century campaigns against monopoly, moreover, there were many antimonopolists who expressed misgivings about the direction of United States capitalism. The muckrakers who had attacked that direction in their exposés of the oil and other industries several decades earlier enjoyed revived popularity, fostered in part by publication of their memoirs and by new works that forwarded many of their perspectives. Popular writers came forward to blame the depression on the failure of industrial and financial capitalists. The most widely read and quoted of the new muckrakers was Matthew Josephson, whose polemic *The Robber Barons* appeared in 1934. His book joined the canon of reformist and radical history, enjoying widespread inclusion on college and university reading lists to the present. Turning late-nineteenth-century business growth into the stuff of ripping yarns, Josephson's business leaders were swashbuckling pirates out to plunder the unwary public; his Rockefeller was just as Ida M. Tarbell had drawn him, and he repeated all the old tales — the widow Backus, the Buffalo refinery fire, the turn of another screw — the usual litany of abuses committed by Standard Oil. But Josephson's target was not so much his despicable robber barons as the system they built, which meant "the misdirection and mismanagement of the nation's savings and natural wealth." It would take the depression to expose "the fearful sabotage practiced by capital upon the energy and intelligence of human society."[1]

If one believed that American capitalists were responsible for the depression, that the economic system they presumably controlled and manipulated periodically impoverished consumers, one could doubt the wisdom of the early New Deal policy, contained in the National Industrial Recovery Act and the Reconstruction Finance Corporation, of government cooperating with business to stabilize and restore the economy. There were many New Dealers with such doubts, and they were ready to adapt old monopoly rhetoric to current conditions. The "expansionists" argued that "stabilization" would lead to higher prices and reduced consumption, thus thwarting industrial recovery; Leon Henderson became a prominent exponent of this point of view. And there might be graver consequences of working with business: former Wall Streeter and expansionist Alexander Sachs warned that stabilization would strengthen monopolies and cartels; Sachs argued that that had happened in Germany and resulted in dictatorship.[2]

As the antimonopoly cry enjoyed renewed popularity, it displaced conservation as the issue of choice for critics of the oil industry. It was not only that oilmen now claimed to support conservation — a travesty to many of their critics — but also that for all the New Deal's talk of conservation, remarkably little had been accomplished. The United States had a secretary of the interior who was an avowed conservationist with solid Progressive credentials, he enjoyed power over the oil industry extending to the wellhead, and the industry was poised to enter the new order preached by Henry L. Doherty. And then the National Industrial Recovery Act was swept away, survived by the Connally Act and the new Interstate Oil Compact, whose combined impact came nowhere near Progressive regulation of the industry. For those who hoped for an oil czar and got former hotel keeper Ernest O. Thompson at the Texas Railroad Commission, this was a frustrating turn of events.

Because Ickes's Petroleum Administration promised much and delivered little, it was a tempting topic for experts, academic and self-styled, to write about, and in the three years following the end of the National Recovery Administration (NRA), René de Visme Williamson, Myron W. Watkins, William J. Kemnitzer, and other critics came forward with analyses of what had gone wrong. Though beginning from different points of view — Williamson from political science, Watkins from economics, and Kemnitzer primarily from old-fashioned republican ideology, these commentators all reacted against one of the directions conservation-oriented discourse on the petroleum industry had taken, specifically the oilmen's linking conservation to industry stabilization. All insisted on a sharp distinction between industry majors and independents and on seeing majors as acting from unified interest. All were highly critical of what Ickes and the Petroleum Administration had done with oil, arguing that it had not served the national interest; all held the traditional Progressive opinion that the petroleum industry was deeply flawed, incapable of acting for the public good without major modification.

Although his professed purpose was the study of government planning rather than the oil industry, René de Visme Williamson focused on the objectives of conservation and stabilization and how they had fared under the NRA; he offered his conclusions in *The Politics of Planning in the Oil Industry under the Code*, appearing in 1936. Williamson, an instructor in politics at Princeton, took much of his evidence from New Deal congressional hearings on oil; he also consulted members of the Petroleum Administration Board, working closely with John Frey; the Petroleum Labor Policy Board, on which George Ward Stocking held a position; and the Bureau of Mines.[3] With these sources, it was predictable that Williamson would arrive at many familiar observations and conclusions, but he went beyond them to argue that conservation would require federal control of every aspect of exploration and production, as well as substantial control of refining and marketing. In large part, his book was an extended brief for Harold Ickes's case for an oil czar.

Like so many oil industry critics of the previous decade, Williamson rehashed the charge that too much competition in the industry led to waste. Too much natural gas was flared off, too much crude was merely skimmed at the refineries, too many gas stations were built, all in the competitive rush for profits. And profits, not real conservation or natural security, were all oilmen cared about.[4] Were national security to be the priority, that could only be achieved by "rigid federal control over every detail of the production process in every well and pool in the United States"; for Williamson, that was "the inexorable logic of conservation." In Williamson's utopian dream of federal micromanagement, there would no longer be waste because all petroleum would be extracted from reservoirs and nothing left underground, a vision betraying his remarkable lack of grasp of industry technology.[5]

Having indicted competition and demanded federal control, however, Williamson did an about-face and took up the old cry that independents were in danger and major companies were the reason why. Here he simply extended the old charges against Standard Oil to all major companies — they were out to end competition (though by his reckoning this should have been a public service), they were out to create chaos and instability, they opposed federal legislation deviously and behind the scenes. In short, they were enemies of the public interest and, especially, of independents, "if there are any independents left in this country."[6] Williamson's uncertainty on this last point stood in contrast to his spirited defense of stripper well operators, whom he saw as the surviving independent oilmen. But what Williamson lacked in logical consistency he made up for in traditional ideology.

The following year, New York University economics professor Myron W. Watkins offered a more sophisticated study of the industry in *Oil: Stabilization or Conservation? A Case Study in the Organization of Industrial Control*. A close personal friend of George Ward Stocking's, Watkins drew heavily on Stocking's *Oil Industry and the Competitive System* in his view of the petroleum industry, particularly its exploration and development sector.[7] Indeed, rather than go through the ritual condemnation of Standard Oil's role in the history of the industry, Watkins simply referred readers to Stocking's work. Unlike Stocking, however, Watkins did not see the industry as dominated by the former Standard Oil companies or see them as constituting a cartel. Instead, he saw all major companies as unified in interest and controlling pipelines, refining, and distribution. Stabilization reflected the desire of major companies to extend managerial order over the "middle" segments of the industry and its unruly extremities in production and marketing, a goal finding expression in the NRA Petroleum Code and its administration.[8]

As Watkins presented exploration, production, and retailing, order was essential to efficiency, and small businessmen could not achieve it, in part because those who looked for oil were gamblers, pure and simple, no different from people who played the horses. Watkins overlooked explorers' consistent efforts to diminish risk and the role stable, predictable prices played in their financial planning. As for production, it was dictated by "reckless gambling, chance and headlong rivalry." In effect, from Watkins's perspective it was normal for oil exploration and production to be out of control, because oilmen were gamblers.[9] Similarly, on the other end of the spectrum, retailers were "an undisciplined lot," but Watkins was more charitable toward them; as a group, they displayed "stubborn independence," thus putting a gendered spin of approval on them by contrast to the reckless gamblers upstream. Unlike Stocking and many other critics of oil, Watkins did not assert that there were too many gas stations.[10]

Overall, conservation was a proper objective of national policy, but one too important to be left to the petroleum industry to accomplish: "The conservation of a nation's supply of this irreplaceable natural resource can never be safely entrusted to the self-seeking, short-sighted stratagems of a special interest group—such as the businessmen constitute." Serving the public interest, which Watkins identified with the interests of consumers and labor, could not be "an incidental by-product of profit making." Significantly, in these last reflections, drawn from the concluding paragraph of his text, Watkins spoke of businessmen, not oilmen, of consumers, not gasoline or fuel oil purchasers: his case study of oil, like John T. Flynn's biography of John D. Rockefeller, which had expressed many of the same opinions several years earlier, was a vehicle for more general criticism of the American capitalist order. What was really needed was not so much conservation as "a scheme of industrial control."[11] Freewheeling capitalism had to give way to something else.

William J. Kemnitzer, a self-professed believer in entrepreneurial capitalism, wanted nothing to do with Watkins's variety of industrial control, and a large part of his *Rebirth of Monopoly*, appearing in 1938, was devoted to showing how industrial control as exercised by Harold Ickes and state bodies such as the Texas Railroad Commission served the interests of monopoly. A graduate of Stanford University, Kemnitzer variously styled himself as an "economic geologist" or a "petroleum technologist," both titles implying expertise. In 1925–26 he worked for Shell Oil, but in 1927 he teamed up with Ralph Arnold to produce *Petroleum in the United States and Possessions*, finally published in 1931, a condemnation of oil imports. During the early thirties Kemnitzer wrote articles for the *New York Times* on the economic condition of the domestic petroleum industry, continuing to blame imports for industry problems at the same time he talked of the "excellent statistical position" of the industry. Despite Kemnitzer's opinion that all the figures showed the industry should be making money, it was not; Kemnitzer decided that production limitation, as well as imports, was to blame. This idea appealed to royalty broker J. Edward Jones, for whom production limitation represented a handicap in selling investors on his projects, and Jones hired Kemnitzer as a consultant, to say virtually the same things Jones said before congressional hearings in 1934. Kemnitzer also prepared charts and other data for Jones, which the broker circulated widely. Later in the thirties, Kemnitzer relocated to Washington and found a friendly ear in Paul E. Hadlick, secretary of the National Oil Marketers Association. Both men became frequent participants in government hearings relating to oil, and Hadlick sent the members of William Cole's House subcommittee copies of *Rebirth of Monopoly*. In short, Kemnitzer launched and built a career from directing public discourse on oil in the interests of promoters, royalty owners, small producers, and independent marketers.[12]

In many respects, Kemnitzer's opinions simply echoed oil's critics of an earlier era, and he openly acknowledged his debt to them. He urged the Cole committee to read Henry Demarest Lloyd; he said not only that he had read Ida Tarbell's book but that she was a "close friend." Not surprisingly, the opening pages of *Rebirth of Monopoly* covered the sins of Standard Oil, drawn from Tarbell, the House investigation in 1889, and the Corporation Commission's report of 1906–7. Like Tarbell, Kemnitzer played games with numbers. He argued, for example, that in 1934 there was no oversupply of oil on U.S. markets by making no allowance for illegally produced oil in his production data. Similarly, in 1939 he could show that Pacific Coast consumers paid far higher prices for gasoline than their Atlantic Coast counterparts by including Nevada and Arizona in "Pacific Coast" and omitting California. Like Tarbell, Kemnitzer was also a fierce advocate of independent oilmen, whom Kemnitzer saw as battling monopoly in the form of major oil companies.[13]

Like Tarbell but unlike more recent critics, Kemnitzer thought competition in the oil industry let everyone prosper and brought efficiency to all levels. Major company efforts to thwart competition created problems. True, the Standard Oil octopus had been dissolved, but in its place were twenty major companies that, because of interlocking directorates, acted "as one giant organization." To support this argument, Kemnitzer offered readers a two-page diagram showing major company interlocking ownership; he did not guarantee that it was correct in all particulars but stated that "data were obtained from sources believed to be reliable."[14] Major companies could not compete efficiently with independents, who could find petroleum and get it to market more cheaply than majors, so they began to talk about conservation in order to dry up independents' supplies of oil. In order to dupe the public into thinking the United States was running out of oil, they organized the American Petroleum Institute (API) to feed the public propaganda and found some oilmen to pose as independents in favor of conservation, stooges ready to repeat major company lines. Comparable to a fifth column, their representatives infiltrated scientific groups and chambers of commerce to make them, like the Federal Oil Conservation Board and the United States Geological Survey (USGS), into major company tools. In Kemnitzer's eyes, virtually any oil industry "expert" other than himself was the pawn of major company interest—thus government oil investigations never got anywhere.[15]

When it came to the usual conservationist topics, Kemnitzer was less paranoid but grandly inconsistent. On the one hand, he argued that the United States was not running out of oil, that there was at least a five-hundred-year supply. Production limitation in the name of preventing waste was totally unnecessary and merely a camouflage for price-fixing. The more oil refined and sent to the consumer, the better. On the other hand, Kemnitzer advocated setting aside federal naval oil reserves; as he put it, "Every drop of oil in government reserves should be guarded jealously and every means should be taken to prevent commercial exploitation by private interests." And even as he insisted that conservation was an idea "foisted" by major companies onto the public, Kemnitzer called himself "an ardent conservationist." One can only assume that while Kemnitzer saw conservation as flimflam, it was flimflam he knew the public respected.[16]

Just as he differed with many industry critics on the value of competition, so also Kemnitzer differed on the matter of federal control of oil. Making oil a public utility would simply play into the hands of the special interests, major oil companies, who would then use corruption of government to get what they wanted. Instead, Kemnitzer wanted to turn the clock back and renew vigorous antitrust action. He wanted to repeal the Connally Act, end the Interstate Oil Compact, and see divorcement of pipelines and marketing from major companies. All this was necessary if one was to prevent the American people from becoming "serfs of the established monopoly" and preserve democracy.[17]

Behind Kemnitzer's alarm with monopoly was profound disquiet with the economic realities of modern life, or as he told a congressional committee, "the corporate life that is exploiting us." "Ordinarily," he said, meaning ideally, "the production of raw materials and the marketing of manufactured products of an industry will be in the hands of thousands of operators." But in modern industry vertical and horizontal integration destroyed this happy state; the more it proceeded, the more it concentrated economic power in the hands of those who owned giant enterprises. Ultimately it would lead to monopoly in everything. For that reason, beyond a certain point, integration must be stopped. To Kemnitzer oil had reached that point, for in oil one could see a greater concentration of wealth and power than ever existed in any industry at any time. From his perspective, and that of the independent marketing association that hired him, oil represented an exemplary target. It demonstrated what was wrong with the modern economic order, in kind and especially in degree.[18]

If one believed the NRA critics' charges that its codes had been used by big business to promote monopolistic objectives, once the NRA ended, it made sense to take action against whatever monopoly one could find. For that matter, NRA or not, the swelling antimonopoly cry of New Deal critics such as Idaho senator William E. Borah, who told NBC and CBS radio network audiences that the depression had been caused by monopoly and blamed the New Deal for not acting against it, made it politically expedient for the Roosevelt administration to show some antitrust activity. In 1936 both Republican and Democratic parties contained antimonopoly planks in their platforms; if there was, as Borah charged, "some great evil lurking in our entire economic system," Republicans and Democrats alike were out to exorcise it.[19] As Ellis W. Hawley has pointed out, ideal targets for New Deal antimonopoly action were widely hated groups, especially when one segment in the group was trying to get the better of a rival segment,[20] a situation historically recurrent in oil. No sooner had party platforms been aired than Roosevelt's Justice Department pursued the oil industry in what became known as the Madison case.

Justice's launching the Madison case clearly reflected ideological tensions among New Dealers, for its highly traditional economic perspective was at odds with what Harold Ickes's Petroleum Administration had done under the NRA. The Petroleum Administration's own variant of official economics, also inherited from Progressives, assumed that greater efficiency led to higher margins, which led to greater production, in turn leading to lower prices, increased consumption, and higher employment. Thus, under the NRA, Ickes's Petroleum Administration tried to address what it saw as retail competition run amok, in the form of gasoline price wars. Such price wars had been deplored by many industry participants in the twenties and early thirties, since, like the overproduction of crude that eventually led to gasoline at distress prices, price wars made it hard to make a profit. As with too much crude on the market, the solution to the problem of too much gasoline was to cut the amount reaching the market. To reach this goal, Ickes divided the nation into refining districts and apportioned gasoline to meet estimated demand among the districts in the form of quotas, set below preregulation levels. Tank car stabilization committees then apportioned quotas for the districts among their refiners. In order not to burden independent refiners, whom most federal investigators saw as "squeezed" by larger competitors, however, the Petroleum Administration got major companies to agree to a special concession: rather than cut back independent refiners' production as much as their market share might dictate, major companies would buy gasoline from independent refiners, being paired with these "dancing partners" by the regional stabilization committee. Allowing that everyone obeyed regulations—which by no means everyone did—the supply of gasoline would thus balance demand, prices would rise, and there would be no surplus gasoline to fuel destabilizing price wars.[21]

This pooling and marketing agreement was not ideal from the point of view of participants. Major companies had to buy part of what they were capable of producing, thereby lowering levels of plant utilization, and independent refiners also had to accept diminished production. Somewhat higher gasoline prices, however, brought most refiners to support continued limitation of production after the end of the NRA in 1935. The other attractive aspect of NRA regulation was that it offered protection from antitrust prosecution. Quotas, after all, had been created under federal direction. Ickes's program thus extended stabilization to the refining sector of the industry with few objections from independent refiners. But one segment of the industry, independent jobbers and marketers, offered noisy opposition from the beginning, because cutting the amount of distress gasoline on the market raised their costs dramatically. Independent spokesmen such as Paul Hadlick lost no time in complaining to congressmen and bureaucrats. By 1936, Roosevelt's Justice Department was listening.[22]

Choosing to launch its attack in La Follette Country, late in July 1936, the Justice Department led a Madison, Wisconsin, grand jury to indict twenty-three oil companies, three trade journals, and fifty-eight individuals for violations of the Sherman Act, conspiring to fix gasoline prices in interstate markets. The indictment of the *Chicago Journal of Commerce*, *Platt's Oilgram*, and the *National Petroleum News* was especially bizarre; the publications aided in price-fixing because they reported gasoline prices set by major companies. The wider implication, that any publication of commodity prices was illegal, brought strenuous objection from trade and financial press. As a result, these charges, with those against seven companies and twenty-eight individuals, were eventually dismissed, but sixteen companies and thirty individuals were ultimately convicted and fined. In vain did oilmen argue they were only doing what Ickes wanted them to; in Washington, Ickes remained silent.[23]

The Roosevelt administration had even more incentive to pursue monopoly when the economy dipped into serious recession in 1937. No matter that administration tax policies may have done more to set back recovery than anything else; New Dealers needed to shift the onus of downturn to some other quarter. There was, of course, the time-honored device of blaming the rich for the misery of the poor, and that avenue was followed by Harold Ickes and Assistant Attorney General Robert H. Jackson.[24] Picking up on a best-seller, they condemned the power and dangers of concentrated wealth as portrayed by Ferdinand Lundberg in *America's Sixty Families*. Lundberg argued that the United States was no longer a democracy but was run by a small group of wealthy families, most prominent among them the Rockefellers; these families had thrived during the depression while many other Americans were "reduced to beggars." Now the rich were out to do in the New Deal — itself not as free of their control as it ought to be — and Lundberg concluded, "The danger of dictatorship of the Right was never more real than at the present moment."[25]

New Dealers such as Leon Henderson who subscribed to expansionist economics, however, had no need of Lundberg to explain the downturn. An overweight, rumpled, cigar-chewing Swarthmore alumnus, Henderson was an economic consultant for the Russell Sage Foundation and Pennsylvania governor Gifford Pinchot before he joined the NRA and, subsequently, the Works Progress Administration, as an economic adviser. Launched on his bureaucratic career, he ingratiated himself with key insiders in the Roosevelt administration, especially Thomas G. "Tommy the Cork" Corcoran and William O. Douglas; in 1939 he would advance by taking Douglas's place on the Securities and Exchange Commission. By the time he did so, however, he had been widely quoted on his views on the prevailing danger of monopoly. Proud of predicting the 1937 downturn, Henderson blamed big business for it; big business kept production low in order to keep prices high, thereby hurting smaller competitors and destroying consumer purchasing power, needed to sustain recovery. Not a rigorously logical thinker, Henderson did not explain how high prices hurt small business or how price cuts, the remedy for recession, could help them. He focused instead on the consumer: as he put it, "The man in the street who buys bread and automobiles is the man who keeps the nation's industries rolling." Lower prices would enable consumers to buy more and prompt industry to produce more. But monopoly, driving smaller competitors out of business and fixing prices at ever higher levels, was in the way of this objective. To Henderson, high prices and prosperity were incompatible.[26]

With the ideological blessing of expansionists such as Henderson, Justice stepped up its antimonopoly crusade, and in 1938 it brought in Yale Law School professor Thurman W. Arnold to take special charge of prosecutions. A Wyoming rancher's son, Arnold went to Princeton and Harvard Law School, returning to practice in Laramie in 1919. During the next decade, he tried politics, serving as mayor of Laramie and as a state legislator, and he taught law at the University of West Virginia and Yale Law School, an unusual progress attributable to Yale friends. At Yale he cultivated a colorful persona. He brought his terrier to classes and addressed his rhetorical questions to the dog rather than his students, interrupting rambling lectures with the interjection, "Now what the hell am I supposed to be talking about?" Yalies responded by dubbing him the "fifth Marx brother" and calling his classroom "the cave of the winds." Notwithstanding his unusual behavior, Arnold's opinions pleased New Deal insiders Thomas Corcoran and former Yale Law School professor William O. Douglas, as well as determinedly agrarian senators such as Joseph C. O'Mahoney of his home state and George Norris of Nebraska.[27]

By the time of his appointment, moreover, Arnold's book *The Folklore of Capitalism* had made him a celebrity. In a thoroughly modern and cynical tone, Arnold told its readers that existing federal antitrust laws were merely "a great moral gesture." For that matter, he dismissed the Constitution as "only a protection against unholy desires." Big business was a fact of modern life, and society needed what it could deliver. That was why new laws and regulations were needed; existing laws only seemed to, but could not, control big business, and while on the books, they were "an effective moral obstacle" to "practical regulation." That practical regulation was what Arnold wanted from government.[28]

Arnold's skill in manipulating discourse makes it difficult to sort out opportunism from idealism. Like Harold Ickes's press releases, Arnold's public pro-nouncements often seem like what the Roosevelt administration wanted said but only a self-styled political enfant terrible could say. Behind the cynical and sophisticated tone of *The Folklore of Capitalism*, however, is nonetheless a tre-mendous amount of normative discourse in aid of a new and enhanced role for government, the kind of New Jerusalem ardent New Dealers saw within reach. Arnold longed to see a society in which principles and ideas — new ones rather than old ones — were "more important than individuals." He wanted to see government get things done, build a new order. In that respect, he could admire Stalin. Admittedly, the Soviet leader's purges were a "failing in organi-zational methods," but Stalin's other "techniques" were bringing about real change. It was time for Americans to embrace change, to cease measuring the New Deal against old, outmoded notions of individualism and "tolerate ex-periment as something essential in meeting changed conditions." Principles had to be "molded to organizational needs,"[29] and not vice versa. Such senti-ments could be taken as both idealistic and thoroughly up to date.

In the context of the role Arnold would take at Justice, there may also have been more than the obvious grandstanding that his critics picked up. In *The Folklore of Capitalism*, Arnold saw the working out of conflicts, "dramatic contests of all sorts," as giving unity and stability to government.[30] With that reasoning, while antitrust laws might be ineffectual, rigorous prosecution of them might serve a highly positive purpose, as might inflammatory tirades to the press and congressional committees. Arnold would draw the opponents of the new order into battle, thus strengthening that order whether or not the battle ended in his favor. In any event, within a year of his appointment, Arnold had some three hundred lawyers working under him in the antitrust division and hundreds of organizations and individuals under investigation.[31]

With respect to oil, the most notable Justice action was the so-called Mother Hubbard case of 1940, in which Justice brought charges against twenty-two integrated oil companies and the API on familiar grounds of fixing prices and restricting competition. Justice also made an attempt at pipeline divorcement, bringing suits against the Great Lakes, Phillips, and Stanolind pipeline sys-tems in the same year, for violations of the Elkins Act.[32] For the Roosevelt ad-ministration's crusaders, oil was a convenient whipping boy but by no means an exclusive one. Action against oil was part of a broader campaign that in-cluded such diverse targets as motion pictures, steel, aluminum, glass, insur-ance, banking, and the dairy industry. With the coming of war, action was suspended on the Mother Hubbard case and the Elkins suits, settled by con-sent decrees because of pressure on Justice from the National Defense Ad-visory Commission.

A month after Arnold joined the Justice Department, Roosevelt came forward with the script the antimonopoly zealots had been demanding. The president told Congress that the nation suffered from a concentration of private economic power, that competition was disappearing, and that existing antitrust laws appeared inadequate to deal with the current situation. He then called on Congress to make a thorough study of the concentration of economic power, and, by joint resolution, it established the Temporary National Economic Committee (TNEC) to do so. Committee membership included Senators Joseph C. O'Mahoney (chair), William E. Borah, and William H. King of Utah; Representatives Hatton W. Sumners of Texas, Claude Williams of Missouri, and B. Carroll Reece of Tennessee; Thurman Arnold of the Department of Justice; Jerome N. Frank of the Securities and Exchange Commission (with Leon Henderson as alternate); Richard Patterson of the Department of Commerce; Isador Lubin of the Department of Labor; J. J. O'Connell Jr. of the Treasury; Garland S. Ferguson of the Federal Trade Commission (FTC); and James R. Brackett as executive secretary. Of the committee members, O'Mahoney, Borah, Arnold, Henderson, and Ferguson were ardent antimonopolists. With respect to oil alone, twenty-five days of oil hearings and forty-eight witnesses produced 3,116 pages of testimony about petroleum accompanied by 986 charts, graphs, and tables.[33] From the point of view of what was actually said, however, there was little new, and those who spoke were the usual spokesmen.

With its nebulous mission, the TNEC, or as journalists dubbed it, the "National Monopoly Committee," could look into virtually any area of the economy that might seem tainted by monopoly, and before it got around to investigating the oil industry, it looked at patents, life insurance, beryllium, investment banking, and the construction industry—the last suspected of sabotaging economic recovery with high prices. The TNEC's investigation of glass unearthed monopolistic design in the sale of milk bottles, as well as of the milk in them, and received ample coverage in newspapers as a result. The hearings generally resembled those of earlier decades, though the staff of the TNEC attempted to fit current data into a boilerplate script. By the time hearings began on oil in late September 1939, the committee had received answers to a lengthy questionnaire it had sent to forty-nine larger oil companies, and the data it received were analyzed for it by John Ise, a like-minded critic of the industry.[34]

Leading off the oil hearings on September 25, Ise assured committee members that if they were looking for evidence of concentrated economic power, the oil industry was certainly the right place to begin. He began his testimony by pointing out that the growth of the oil industry made possible the widespread use of the internal combustion engine—which, used in automobiles, trucks, and tractors, destroyed America's self-sufficient farming communities, brought farmers into national markets, threw them into debt, and effectively ruined them while "billions pile up in the great commercial and financial centers." In effect, oil was an accomplice to the ruin of the yeoman farmer. This dire adaptation of traditional agrarian ideology, once used against the railroads, set the tone for the rest of Ise's testimony, which reiterated what he wrote in 1926, with emendations based on the work of Myron Watkins. The United States was running out of oil, production was hasty and wasteful, independents were in trouble. Pipelines were monopolistic, as were crude purchasing, refining, and gasoline marketing. Yet, having said the last, Ise asserted that marketing was competitive enough to be carried on at a heavy loss, which brought Congressman Sumners to ask why companies would be in retailing if it meant loss. Ise explained this as a miscalculation on the part of integrated companies, leaving the survival of independent retailers something of a mystery. He could not tell Sumners what percentage of market share independent distributors had and, more embarrassing, could not explain what he meant when he used the term "independent."[35]

Much of Ise's testimony was imprecise and inconsistent as well as unoriginal, but one lucid segment of it expressed an idea about the industry, at variance with oil industry discourse, that would be important during World War II and thereafter. According to Ise, it was a peculiarity of the industry that the supply of oil showed "little response to price." Whereas in other industries high prices encouraged greater supply, in oil, discoveries were "the result of chance." Seeing oilmen in the old stereotype as gamblers, Ise reasoned that since a high crude price could not guarantee discovery, it would have no effect on the willingness of oilmen to seek new reserves. Thus, supply could be high when prices were low—as, in fact had been true for much of the decade.[36] He concluded that policies aimed at raising the price of crude would increase consumer cost without beneficial effect. This belief would dominate policy making in Washington for decades thereafter, but it would be especially evident at the Office of Price Administration during World War II.

As the hearings proceeded, witnesses advanced their usual arguments. The cry of "independents in danger" surfaced with respect to the looming elimination of independent producers, refiners, jobbers, and retailers; major companies were "squeezing" independents by, among other things, pipeline monopoly, fixing prices, and restricting supply through proration programs. Criticism of proration offered one of the few relative novelties in discourse; Fort Worth lawyer Karl A. Crowley thought Texas proration showed that the Railroad Commission was run from 26 Broadway, the headquarters of the Standard Oil group until dissolution, while Eastern States Petroleum's Louis V. Walsh thought it had eliminated all incentive for Texas producers to develop new property. Witnesses from the USGS and Bureau of Mines repeated the traditional positions of their agencies, warning that the United States was running out of oil, had been wasteful with what it had, and needed to bring all oil fields under unit operation. James A. Horton of the FTC aired his agency's allegations of retail price-fixing on the part of major companies. Promoting his agency's position, he stressed the need for a thorough and complete investigation of industry marketing by the FTC.[37]

Familiar charges elicited customary responses from the usual industry spokesmen. Most prominent were W. S. Farish and J. Howard Pew, both of whom were usually called on to give major company positions on petroleum issues; Russell Brown testified for the Independent Petroleum Association of America. That perennial witness for the independent refiners John E. Shatford of El Dorado, Arkansas, repeated the opinions everyone had heard before. In short, part of the reason testimony before the TNEC included familiar themes was that many of the same people who testified regularly on oil-related questions were present once again.

Undoubtedly, oilmen wanted to present an image of oil as a good corporate citizen, and one of the themes they used to that end was conservation. Here what they had to say often presented considerable contrast to what they had said in the twenties or, in J. Howard Pew's case, as recently as five years earlier. Thus, the same man who in 1934 said that he did not want a nurse for his business, referring to proration plans, now praised the Connally Act and the Interstate Oil Compact and talked about the constructive leadership the industry had taken in conservation. Everett DeGolyer of Amerada talked about reservoir management and how proration programs worked to conserve the nation's oil in the public interest. W. S. Farish went so far as to give a brief history of conservation in the industry, admitting that before the mid-1920s the industry had been wasteful — "for the lack of good science." Now oilmen believed in a conservation that meant getting the most oil out of the ground at the least cost: "The concept . . . embraces economic aspects as well as considerations of physical waste." It would be effected best by state-directed programs.[38] There had thus been an industry change of heart, and, if one believed Farish, the industry was now the wise steward of America's bounty. More to the point, there was no reason, on grounds of conservation or anything else, for the federal government to intervene in industry affairs.

In all, the TNEC investigations had little effect on either the oil industry or oil-related discourse, so for antimonopolists the TNEC was thus a disappointment and a chance to say "I told you so." As the TNEC began its investigation, John T. Flynn, now writing for the *New Republic*, warned that business would be "pulling strings" to keep it from doing much; he thought every member of the TNEC ought to read Kemnitzer's *Rebirth of Monopoly* at least once. A year later he found alarming signs that, rather than pursuing monopoly, the Roosevelt administration was lapsing back to allowing business to work against competition, as it had in the NRA. Far from advancing to a new order, the TNEC seemed to be heading back toward an old one. Delighted to see the Roosevelt administration turn to trust-busting, the *Nation* nonetheless expected it to move on to much more stringent business regulation. In 1939 it reported on the pipeline divorcement bill in the Senate, telling its readers that pipelines permitted the growth of a new oil trust taking the place of the Standard Oil trust dissolved in 1911, and observed that since "oil can actually be shipped more cheaply by rail than by pipeline," all pipelines did was sustain monopoly. But neither divorcement nor the TNEC revolution came about, and when the TNEC wound up its activity, I. F. Stone disgustedly called its proposals "feeble and equivocal." The TNEC had been careful not to look far below the surface of business activity, and its work would "provide no beacons for the future." Leftist sophisticates said that they had not expected it would. In a July 1939 article subtitled "A Study in Frustration," Dwight MacDonald described the TNEC as the arena of contest between greater and lesser bourgeoisie, its lack of action as reflecting "the agonized indecision of the middle class faced by the disintegration of the capitalist order." With capitalism in its death throes, what the TNEC did or did not do was beside the point, but it could not dare to do much.[39]

For those less disappointed and pessimistic, the exhibits and testimony introduced during the TNEC hearings were potentially valuable; as a journalist for *Time* commented, its study could provide "a factual basis on which U.S. business problems will be approached." In that regard, the hearings were but a skirmish in the ongoing battle for control of business-related public discourse, whose next episode involving oil took place with the publication of TNEC Monograph No. 39, Roy C. Cook's *Control of the Petroleum Industry by Major Oil Companies*, in 1941. A subordinate of Thurman Arnold at the Department of Justice, Cook modestly styled himself "Expert, Antitrust Division." His monograph was in fact a research project in economics at George Washington University, and when it appeared, the TNEC and the Department of Justice denied that Cook spoke for them. But it nonetheless appeared as one of the official monograph series, much to the chagrin of industry leaders such as Pew and Farish. Because the study had not originally been commissioned by the TNEC, Pew and Farish succeeded in having the committee publish their refutation of Cook, *Review and Criticism on Behalf of Standard Oil Co. (New Jersey) and Sun Oil Co., of Monograph No. 39*, more easily referred to as TNEC Monograph No. 39-A.[40]

Since Cook drew heavily on the works of Tarbell, Stocking, Ise, Kemnitzer, and Watkins, as well as the Corporation Commission's report and the FTC investigations of 1916 and 1921, what he wrote was thoroughly representative of orthodox traditional political and anti-industry discourse.[41] As Cook saw it, control of oil by major companies was as complete in 1941 as under Standard Oil before dissolution. With their tool, the API, majors acted as a unit; and there was little opportunity for the small nonintegrated company to survive. Majors worked together to keep production down, crude prices high, and independent refiners "squeezed." In the field majors kept independents from drilling wells and producing them; conservation measures were merely devices of price manipulation. Majors' control of pipelines kept "most of the independents from using them." Majors fixed both crude and product prices to eliminate independent refiners and jobbers. Indeed, so grim did the majors make life for independents that Cook himself wondered, "After reading this report, one may ask how does the independent exist in view of all the controls exercised by the majors." Given Cook's presentation, this question was a conundrum. Cook's remedy was regulation of the whole industry as a public utility.[42]

That conclusion and its position in the TNEC literature were ominous, so quite apart from what they might have felt as a need to correct distortion, major company spokesmen had strong motivation to contest Cook's position. W. S. Farish and J. Howard Pew put their names on the refutation of Cook, but the work really amounted to a major company attempt to set aside decades of industry-hostile criticism that Cook rehashed. Farish and Pew began by attacking the idea, assumed by antimonopolists, that one could speak as though nonmajors were a homogeneous group opposed by some more cohesive entity. They pointed out that there were a great number of independents, some of them operating integrated oil companies. The majors did not act as a uniform group and did not pursue identical business strategies; indeed, they could hardly be said to represent a concentration of economic power when critics agreed there were as many as twenty of them. If the majors were integrated, that was because integration was efficient, "the essence of mass production," not because it was a means to achieve monopoly of power over prices. As for the API, it was open to anyone who wanted to pay $12.50 to get its publications; how could that represent an exclusive tool of major companies? In general, Farish and Pew disputed Cook's vision of majors at odds with independents: instead, they argued, "There is considerable cooperation between them and each group realizes that depression and prosperity will treat them alike."[43]

In their attempt to set aside the traditional view of relations between majors and independents, Farish and Pew also tried to refute old arguments about price squeezing and pipelines. On the former they tried logic: How could major companies at one and the same time be keeping crude prices low to plague independent producers, high to squeeze independent refiners, and product prices higher to abuse independent jobbers? Farish and Pew reminded readers that there were many pipelines, that they shipped oil for independents, and that their prices were regulated. Moreover, contradicting Cook's depiction of independent refiners as cut off from both crude and urban markets, Farish and Pew pointed to independent refiners who bought crude from majors in order to serve urban markets in which they were competitors; pipelines were no handicap to them. Cook had made the mistake of basing his observations on pipelines on the Bureau of Corporation's investigation, and, they argued, "Obviously, evidence of 37 years ago cannot be used to support allegations of monopoly today."[44]

As in the TNEC hearings and the earlier Cole committee investigation, industry leaders returned to the issue of conservation, waving a flag they had happily appropriated as their own. Conservation, they contended, meant wise use of reservoir energy, and because proration served that objective, it was no mere device for price manipulation. But they were not afraid to stress their economic understanding of conservation: "It makes little difference whether the preservation of reservoir energy is justified by calling it prevention of physical waste or prevention of excessive cost of recovery. The economic advantage of preserving reservoir energy is so clear that it is not open to effective challenge." With this they also took moral high ground by charging that Cook implied that conservation was unnecessary.[45]

In all, TNEC Monograph No. 39-A amounted to a readable refutation of Cook's traditional opinions, but if Pew and Farish thought their effort could amount to having the last word, they were quickly disappointed. Although the TNEC agreed to publish their refutation, it allowed Cook his rejoinder in the same volume. He simply reasserted his main points, brushing objections aside.[46] Accordingly, oilmen made another effort; the executive committee of the API published statements given the TNEC by witnesses hostile to the antimonopoly stance, introducing them with its own picture of conditions in the industry and with a digest of the hearings. The justification it gave was that the TNEC's information was "readily available."[47] If the TNEC's proceedings really were going to be an information base for the future, this was a reasonable strategic step, especially in the light of the existence of Monograph No. 39 and Cook's argument that oil might be treated as a public utility. The API's version of the TNEC, however, was no more scintillating than the original hearings. More to the point, when it appeared in 1942, the industry faced far graver challenges than trying to reshape discourse generated by the TNEC.

In the antimonopolist crusading of the late thirties, then, conservationist discourse, although continuing, played a distinctly secondary role in the discussion of petroleum and, when surfacing, was as likely to be used by industry spokesmen as their critics. Old questions came up from time to time. In 1935, for example, that geological prophet of doom L. C. Snider was joined by

Benjamin T. Brooks in predicting that the United States was running out of oil; *Time* and *Literary Digest* both carried their prediction of shortage in five to eight years. The following year, the Department of the Interior's *Report on the Cost of Producing Crude Petroleum*, originally ordered by Harold Ickes as Petroleum Administrator, noted that the United States would run short of oil before other countries. These gloomy forecasts were reinforced in 1939 by the report of the Energy Resources Committee of the National Resources Committee. The committee's findings are the best example of traditional conservationist discourse continuing in nonindustry channels in the late thirties.[48]

Like earlier conservationists, the Energy Resources Committee warned that oil reserves were "strictly limited," they were being used up faster than in other countries, and shortage, accompanied by higher prices, was likely in a decade or two. There had been "grave waste" in the past, and, though progress had been made toward its elimination, not enough had been done. Proration plans by states had not necessarily ruled out waste, and unitization of production had not been made mandatory. Citing elsewhere such industry commentators as Watkins and Kemnitzer, the committee observed that "an industry which is harassed by chaotic conditions of competition can hardly be expected to develop a broad and socially constructive program of conservation." The committee recommended creation of a federal board that would make regulations for oil and gas production, assisted by an advisory planning group and studies by the best technical experience available. There also needed to be a comprehensive national energy policy. In its absence and without federal regulation, there would emerge "an insistent and eventually irresistible demand for public ownership and control." That, of course, was what Harold Ickes had threatened the petroleum industry with in 1934.[49]

Of course the committee could not foresee that federal officials would be in a position to do much of what it recommended little more than two years later. There would be regulation of the petroleum industry from drilling rig to gas pump, and a plethora of advisory committees would join bureaucrats in carrying out regulation. The humblest poor boy prospector and the major company refiner alike would be filling out quadruplicate forms for review by federal experts in Washington. Competition of the marketplace would give way to price ceilings and rationing. It would be a planner's dream come true. But whether the ensuing planning and regulation would spring from any consistent or practical comprehensive energy policy was quite another matter.

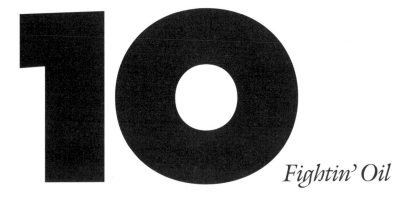

## *Fightin' Oil*

When federal officials constructed and implemented policy toward the petroleum industry in World War II, they made a number of assumptions derived from ideas that were far from new. Falling back on the old Progressive faith in government by experts, they assumed that successful mobilization of the industry would require management of all its sectors from Washington. Drawing from the experience of the previous world war, however, they assumed that management would best be effected by federal officials working with industry experts serving in advisory capacities; they followed the neo-Progressive model promoted by Bernard Baruch, albeit on a far grander scale. Once experts and advisers were in place, they chose to act on a wide variety of ideas long common among scientists, economists, and, especially, federal bureaucrats: fear of shortage; need for prioritized use; requirement of uniform application of rules on well spacing and production rates; compulsory unitization; need for central decision making accompanied by voracious fact-finding; and, later in the war, the desirability of a government-owned oil company to acquire foreign reserves. When men in Washington acted on these ideas, they assumed ideas could be translated into practical operations with efficient results. It was only reasonable to assume that if one framed federal policy to eliminate all the features of the petroleum industry critics had long identified as undesirable and followed the critics' recommendations, one would improve the industry's efficiency — and, hence, serve the defense effort well. No one reckoned that a mixture of old ideas might not add up to coherent and workable policy or that bureaucrats and experts might differ among themselves. Because no one foresaw such problems, Washington officials ended up with regulation that was at odds with economics and operations — and, by the same token, did not produce desired results. Policy shortcomings then required strenuous discursive efforts directed at shifting responsibility and covering up disappointing results.

As the United States drew closer to war, the Roosevelt administration began to organize bodies directed toward rationalizing a defense effort. It set up the Office of Production Management (OPM), which included a petroleum section, and, in April 1941, it established the Office of Price Administration and Civilian Supply (whose functions would later be subsumed under the Office of Price Administration), headed by Leon Henderson. Inclined to see "profiteering" at every turn, Henderson told reporters that to have an effective defense program all prices ought not go any higher — especially prices on basic commodities. This philosophy would guide the OPACS/OPA even after Prentiss Brown replaced Henderson as OPA head in January 1943, and it would have a profound effect on the oil industry. But of more immediate importance was Roosevelt's creation of the Office of Petroleum Coordinator for National Defense (OPC, which became the Petroleum Administration for War [PAW] in December 1942) on May 28 and his appointment of Harold Ickes as petroleum coordinator. Oil was thus one of the first industries to face wartime control.[1]

Ickes's appointment, as he himself would stress two years later in *Fightin' Oil*, did not elate oilmen. After eight years of participation in industry affairs, most oilmen had reason to dislike him for one reason or another, recently for his leaving the industry in the lurch in the prosecution of the Madison case. Now, without naming names, the *Oil Weekly* expressed the editorial hope that the growing crisis would not be "prostituted by those in political power to retain their power, to increase it, or to misuse it." Its Washington columnist, B. F. Linz, told readers that some in Washington said the administration was now going to assume a control of the industry it had wanted for eight years and could not get. In fact, Ickes would ostensibly fill a role much like the one he had under the National Recovery Administration, but Roosevelt did not give him compulsory authority; he could persuade, but he could not command. That limitation meant he would have to cooperate with oilmen far from thrilled to work with him if he was to accomplish anything. To organize such cooperation, Ickes divided the country into five regions, organized a regional committee in each one, and appointed scores of oilmen to fill them. On the national level, he created a Petroleum Industry War Council as an advisory body, again composed of industry representatives. Thus, ironically, the Progressive vision of regulation of oil through a federal office would be realized:

but through cooperation with the very people conservationists and others had been accusing of malfeasance for years. An indignant old Progressive, Iowa senator Guy M. Gillette, observed that of 205 persons initially named to regional committees, 134 had either been convicted of violating antitrust laws or were being prosecuted for doing so.[2]

Having to work with droves of oilmen, however, was not the only problem Ickes faced. Much to his chagrin, federal direction of oil was not concentrated in his hands but shared out among more than three dozen agencies, most important among them the War Production Board (WPB), the Office of Price Administration, the Office of Defense Transportation, and the Federal Power Commission. None of the other agencies yielded an inch to Ickes, and his relations with them soon became more confrontational than cooperative, a situation not improved by his attempt in 1942 to take all jurisdiction over oil away from the other departments and concentrate it in his own hands. Ickes's constant scrapping with other agency heads made it difficult for him to get constructive responses to urgent problems the petroleum industry would face during the war. Last but not least, Ickes also had to work with state conservation bodies where they existed, and here his past acrimonious relations with state agencies such as the Texas Railroad Commission were no asset. Ickes could not compel state officials to follow his rules, and as William R. Childs has pointed out, state officials tended to continue doing what they had been doing before the war, while condemning attempts to meddle on the part of Washington with the old states' rights rhetoric they had used during the thirties.[3]

Within a few days of becoming petroleum coordinator, Ickes began to talk about the imperative need for gasolineless Sundays, World War I–style, in order to conserve short fuel supplies on the East Coast. According to the coordinator, the source of the problem was wartime diversion of tankers to British use; rather than take Gulf Coast petroleum and products to East Coast refineries and markets, the tankers, fifty of them at first and up to one hundred by July 1941, were put to carrying Caribbean oil to Britain. Ickes said he had seen that such diversion could cause supply problems and had recommended construction of a pipeline to carry Texas crude to eastern refineries. Until a pipeline was built and operating, Ickes worried about shortage loudly and frequently, setting the stage for the East Coast gasoline crisis of late summer 1941.[4]

Ickes's cries that the East Coast would run out of gasoline were greeted with skepticism. New Yorker Robert Moses said the whole oil emergency situation was Ickes's concoction, "sprung on the public by an arbitrary fellow in Washington without study or investigation." In Kansas, Alf Landon told reporters that gasolineless Sundays were unnecessary; if there was a shortage, it must have been created artificially by the administration's tanker loan. The *Oil Weekly* warned that gasolineless Sundays would hurt public morale; as for a shortage, its feature writer Leonard M. Logan noted on August 18 that there had not so far been any material lagging of the movement of oil to the East Coast and fuel stocks were not much different from levels of the previous year. Newspapers were dubious about how grave the situation might be; Raymond Moley thought shortage was surely not as serious as Ickes suggested, while *Time* told readers, "This whole situation is fantastic."[5] One can wonder if decades of Washington's crying wolf about oil shortage now took a toll.

Skepticism turned to outright hostility and resistance when Ickes went beyond warnings. He prevailed upon retailers to close East Coast filling stations between 7 P.M. and 7 A.M. beginning August 1, and, through Leon Henderson, who had power to effect it, he ordered a 10 percent cut in supply of gasoline to retail outlets from August 15 through September. Consumers responded to the gasoline curfew by topping off tanks during daylight hours; as a result, during the initial week of the cutback East Coast gasoline consumption rose. The president of New York City's borough of Queens called the curfew a "crackpot" idea and urged citizens to ignore it in the interest of the defense effort. Newspapers continued skeptical: the *New York Daily News*, for example, headed an editorial "Is the Gas Crisis Real?" and doubted there was a genuine tanker shortage. Frustrated in their attempts to investigate matters, because the administration censored relevant information on the ground of military expediency, journalists reported rumors to the effect that the British were using the tankers for petroleum storage because tankers already acquired from Norway and the Netherlands had remedied the deficiency of ocean transport. Such tales led the *Oil Weekly* to taunt Ickes by editorializing against withholding information from the public, on the grounds that it might lead people to see a real emergency as phony.[6] What stood out in journalists' coverage generally was the inclination to blame disruption on Harold Ickes rather than the oil industry.

Ickes's consumption curtailment measures also drew strong reactions from Congress. Ohio senator Robert A. Taft saw them as designed simply to stir up prowar sentiment. Florida senator Charles O. Andrews decided that the petroleum industry needed to be taken from Ickes and placed under congressional control. Responding more directly to the gasoline situation, Senator Francis Maloney of Connecticut launched an investigation of the shortage, explaining to the press that Ickes meant well but, "in his conscientious effort, he is inclined to be dramatic." Stung by suggestions that petroleum shortage had been trumped up by his agency, Ickes's second in command, Ralph K. Davies, suggested to Maloney's investigation that the critics of curtailment were dupes or worse; as he put it, "One cannot but wonder if there does not lie behind these efforts to promote confusion and create dissension, some purpose, sinister and planned. Is it possible that we have here an attempt at sabotage?"[7] Such public paranoia would also surface in Thurman W. Arnold's crusade against Standard Oil of New Jersey and other corporations some months later.

Ickes, erstwhile journalist and unretiring scrapper, would not let either columnists or congressmen have the last word on the shortage of 1941. Periodicals such as *Collier's* and the *Saturday Evening Post* frequently ran his articles; now he gave *Collier's*, which had just run a lengthy biographical feature of him, his version of the crisis. He argued that there was a real shortage — existing not only on the East Coast but also in the Midwest and Far West, "though the public hasn't begun to notice it yet in those latter areas." Supplies were not adequate for projected needs, and, making a safe assumption, he told readers, "No man can state definitely that the shortage will not become worse." Casting himself as a persecuted visionary, a role he liked to assume, Ickes told readers how he had pushed for a pipeline to supply the East to no avail. Here he shared out blame between the railroads, which resisted competition from pipelines, and the major oil companies, which he said did not want "small Southwestern producers to get their competitive products into the rich Eastern market" — a creative application of traditional antipipeline rhetoric. As to the criticism and ridicule that followed his proclamation of shortage, it was a price he would pay: "When I became Petroleum Coordinator, I knew that my actions would be misrepresented by those who had their reasons for so doing."[8] In 1943, publication of *Fightin' Oil* gave Ickes yet another chance to keep his version of 1941 in circulation, with even greater emphasis on his position as a misunderstood, maligned Cassandra; when it came to controlling discourse, Ickes was a man who just did not quit.[9]

Whether or not the gasoline shortage of 1941 was phony, people were accustomed to assuming that when it came to oil, things were never what they seemed. It was usual to mistrust things done by the oil industry, regardless of circumstances. As another specimen of the running-out-of-oil cry that Americans had been listening to for decades and had learned in the past was inaccurate, the crisis of 1941 seemed to prove that shortage was contrived, an opinion that would receive widespread circulation during the 1970s. More immediately, negative reactions in 1941 produced problems when mandatory rationing was imposed in mid-1942. Ickes at first shifted focus from gasoline by claiming that there would be a fuel oil shortage in the coming winter if easterners did not conserve gasoline and convert from home heating oil to coal. Ickes also appealed to patriotism, declaring, "No patriotic American can or will ask men [on the remaining tankers bringing in oil] to risk their lives to preserve motoring-as-usual." A motorist filling up his gas tank to go on a picnic, moreover, might make it impossible for a defense worker to get to a job. Indeed, "The greatest and most vital contribution that the individual American can make to the winning of this war is his self-denial as to gasoline and heating oil." Here was conservationist prioritizing with a vengeance. Even under wartime conditions, however, there were doubters: columnist Paul Mallon and Republican oilman Walter Hallahan questioned the need for gasoline rationing before it was imposed; Ickes responded to Mallon by calling him a "professional traducer."[10]

The gasoline rationing established in mid-1942 applied only to the East Coast, but there were federal officials, chief among them Leon Henderson, who thought it ought to be extended to the whole nation — not so much to save gasoline as to save rubber. The shift of attention to a rubber shortage, however, did not let the oil industry off the hook. The targets of critics included the Reconstruction Finance Corporation's Jesse Jones, of Houston, but chief among them was the Standard Oil Company of New Jersey (SONJ).

In 1927 SONJ entered into a twenty-five-year contract with the German industrial group Interessen Gemeinshaft-Farbenindustrie Aktiengesellschaft (IG), commonly referred to as I. G. Farben. As the historians of SONJ point out, at this time IG was the most advanced industrial group in chemicals research, and SONJ was especially interested in its petrochemicals research bearing on the use of the hydrogenation process to convert coal to gasoline. Among other features of the agreement, SONJ and IG entered into patent sharing. It was one thing to enter into an agreement, another to make it work, and neither side was entirely comfortable with the bargain. The companies modified arrangements in 1929 and again in 1930, when a jointly owned company, Jasco, Incorporated, was organized; IG assigned to Jasco world rights to such processes as paraffin oxidation and Buna rubber manufacture. The year 1930 was not a propitious time to work on synthesizing gasoline, but SONJ went ahead with the hydrogenation research that would lead to the development of 100-octane aviation fuel, toluene, and the production of fluid catalytic cracking, all spinoffs of research that originated with IG. From 1933 onward IG worked at improving its synthetic rubber beyond the inferior quality of Buna, with considerable success. Here, however, IG kept the patents on its improved products, and though SONJ wanted its German partner to test them with American tire companies, the German government blocked that plan until late in 1938. Finally in 1939, IG turned over its Buna rubber patents to SONJ, and SONJ tried to interest the federal government in subsidizing an experimental synthetic rubber program. Subsidy would be necessary because the low price of natural rubber made synthesization unprofitable. In the face of low natural rubber prices, however, the government was not interested.[11]

Outbreak of war in Europe brought knotty problems to an already complex and uneasy international business relationship, problems that escalated once the United States was actually at war with Germany. Thurman Arnold began SONJ's domestic tribulations when he sent his antimonopoly squadron to investigate SONJ's connection with IG; after looking through thousands of documents, they worked up an antitrust case against SONJ based on the patents-sharing agreement. Some of SONJ's directors wished to fight the suit, but a majority, preoccupied with wartime operational problems, supported a negotiated settlement with Justice, which was signed on March 25, 1942. Under it, SONJ not only broke all connection with IG but also released two thousand SONJ-IG patents. The next day Thurman Arnold told Senator Harry S Truman's Special Senate Committee on National Defense that SONJ's relation with IG was the only reason the United States faced a rubber shortage. Standard Oil of New Jersey, he claimed, had given synthetic rubber technology to the Germans and denied it to American manufacturers.[12]

The focus of oil-related discourse thus shifted to another round of Standard Oil–baiting on the part of Washington antimonopolists, with full guarantee of headlines in the press. Arnold made headlines by claiming that SONJ's agreement with IG "absolutely stifled" development of synthetic rubber production in the United States, that Standard Oil helped the Nazis in 1938 by building a gasoline plant to supply the Luftwaffe, and that until he stopped it, SONJ was getting ready to aid Japan, too. As days passed, Arnold's attack expanded: Standard Oil was charged with having supplied the Axis with gasoline by selling it to German and Italian airlines in Brazil against State Department admonitions. Newspaper headlines read: "Arnold Says Standard Oil Gave Nazis Rubber Process," "Arnold Accuses Standard Oil of More Nazi Aid," and "Would Prevent in Future Such Deals as Standard Made with Nazis." To the casual reader, Standard Oil was a Nazi collaborator. But that was not precisely what Arnold was getting at. In baiting Standard Oil of New Jersey, Arnold was still pursuing his antimonopoly crusade, but by a new avenue. He saw relations between SONJ and IG as a compelling example of the evils of international cartels and how they subverted the defense effort in the interest of monopoly. For Arnold, as for so many crusaders before him, Standard Oil was to be an "a great educational lesson to the American people." The company in effect had the role of scapegoat, both for the evils of international big business and for the nation's lack of adequate preparation for war. But while Standard Oil might, in Arnold's mind, be primarily an example of a general social problem, others had a simpler perspective. Senator Harry S Truman said, "I think this approaches treason."[13]

The newspapers also carried SONJ's answer to Arnold, Standard Oil's spokesman being its CEO, W. S. Farish, and Farish gave vigorous challenge to Arnold's allegations. He denied that Standard Oil gave rubber processes to the Germans that it kept from Americans and asserted that overall Standard Oil got more from IG than it gave, an assessment with which later historians have generally agreed. It had given the U.S. government its toluene formulas, essential to the explosive TNT; it had declined to build hydrogenation plants in France, nor did it sell gasoline to Axis airlines against the State Department's wishes. More to the point, Farish reminded reporters that when SONJ began its association with the German group, the Nazis were not in power and Germany was not at war with the United States. In all, he claimed that Arnold's charges had no foundation. Even so, the papers reported Senator Joseph O'Mahoney's rejoinder: "You are bound by two loyalties; one a loyalty to I. G. and its world cartel and the other to the United States and its world policy."[14]

Arnold reiterated his charges in June when the Truman committee issued its report, and Washington senator Homer T. Bone's Patents Committee went over much the same ground in July, so the question of Standard Oil's relations with IG and their effect on defense preparedness stayed in the air during much of the spring and summer of 1942. Indeed, so greatly had the reputation of the oil company and its executives suffered that some four hundred persons attending the annual meeting in June demanded assurance that no board members had benefited personally from association with the Germans. Farish, Walter Teagle, Orville Harden, F. A. Howard, and Frank Abrams all swore they had not—which the New York Daily News headlined, "Farish Denies Nazis Paid Him."[15]

Some journalists, on reflection, decided that Standard Oil was accused unfairly. The New York Times, for example, thought Arnold's charges without foundation; the Journal American said his efforts were "more helpful to Hitler, Hirohito, and Mussolini than to the American people." Time called the assault on Standard Oil a "smear." By contrast, however, syndicated columnist Drew Pearson, ever on the hunt for a conspiracy, charged that SONJ was "secretly used by the Nazis to further plans for world conquest." Longtime critics of Standard Oil at the New Republic and the Nation made the most of its German connection. In "Standard Oil: Axis Ally," Michael Straight told New Republic readers that officials of Standard Oil were "members of a conspiracy with a Nazi corporation" and that there were similar conspiracies on the part of Alcoa and General Electric. In similar manner, emphasizing the theme of conspiracy, Guenter Reimann described Standard Oil's relations with IG as "shrouded in secrecy" and claimed that Standard Oil got "privileged treatment from the Nazi government." At the Nation, I. F. Stone asserted that Standard Oil was much more inclined to cooperate with the Nazis than with its own government and that the State Department was covering up for it on the matter of gasoline sales to Axis airlines.[16]

While Thurman Arnold and supportive journalists were busy baiting Standard Oil, Harold Ickes was trying to cooperate with the petroleum industry in the war effort. In theory, he might now direct industry operations from the wellhead to the gas pump, but in fact, even after Roosevelt gave him independent authority as petroleum administrator for war in December 1942, he was far from being an oil czar. Nonetheless, Ickes took the opportunity to push conservationist principles of oil field management. As early as August 1941, the petroleum coordinator's office told oilmen that "unnecessary" drilling in proven fields should cease and drilling in new fields should be on the widest spacing possible. The immediate reason given for the directive was the need to conserve iron and steel, both in heavy wartime demand, but it also served to curtail small-tract or town lot drilling, long the bane of good conservationists. Of course, good conservationists urged these things to keep more oil in the ground, not to produce it in a hurry. On December 23, 1941, the OPM and OPC issued Order M-68, which made 40-acre spacing for oil wells and 640-acre spacing for gas wells a condition of obtaining material. The agencies would give materials requests for wildcat wells priority; these were defined as wells not less than two miles from production. To get material for wells on the pump, there could be an average of no more than one well per 10 acres on the tract being produced. In January, these rules were amended to require unitization around the 40-acre area of any new oil well drilled. All this fit conservationist logic. At the marketing end of the industry, the OPM and OPC prohibited construction of new gas stations unless they could be shown to promote the war effort, thus tackling another venerable conservationist complaint. Ickes also took over allocation of petroleum production among the states; to prod all states into passing their own conservation statutes, he could cut back the allocation of states that had not passed such laws, giving additional amounts to states that had, such as Texas.[17]

Although it was possible simply to use production allocation as a punitive device, doing so did not address questions of allocation among states that had regulatory commissions or allocations to fields within states. Of course, such questions were intensely political, both because they invaded the jealously guarded turf of state regulatory bodies such as the Texas Railroad Commission and because how they were answered determined how much oil the individual producers could sell. What the PAW needed was some allocation method that would not seem arbitrary, ideally something apparently grounded in conservation and engineering. For its purposes, the emergent concept of "maximum efficient rate of production" (MER) was exceptionally useful. In 1938 the American Petroleum Institute's Special Study Committee on Well Spacing suggested that it might be possible to find a maximum rate of production for each oil field that would result in maximum ultimate recovery. The committee was, in effect, trying to work with the older notion that flat-out production rates resulted in lower recovery and come up with a concept defining an optimal rate of production. By 1941, the API had come to think that there might be no one optimal rate for a whole field, but it still thought one could tie recovery to rate of production, albeit with a range of rates within a given field. Even so, it hedged its speculation on maximum efficient rates with regard to market demand; if an optimal rate resulted in production beyond market demand, then field production should be slowed below what would result from MER. Engineering would not take precedence over market economics.[18]

However dubious the logic surrounding MER might be, the concept itself was very useful because it could justify varying allocations between fields on some other basis than grade of crude production or mere proximity to markets. But it was one thing to recognize the usefulness of MER and another to assume one could use it as a basis for decision making in a crisis. Ideally, to calculate MER, one would have complete information on all wells in a field and all producing formations. That meant to establish the MER for some of the older fields in a place such as Texas — fields like Spindletop, West Columbia, Humble, or Orange, all of which had been reentered many times — one would need information on what every operator had done over as much as four decades, and on multiple producing horizons: a truly daunting task, but at least there was the possibility of obtaining some information dating from after 1919, when the Texas Railroad Commission (TRC) began to gather it. But what did one do in California or Illinois, where there were no state agency records on which to fall back? As the Production Committee for District 2, which stretched through the middle of the country from Michigan and Ohio to Kansas and Oklahoma, recognized, no complete list had ever been compiled of the thousands of companies operating in the Midwest or the hundreds of thousands of wells drilled. There was no comprehensive collection of field data for the whole district and no correlation of information on individual fields.[19] Yet the committee and its engineering subcommittees not only had to come up with such data but they also had to let their engineers process the information to arrive at MERs and field rules: all in the age before the computer and rapid computation. The district committees calculated MERs, but how they did so is a mystery. Indeed, in 1948, when one scientist tried to determine how the PAW had set MERs in Texas, he was told that the figures were "restricted information."[20] No wonder the PAW did not look for challenges to the Texas Railroad Commission in how the TRC managed Texas oil fields. By contrast, in California, the PAW ended up allocating production on the basis of its own MERs.

Since MER continued to be a staple in oil conservation discourse after the war, it is worth making several more observations about it. How much oil is recovered from a reservoir is a matter not only of well-by-well engineering but also of enhanced oil recovery technology and the economics of its application. After World War II, with greater understanding of reservoir engineering, scientists demonstrated that in dissolved gas drive reservoirs rate of depletion did not affect the percentage of oil recovered; wells produced at differing rates in such reservoirs ended up recovering the same proportion of oil.[21] More to the point, however, with improved recovery technology, one can go back and recover many more barrels of oil left in the ground, a possibility the conservationists of the twenties overlooked, understandably, in the infancy of enhanced recovery technology. Indeed, depending on oil prices, one may have a better financial return on what is produced secondarily or thereafter than on discovery, if that discovery takes place when prices are low. Such a consideration makes it harder to label oil left in the ground after development as "wasteful," as early conservationists did.

If arriving at MERs for all oil fields in the country represented a staggering amount of engineering, adding to that increasing unitization and secondary recovery in all oil fields, also PAW objectives, took the amount of engineering beyond the realm of the possible. For unitization, whether of a new or old field, engineers had to determine the number of wells necessary for optimal development; the appropriate extent of the unit; the best location for wells; and the best way to conserve reservoir energy. How did one do this in fields already drilled, carved up among large and small producers, where the principle had been "Devil take the hindmost?" To apply secondary recovery, one would have to locate not only all producing but all abandoned wells and then work out the engineering for each project. For that matter, to drill injection wells for water flood recovery, one would have had to have gotten the materials from the WPB — which was niggardly even about oil wells, let alone ones to inject water! In short, as the author of the District 2 PAW report admitted, little was done with either unitization or secondary recovery during the war.[22] One might be skeptical about the amount and quality of engineering done by the PAW on other terms, as well.

Enforcement of PAW regulations in effect depended on the Office of Production Management, later the War Production Board. Oilmen sent in their plans and requests to Ickes's office; if PAW approved, requests for materials went to the WPB. Denial by either the PAW or WPB meant suppliers could not provide operators with materials. Companies, moreover, could not evade rules by drawing on materials in inventory; they had to provide detailed inventories of materials on hand to Washington. In fact, they were obliged to return many other questionnaires about operations to Washington, so much so that in March 1943 an indignant Henry M. Dawes of Pure Oil claimed that filling out government questionnaires cost his company 8 to 10 percent of net earnings; as far as he was concerned, the government was gathering "a mass of uncoordinated, immaterial and irrelevant data." Smaller operators faced a far greater financial burden in handling paperwork. Yet all this activity was in the best tradition of informed Progressive government, in which policy followed intensive data gathering.[23]

In operational terms what government needed was ever higher levels of oil production to meet higher demand generated by war needs, and according to conservationist discourse, the kind of steps Ickes took should have done the job. In fact, after only three months of regulation he had to give up his goal of thirty thousand wells producing 3.6 million barrels of oil a day. Drilling fell in February, fell further in March, and stayed low in April. Continued transportation bottlenecks caused production and refining cutbacks. By midyear drilling had declined by half from July 1941, and well completions were down by nearly two-thirds; in Texas new production was down 75 percent, and Texas produced more than one-third of the nation's oil. Nationally, the production of all petroleum liquids in June 1942 was 10 percent below the June 1941 output of crude oil alone. Clearly things were not going according to plan. What went wrong?[24]

Answers gradually emerged in trade journals and thereafter appeared in wider circulation. To begin upstream, Ickes's blanket regulation on well spacing and wildcatting did not fit operational realities. How wells might be spaced to deliver maximal yield varied from reservoir to reservoir, and different areas of the same field might require different spacings. Depending on geological structure, optimal spacing varied. Wells in water-driven fields could often be spaced more widely than wells in gas-driven fields; in low-porosity areas wells needed to be closer than in high-porosity areas. In short, there was no specific figure — such as forty acres per oil well — that was valid for all fields, and to impose one courted production decline. The same held true of Ickes's definition of a wildcat well. Depending on structure or pay depth, a wildcat well might be much closer to established production than two miles. In exploration terms, complying with regulation meant the wildcatter could not pursue attractive prospects on relatively accessible leases. Nor could wildcat extensions of fields qualify under Ickes's rule. Thus the wildcat success rate fell 2 percent during the first six months of regulation. After a year, geological and operational discourse edged out conservationist orthodoxy in Washington, and Ickes's office began to grant frequent waivers of spacing and other regulations.[25]

Difficulties obtaining both materials and manpower also held back exploration and development. Delays in obtaining materials and equipment, in particular, held back additions to reserves, for the best chance for significant additions, as the experience of the later forties would show, lay in drilling deeper — and that required amounts of pipe and heavier machinery that wartime regulation made difficult to obtain. Wartime regulators made it especially hard to get new, heavyweight well casing, for they urged that casing programs be curtailed to a minimum and that lightweight casing be substituted for heavier when possible. This use of less durable materials made it harder to drill without mishap, especially to greater depths; it also set the stage for future problems of an environmental nature from casing failure. Then, even if his project fit federal guidelines on well spacing and operation, an operator had to send Washington multipage application forms—four copies of all papers—and accompany them with plats, field maps, inventory statements, and notarized declarations. Submitting all this paperwork not only slowed down operators, especially small ones, but also put an added cost on operations while prices of crude were frozen. Once the operator sent his forms to Washington, both the OPC/PAW and OPM/WPB reviewed them, meaning further delay, adding costs for operators of all sizes. In all, the way regulation structured operations guaranteed that they would not be speedy and that exploration would be more costly; there was always the possibility applications would be turned down and operations could not proceed at all. If operators navigated what many of them saw as a bureaucratic minefield, as they waited on equipment and supplies, they watched the draft take young workers, the backbone of the industry's drilling force; other workers left the oil field to enter better-paid defense industry jobs. Operators increasingly hired high school boys too young for the draft, middle-aged men, and others whose physical impairments exempted them from military service. As most drillers soon learned, they acquired a less efficient workforce. That further slowed the pace of operation and drove up costs.[26] Unless petroleum prices rose at a pace comparable to cost, there was little incentive to prospect and drill.

Prices were the largest barrier in the way of increased production. While operators' costs rose, prices did not rise with them. In fact, when the OPA's Leon Henderson froze all petroleum and products prices in October 1941, crude oil prices were only at 1937–38 levels, in real terms less than 60 percent of 1926 levels. Despite pleas from Ickes and oilmen, Henderson proved obdurate. Although he wanted enhanced reserves, he thought there was a perfectly adequate "stockpile" of oil to meet war need, and he could not see any reason to add to consumer costs just to encourage wildcatting. As far as he was concerned, oilmen were reaping unearned profits, and there was no reason to increase them.[27] Henderson's attitude reflected not only the old view of the oil industry as vastly profitable but also the turn-the-valve concept of petroleum reserves: if one already had them, it would cost nothing in time or additional capital to produce them. More generally, Henderson's position reflected the idea that prices had little to do with supply. And if one reflected on the experience of the 1930s, that seemed to be axiomatic.

According to industry commentators, the administration's price policy had peculiar and unforeseen effects on refining and regional supply problems as well as on wildcatting and additions to reserves. Writing for *Oil Weekly* in June 1942, Leonard Logan argued that transportation bottlenecks caused East Coast refiners to curtail production by as much as 30 percent and that a few had even shut down plants. When they could get crude, however, they paid higher prices to bring it in by rail, a cost not offset by price increases. In the face of shrinking margins, refiners were likelier to recoup costs by producing gasoline, which sold for more than fuel oil. But that shift created the possibility of a serious East Coast shortage of heating fuel — as, indeed, there was during the winter of 1942. Ickes understood this and continued to push for higher prices, but he met effective resistance from the OPA. The Progressive model of enlightened government by disinterested managers did not foresee managers' policies being at odds with one another.

Caught in the policy deadlock, oilmen filled industry journals and the records of congressional hearings with complaints. They argued their position with many of the themes familiar in industry discourse: stripper wells, independents, exploration, conservation, and future oil supply were all in jeopardy, above all because of low prices. Secondarily, they complained about red tape and materials and manpower problems. Most of them who did more than describe their problems traced them back to ignorant and arbitrary bureaucrats in Washington, particularly at the OPA. Thus, in May 1943, Independent Petroleum Association of America (IPAA) president Frank Buttram decried "the socialist approach evident in some bureaus." By that time, Ickes himself jumped on the anti-OPA bandwagon, telling the House Small Business Committee that the agency was inept and its policies were actually increasing monopoly by driving independents out of business.[28]

Oilmen were far from happy and oil problems far from resolved in 1943, and Harold Ickes knew it — which may have been why he embarked on a grander attempt to shape discourse than broadcast speeches and periodical features. Ickes gave readers *Fightin' Oil*, a book devoted to showing the public how well he and his agency managed wartime problems. The work's unequivocal purpose was to control discourse. Too much appearing in print, he argued, was contradictory. He would "clear up a few issues that seem to have too many people confused" and at the same time let the public know what his agency was doing. He hoped this would lead the public to take seriously only statements from authorized sources such as the PAW and "dismiss another as hearsay."[29]

Of course, Ickes once again assured his readers that the "phony" eastern gasoline shortage of 1941 had not been phony at all. Still, challenges had been met, and a "transportation miracle" had taken place to meet them. True, rationing was necessary on the East Coast and would probably continue to be so. He, Ickes, had urged voluntary cutbacks in consumption, but to no avail.[30]

Turning to his agency, Ickes admitted that oilmen had not initially been overjoyed at the prospect of working with him. Nonetheless, he had gotten them organized before any other industry was comparably coordinated. Cooperation had been splendid, the single most important civilian contribution to the war effort, demonstrating that "government and industry can work together, although it has taken a war to prove it." Presumably the skeptics to whom this possibility had to be proven were oilmen, for Ickes stressed that the basis of the cooperation was partnership: "No dictatorship exists or impends." The latter phrase suggested that a few oilmen may still have been skeptical.[31]

Ickes praised oilmen, but he argued that without government direction of oil, an effective war effort would have been impossible. Take transportation problems, for example. Ickes assured readers, "Had it not been for the most careful planning and the skillful execution of the plans, complete chaos would undoubtedly have resulted." On its own the industry could not have worked out a way to get petroleum to the East Coast. More generally, in wartime, "The requirements of the Nation must be ascertained by this central governmental organization; they must be interpreted to the various units of the industry; the necessary allocation of materials must be arranged." Here was a not unfamiliar bureaucratic assessment of agency performance offering sharp contrast to the articles in oil industry journals, and, for that matter, what newspapers reported as happening. With respect to upstream developments, Ickes assured readers that well spacing regulations ensured proper distribution of scarce supplies. They also prevented wells from being drilled "virtually on top of one another as too frequently had been the competitive practice." Rather than acknowledge the burden paperwork placed on oilmen, Ickes stressed Washington's heroic processing of required forms; his agency received thousands of applications for materials, each "subjected to the most searching study," indeed a "tremendous job." But doing the job meant the industry had saved hundreds of thousands of tons of steel and other materials critical to the war effort. He did not add that the savings came at the cost of exploration and development.[32]

In the latter part of *Fightin' Oil*, Ickes dwelled at length on the problem of future oil supply, raising the old fear of running out of oil. As matters now stood, serious shortfall of production was likely in only two years, and "beyond that, the future is not predictable with any confidence." That meant conservation was essential. It implied two other matters that Ickes did not go into at length. First, having raised the possibility of shortfall and suggesting it was caused by negligence on the part of industry and the public, Ickes could imply that if the United States ran short of oil to fight the war, it was not the PAW's fault. Second, Ickes's gloomy reflections set the stage for acquisition of foreign petroleum reserves, a project in which he had keen interest.

Although Ickes left an extensive justification for acquiring overseas reserves out of *Fightin' Oil*, he had no hesitation in presenting it to a potentially wider readership in the form of features for *Collier's*. For example, he raised the alarm of running out in "Hitler Reaches for the World's Oil" in August 1942. Echoing what Mark L. Requa said many years earlier, he feared what lack of oil would do to American living standards. Just as it was essential to keep world oil out of Axis hands, the United States had to look ahead to a search for new reserves. For the next two years, he repeated his warnings; as he said in 1944, "Maybe there is enough undiscovered oil in this country to last us indefinitely, but we don't know whether there is or not." So the nation needed to look beyond its boundaries to the rest of the world, and it needed to get "an equitable share" of world oil for itself. To reach that objective Ickes favored a government-owned oil company that would claim a share of Near Eastern oil.[33]

Ickes's perspective was shared by the State Department's Herbert Feis, who in 1944 presented many of the time-honored arguments from running out of oil in his *Petroleum and American Foreign Policy*. Certainly the resolution of American reserves problems was too important to leave in private hands. As Feis saw it, private companies might simply use up overseas reserves without setting anything aside to meet future national defense needs; they might, as Josephus Daniels once feared, overcharge government for oil. Clearly, there needed to be a government-owned oil company to sustain national interests; its mere existence would "persuade or compel private companies to furnish adequate supplies at fair prices."[34]

By the time Feis aired these opinions to the wider public, the kind of body he and Ickes envisioned existed. Acting on the recommendation of the Committee on International Petroleum Policy, which Feis chaired, Congress created the Petroleum Reserves Corporation (PRC) in June 1943, and Ickes was its head. Part of the reason Feis jumped into print, however, was to counter arguments from oilmen who decided they did not favor the new creation. As a number of scholars have described, industry members and state conservation agency officials worked up strong opposition to the PRC by December 1943, and the API and the IPAA alike condemned it. Industry opposition mounted when Ickes proposed that the PRC take on construction of a pipeline from the Persian Gulf to the Mediterranean in February 1944; his own Petroleum Industry War Council (PIWC) condemned the plan, just as it had rejected a government-owned company attempt to purchase Arab oil.[35] On the question of foreign reserves in the hands of a federal government body, it was obvious that the wholehearted cooperation of government and industry, of which Ickes said so much, ran aground.

One group about which Ickes and successive authors talking about oil and the war said little was that of state agency officials, and with good reason, for if one looked to paint the view of wartime regulation in the rosy hues of harmony and cheerful cooperation, one would not find it easy to do. Not that state officials refused to cooperate with Ickes, but neither they nor the officials at the Interstate Oil Compact Commission (IOCC) forgot that they were dealing with a man who had said he wanted oil to become a federally managed public utility. As William Childs has pointed out, during the war, state agencies such as the Texas Railroad Commission simply regulated as they had been doing with little attention to Washington, save to join with industry members to protest some aspect of Washington they disliked—gasoline rationing, for example, or price ceilings. Ickes did not have power to force a state such as Texas, with its own conservation agency, to accept his rules, so TRC business could be conducted as usual. But when Ickes launched his PRC project, state and IOCC officials lost no time in attacking the plan for a nationalized oil company to begin exploiting foreign oil, doing so with a barrage of states' rights rhetoric of the sort common a decade earlier. In 1944, the TRC chairman, Beauford H. Jester, said that Ickes and his Washington colleagues were out to destroy state controls over oil, that state regulation would not be regained after the war. Similarly, Ernest O. Thompson of the TRC said the PAW ought to be terminated: "We can handle the oil-well spacing, drilling and producing problems within our sovereign state." He added, "We are fearful that the federal government is attempting to set up a super-duper oil and gas conservation and production regulatory bureau under the guise of war emergency that will not liquidate itself when the emergency is over." Before the end of the year, the IOCC called for an end to the PAW's drilling, spacing, and production regulations.[36]

At this point one might well wonder what wartime federal regulation of petroleum accomplished. What kind of overall record did the PAW really have? Was it a model of government-industry cooperation? How well did it work? Since a thorough assessment of federal regulation of oil in World War II would require a volume of its own, the following observations are but a beginning. In general, notwithstanding "spin" control documents by Harold Ickes and others, federal effectiveness was uneven, and failure was more common than success.

On the plus side, Ickes did get oilmen to work with him on a myriad of committees great and small. This was the cooperation Ickes stressed and that Gerald Nash sees as the key to wartime achievement. But, as John G. Clark has pointed out, oilmen had little choice but to go along with a system imposed on them; Ickes might talk cooperation, but he believed in coercion, and oilmen knew it.[37] Certainly as the war went on, the cooperative spirit wore very thin, not only among oilmen but also among state authorities.

Perhaps the most significant project the PAW pushed through was the construction of Big Inch, the crude oil pipeline from Longview, Texas, to Philadelphia and New York, and Little Inch, the products pipeline from Beaumont, Texas, to Linden, New Jersey. So vast a project would not have been attractive to private industry at the time, and it was unquestionably a strategic and economic asset. With respect to other transportation problems, the PAW may have helped resolve some bottlenecks, though it may have created others. The PAW did press refiners to produce 100-octane aviation fuel. On the other hand, the PAW insisted on keeping all refineries operating regardless of whether they could or would convert to making high-priority products.[38] Clearly such a policy owed more to traditional thinking about small refiners than to wartime efficiency.

As the PAW stressed, the oil needed to fight the war was produced: the United States did not run dry. But was this because of the PAW and other federal agencies or in spite of them? Throughout the war, one of the main barriers to getting things done, as Ickes argued, was the sheer number of federal agencies with decision-making power relating to oil industry operations. As Clark has noted, FDR created a "witches' brew of agencies," all seeking to enhance their own power at the expense of other agencies.[39] Far from much-vaunted cooperation, relations between these agencies tended to be confrontational. Leon Henderson and Prentiss Brown at the OPA and Donald Nelson at the WPB seem to have reached a point in dealing with Ickes at which the PAW head's wanting something was sufficient reason to oppose it. But even without intra-agency scrapping as a brake on action, to have to get approval from at least two federal agencies before beginning to drill a well slowed even the most ambitious oilman.[40] If one valued getting more wells on line quickly, this was not the way to go about it. Indeed, if one valued efficient operation, giving operators increasing quantities of paperwork to fill out was counterproductive.

With respect to both producers and refiners, what federal regulations did tended to raise operational costs and reduce efficiency, but especially with regard to exploration and production, federal regulation was at best beside the point and at worst totally misguided. The kind of regulation spelled out in Order M-68, more aimed at conservation of steel than oil production, invaded areas of operation that state agencies such as the Texas Railroad Commission already handled and continued to regulate during the war. Telling such bodies how to allocate production was unnecessary, even if directives were only guidelines. Moreover, the PAW's regulatory objectives of drilling fewer wells, spacing them farther apart, and unitizing new pools all reflected conservationist discourse aimed at keeping more oil in the ground for the future, not producing it for wartime consumption. Spacing rules reflected lack of familiarity with field operations, and the PAW was obliged to back away from them long before the war ended. As for the PAW's goals such as more unitization and additional secondary recovery, its assessment of what it could do was unrealistic not only in terms of field conditions but also in terms of manpower and time available.

To be fair to the PAW, the biggest brake on the drilling of more oil and increasing oil production was the OPA. Adamantly holding crude oil prices to already low October 1941 levels harmonized with old antimonopoly ideas about high petroleum prices hurting consumers, but in every other respect it was counterproductive when all industry costs rose. Such policy already worked against additions to reserves. In fact, the OPA proved willing to give some ground on price only to stripper well producers. The OPA's price policies may also have contributed to spot fuel oil shortages, though to what extent is arguable.

What kind of bottom line did the PAW have with respect to well completions, wildcatting, and production? Looking at District 2, oil well completions fell from 6,797 in 1940 to 4,277 in 1945. In the same span of years, total well completions fell from 10,985 to 7,255. Wildcatting figures were somewhat better, rising from 1,575 in 1942 to 1,900 in 1945. But total crude oil production for District 2 declined from 1,102,000 barrels per day in 1940 to 947,000 barrels per day in 1945, at the cost of producing fields such as East Texas wide open.[41] Total national production did rise from 1,353,200,000 barrels in 1940 to 1,713,700,000 barrels in 1945. On the other hand, total national additions to reserves dropped sharply between 1940 and 1943, returning to 1940 levels only in 1945. The production record was acceptable, but the record on additions to reserves was poor.[42]

Notwithstanding the federal government's undistinguished performance in directing the oil industry in wartime, if one wanted continued federal direction of the industry during peace, or if one wished to build support for the Petroleum Reserves Corporation's acquisition of foreign reserves, it was highly useful to stress how well things had gone. Ironically, from the point of view of the industry it could seem equally essential to stress success to avoid antitrust action of the sort Thurman Arnold preferred. Because emphasis on success was useful to so many groups, it is not surprising that assessments of the war experience were highly upbeat in tone. The most widely read and still most often cited history of oil during the war was John W. Frey and H. Chandler Ide's official history of the PAW. Their message was that, under the direction of the PAW, oil won the war. As they took up successive areas of industry operation, Frey and Ide stressed how essential federal direction had been to success. With respect to exploration and production, for example, the PAW kept wheels turning and airplanes flying by developing "with painstaking study" long-range programs of exploration, development, and production that made the "most efficient use" of resources. Imposing wider well spacing "fostered principles and methods . . . of lasting benefit." True, not everything went smoothly; hopes for many prolific discoveries went unrealized. But, nonetheless, government and industry got oil where it needed to go "with a minimum of regulation and a maximum of cooperation" and did so "without practically anything but courage, determination, resourcefulness, skill, and the willingness to put the national welfare ahead of the individual interest."[43] Or, against all odds and with raw guts, one might say to this crusaderlike list of masculine virtues: here was language more descriptive of frontier lawmen than bureaucrats and businessmen, a gender spin on the situation. These heroes had put national welfare ahead of individualism — something conservationists and collectivists had urged for decades.

Frey and Ide made it clear that heroics in production had counterparts in every other sector of the wartime oil industry. Completing Big Inch and Little Inch was "a saga of industrial achievement unexcelled" by anything else in industry. Refining was the story of "miracles from molecules," successful "because of the close government-industry cooperation." When problems arose in distribution and marketing, as in the fuel oil crisis of 1942, oilmen worked twenty hours a day to avert "stark tragedy."[44] Thus, in language reminiscent of what moviegoers heard in wartime newsreels, Frey and Ide made what the oil industry did on the home front seem like battlefield heroics.

Heroic hyperbole certainly made Frey and Ide's bureaucratic history more readable than most, and when it came to credibility, the authors were apparently ready to admit that the PAW had not solved all problems, a seeming candor that gave their work an air of balanced, impartial appraisal. At the same time, however, they either omitted or carefully played down embarrassing instances in which government and industry were seriously at odds. Thus the Petroleum Reserves Corporation was buried in chapter 16 and strenuous objections to it from industry in appendix 8. Many readers would not have read that far, to see PIWC condemn state-owned oil companies, oil hoards on public lands, federal control of domestic production, or any further federal direction of the industry.[45]

In the end, Frey and Ide left readers with three main points they argued wartime experience confirmed. First, the domestic oil supply was limited, and war showed the necessity of preventing any future "oil famine" in the United States. Here was the old running-out-of-oil theme, reasserted in a timely way to support acquisition of foreign reserves. Second, the war showed "the effectiveness of the cooperative approach to Government-Industry relationships," and in that respect the PAW was a model for the future. But, third, the PAW's record showed that all authority over oil ought to be concentrated in a single government agency; overlapping jurisdictions did not work well. That, of course, was what Harold Ickes had been saying for years.[46]

Not content with one overview history of the PAW, Ickes intended that each district should have its own history, a goal only realized by journalist and public relations consultant D. Thomas Curtin's history of District 2, *Men, Oil, and War*, published in 1946. By contrast to Frey and Ide, Curtin's work was less hyperbolic, and problems and shortfalls in operations received more candid coverage. Curtin also had less commitment to a spin on events supporting federal direction of oil in peace as well as war; instead he began by emphasizing how different wartime needs were from those of peace. Still, because wartime was so different, the oil industry had to transform its operations and was able to do so "due largely to the most unusual cooperative arrangement which existed between industry and government." To sum up such cooperation, Curtin often used the word "teamwork"; industry and government were a "team." For that matter, sometimes the government component got forgotten; "the men on the petroleum industry team" won "the supply and transportation battle from the Rockies to the Atlantic." Here mixing athletic and military metaphors, Curtin, like Frey and Ide, used gendered language to conjure up a vision of heroic businessmen giving their all for their country and getting, as he put it, "colossal results."[47] Indeed, the title of his book would lead a reader to expect he-man conflict rather than bureaucratic history, surely a marketing tactic.

Curtin gave readers not only teamwork but also, sports-column style, something about individual oilmen working in District 2, offering brief biographical sketches of the members of the general committee that often mirrored Horatio Alger's tales of successful self-made men. Marketing representative Bernard L. Majewski, for example, went from office boy to company vice president. E. J. Seubert went to work for Standard Oil "as a boy," "learned the industry literally from the ground up," and went from mechanic to president of the Standard Oil Company (Indiana). Burt R. Bay of Northern Natural Gas began as a machinist, Phillips Petroleum's president Kenneth S. Adams as a warehouse clerk. Such details showed off oilmen's industrial experience and also played up their energy as successful competitors among businessmen. For that matter, Curtin presented the petroleum industry in general as an arena in which men "pioneered": men "took chances, they risked capital, they worked."[48] Until seeing these activities in tandem with pioneering, a reader might not have seen them as out of the ordinary. In all, Curtin lavished as much favorable light on oilmen as on the agency whose history he wrote.

The editors of *Look* magazine offered an even more glowing account of the industry and men in it in their 1946 *Oil for Victory*. In addition to acknowledging the help of the PAW and PIWC in completing their work, they thanked the

American Petroleum Institute and one of the executives of the Standard Oil Company (New Jersey), which may explain why government did not get as extensive coverage as industry in the teamwork these authors also described. As *Oil for Victory* had it, the crisis of the war obliged oilmen to forget competition and work with government to launch a coordinated effort to win the war. It was the coordination of effort industry needed from government. For the PAW's part, it "leaned over backward not to deal with the industry in a high-handed manner" and took advice "as often as it gave directives." Industry could not have done the job alone; "PAW helped at every turn."[49]

Perhaps so: but the editors made it clear that most credit should go to the industry. War brought "mountainous problems," but, as the authors dipped into gendered imagery, they were met by "the huge, tough, competent petroleum industry," an industry "abounding with technical skill, ingenuity, and plain guts," an industry capable of "mighty efforts." Oil fought wherever the army fought; when the navy fought, "oil was there"; oil fought on the home front, too. Rhetoric changed Big Oil into a militarized version of Superman, a businessman into a warrior. As for oilmen, they were self-sacrificing patriots: "They produced their wells at an uneconomical rate, overran their refineries, and pushed their trucks until pistons rattled and tires collapsed." All this came from people who were "by instinct and tradition keen competitors" but who were ready to surrender individual advantages for the common good.[50]

When authors of *Oil for Victory* went beyond describing the heroic role of oil at war to discuss what oil would be in peacetime, they made it clear that federal regulation should not be part of the picture. The United States had been and would continue to be the greatest oil-finding nation in the world because of "freedom with competition," in contrast to nations in which "nationalistic laws" hampered exploration. Not only was the oil industry highly competitive, but it was also "the most characteristically American of all industries," having developed first and most completely in the United States. Resorting again to gendered discourse, the authors presented the wildcatter as characteristic of American spirit. He was the embodiment of "the enterprising, pioneering spirit, the recklessness and curiosity . . . such a part of American character," the modern counterpart of the clipper ship captain who would "kiss his wife good-bye and come back three years later with a chest of treasure." Warriors in wartime, oilmen were pioneers in peace; either way, the authors of *Oil for Victory* construed oilmen and their industry in terms of masculine heroism. Obviously, in the world of the pioneer there was no need for a federal agency. In fact, as the authors extended their view to an international level, the freedom to develop oil over the world "with only the necessary minimum of government regulation" would be the key to American prosperity and world peace in the future.[51]

With their antiregulatory bias, their vision of a peacetime industry restrained only by the visions of oilmen, one could say the authors of *Oil for Victory* set the Progressive conservationist vision back half a century. At the same time, with their stress on the competitiveness of the oil industry, they set aside antimonopoly anxieties. That left a smaller role for government in industry affairs, which by 1946 many oilmen hoped for; notwithstanding the "spin" the histories of the war years put on industry-government relations, cooperation was problematic, problems numerous, and results often short of goals. When the PAW ended, there were few who lamented its passing, fewer who wanted a peacetime counterpart to it. Far from being a model for the future, the relation between federal government and industry reminds one of Samuel Johnson's cranky verdict on women preaching: " 'Twere not that 'twere done well but that 'twere done at all."

Nor does one have to look for "special interests" ready to sabotage policy makers to explain why the PAW left little enduring impact on public policy or why it did not usher in a new era in industry-government relations. As it was administratively created, sharing power with many other state and federal agencies, the PAW was not constructed to work well. But, more to the point, as a variety of federal agencies including the PAW worked with old ideology about the industry, they came up with counterproductive regulation. One has only to come back to Order M-68, with its conservationist provisions originally aimed at keeping oil in the ground applied to a situation in which maximal production was needed, to see how policy could misfire without the help of sinister forces conniving behind the scenes.

Beyond this, the legacy of decades of public discourse on the petroleum industry emphasized an adversarial relationship between government and industry, the kind of relationship visible in Thurmond Arnold's antimonopolist crusades against Standard Oil of New Jersey; prospects for industry-government cooperation were scarcely enhanced when government suggested that some leaders of the industry were traitors. That Arnold and politicians such as Senators Harry S Truman and Homer T. Bone thought an antimonopoly crusade was appropriate during wartime, moreover, speaks volumes for the strength of traditional rhetoric and ideology independent of operational reality. The legacy of discourse, by 1945, meant that visions of cooperation or even of practical policy making would not be realized when war gave way to peace. But, paradoxically, that same legacy insisted that there ought to be some variety of federal energy policy, that the federal government did have a role to play in energy planning and resolving energy problems. The legacy of discourse did not help much in establishing precisely what the federal government should do. It raised more questions than it answered.

# Conclusion

When one looks at what has been said about the American petroleum industry from a historical perspective, it is possible to make a number of general observations about discourse and ideology. These observations can be demonstrated in industry experience and in industry-directed policy.

Ideology embodied in public discourse created a cultural construction of the American petroleum industry, a body of assumptions about it that came to be repeated often and questioned only rarely. According to this construction, the industry was monopolistic, overpowerful, speculative and risky, conspiratorial, wasteful, disorderly, out to gouge consumers, out to corrupt government, and, in general, a threat to public welfare. Consistent in these assumptions is moral discourse; the assumptions boil down to normative ideas about the industry that may or may not have foundation in any operational reality.

Those involved in public discourse on the industry advanced personal interests as well as ideological positions. Oilmen, journalists, social scientists, geoscientists, state and federal bureaucrats and regulators, politicians and policy makers all had axes to grind and jostled for position at the grindstone.[1] They all sought to control or dominate discourse to their own ends. They spoke out again and again, repetition enhancing credibility. As they did so, however, they often responded as well to broader social and cultural concerns. The discourse about oil was seldom just about oil. It usually encompassed more general questions that involved moral judgments about public welfare.

Over time, discourse on the industry evolved, and it came to have channels emphasizing antimonopoly and conservation. As these channels evolved, so did the cultural construction of the industry, developing to include new ideas tied to new developments. But as channels evolved, they not uncommonly incorporated contradictory ideas about the industry — that it was, at one and the same time, monopolistic and overly competitive, for example. They also included ideas that did not translate into operational discourse or whose main focus was not operational. As a result, policy based on discourse misfired. Alternatively, because of conflicts in discourse, those working to arrive at constructive policy got nowhere.[2]

Now let us translate these general reflections into the history of the industry. In ideological terms, a negative view emerged in the earliest writers' treatment of it. They emphasized that it was speculative and dangerous for investors; that it was disorderly, both in its operations and in the society of those who worked in the oil fields; and that it was dirty — smelly, greasy oil contaminated what it touched. As the industry grew and producers and refiners competed for profits in a highly volatile market, industry participants began to charge one another with various sharp practices usually tied to schemes to "fix" the price of crude oil; talk of conspiracy was common by the early 1870s. By that time oilmen in competing regions had begun to use the press and public assemblies to air their differences and accuse one another of unethical practices; they did so to bring political pressure against competitors. Charges against railroads for preferential treatment of shippers emerged early because of the critical element of transportation cost in calculation of profit. In short, even before the rise of Standard Oil, discourse on the petroleum industry typically included many negative ideas, advanced by those within and without the industry alike.

With the appearance and spectacular growth of Standard Oil, discourse on the industry came to focus on that company, particularly as its competitors saw the value of putting pressure on the firm in both legislative and judicial arenas. As Standard Oil became the focus in discourse, its competitors, and especially the lawyers retained by them, began to charge it with the kinds of unethical practices already current in industry-related discourse: collusion with railroads, fixing crude prices, conspiracy to drive competitors out of business. By the end of the 1870s, Standard Oil's size, prosperity, and efficiency made it vulnerable to the additional charge of monopoly, exploited by Simon Sterne and others in the Hepburn committee investigations. Given this ideological identification, Standard Oil was a public menace, an overpowerful moneyed interest threatening the well-being of many more than its business competitors. And the man behind the menace was John D. Rockefeller, who became the personification of his company.

In the 1880s and 1890s, Standard Oil not only continued to be assailed by competitors, George Rice being a prime example, but also acquired a host of new opponents and critics—journalists, intellectuals, social thinkers, ambitious politicians. In part, this was the result of successful growth and integration, both vertical and horizontal: the more oil fields and markets Standard Oil entered, the broader the range of regional interests it challenged. In larger part, however, it was the result of Standard Oil's coming to stand for a wide range of problems and anxieties Americans of the time experienced in response to economic growth and social change—fears of great wealth corrupting democracy, of irresponsible riches creating social tensions, of increasing wage employment destroying manly independence. Such fears prompted more and grander moral judgments, moral discourse of a sort that required no special knowledge of the petroleum industry to apply. For example, understanding complicated price movements took knowledge of the industry and economics. Understanding "predatory pricing" only took moral judgment. It was easier to talk about Standard Oil as a great evil than as a novel business phenomenon, easier to use gendered imagery to condemn Rockefeller as an effeminate destroyer of manhood than explain his financial success in overseeing the growth of a large firm.

As its opponents used moral discourse to condemn Standard Oil, they regularly repeated not only the same figures of speech, such as predatory pricing, but the same sound bites—"turn another screw"—and the same cautionary tales of Standard Oil's misdeeds, for example, the widow Backus story and the Buffalo refinery saga. The constant repetition of these ideas and stories lent credibility to the negative view of Standard Oil. Ultimately, they became part of a widely shared version of the history of the company and the petroleum industry, a history expounded at length by such writers as Henry Demarest Lloyd and Ida M. Tarbell. That was how, eventually, "everybody" came to know what John D. Rockefeller did. Thereafter, during the first half of the twentieth century, when social scientists wrote about the industry or politicians and policy makers turned their attention to it, the sins of Standard Oil were repeated regularly, like a litany, into text or record. Even as subsequent historians found reason to question the most outlandish tales and judgments, as Tarbell did with some of Lloyd and Rice, and, more recently, as Yergin and Chernow did with Tarbell, they fell back on the same larger body of interested sources to sustain the conventional negative judgment.

As a brief for industry malfeasance, the sins of Standard Oil justified public action and regulation, antitrust legislation, and, in 1911, the dissolution of Standard Oil. Ambitious politicians such as Theodore Roosevelt and Robert M. La Follette were ready to respond to the view of Standard Oil as prime example of the menace of trusts, and after dissolution, federal regulators at the new Federal Trade Commission acted as if Standard Oil continued to be a danger. The ghost of Standard Oil still stalked abroad, perhaps because, having been so prominent a part of public discourse for decades, it was too useful and too difficult to abandon. It was what observers of the industry expected to see.

The dissolution of Standard Oil in 1911 was a triumph for antimonopolists, but there is no strong evidence to show that this action actually enhanced competition by lowering barriers to entry to the industry. As many of the liberal critics of the Supreme Court decree argued at the time, the national monopoly was merely replaced by regionally dominant components of the old company. The *Daily Socialist* (Chicago) put it bluntly: "This is no dissolution." The *New York Press* claimed that it would actually give Standard Oil more opportunities to break the law. The *Journal of Commerce*, among business publications, foresaw increased regional concentration as former Standard Oil components bought out small holders to integrate and grow. Though the Taft administration would take credit for having slain the dragon, many writers doubted that it was dead, and scholars have disputed the efficacy of the decision.[3] Though the decree did not lower barriers to entry, there was little need for it to do so, because unrelated events and developments had already advanced competition. They continued to do so even as the former Standard Oil companies emerged gradually as occasional competitors with one another. Broad changes in the oil industry and in American society accomplished what the court sought to effect.[4]

Most notably, the rapid expansion of oil production in new regions such as California, Oklahoma, Texas, and Louisiana fostered the emergence of new companies, including Gulf and Texaco, and the entry of other competitors, especially Shell. These new companies created functionally integrated operations, exploited regional economic, political, and operational advantages, and competed effectively with the successor Standard Oil companies. Unstable production levels frequently flooded markets with cheap crude, encouraging new entrants, many of whom were adept at using the arena of state politics to secure competitive advantages. The same advantages favored a host of smaller but significant firms, including Sun, Humble, Marland, Continental, Phillips, Cities Service, Skelly, and Southern Pacific. Diversification of products also favored new entrants, as Gulf Coast crude came to dominate fuel markets, challenged only by imports from Mexico and Venezuela, some of which were produced and/or refined by non–Standard Oil companies. Gasoline refined by hundreds of companies that served growing local and regional markets, from the first decade of the century to the 1980s, sustained rivals to the onetime Standard Oil group. In short, changes within the industry and the broader economy accomplished what the legal system did not, lowering barriers to entry and, thereby, increasing competition.

By 1911 discourse on the industry broadened to include a conservationist channel, and new participants, especially federal bureaucrats and geoscientists, came forward to advance their interests through it. Conservationists added to the cultural construction of the petroleum industry. It was wasteful of a vital resource; it was heedless of future need; it was recklessly and destructively competitive; by giving consumers cheap gasoline at thousands of service stations, it encouraged extravagance and materialism. It is easy to see, in this set of ideas, how the new channel of discourse included elements at odds with the antimonopoly channel. Conservationists, however, also developed ideas responding to broader concerns — an overly materialistic society at home, for example, and increasing apprehension about international rivalries and tensions that might challenge America's position in the world. With respect to the latter they responded by advocating naval reserves, "Uncle Sam's oil barrel," an idea that was operationally dysfunctional, and economically imperialistic schemes to save U.S. oil by pumping foreign nations dry first. Both responses presumed what could not be operationally or technologically proven, that the United States was running out of oil.

From 1920 on, it was clear that control of discourse about the petroleum industry was no longer in the hands of industry participants. In fact, once the conservationists emerged, they broadened the onus of public malfeasance to include the whole industry, not just one or some members of it. Put on the defensive, some oilmen first resisted and then began to adapt what conservationists said. When they adapted conservationist discourse, however, they had to try to accommodate industry operations, technology, and economics, and they emerged with different understandings of conservation than what nonindustry conservationists maintained. Their adversaries would see this as cynical self-interest. Then again, some oilmen had little use for any talk about conservation, which they saw as a ploy of larger companies. Instead they continued to frame industry issues through antimonopoly discourse. This explains how, having lost control of discourse, industry participants did not regain it. Industry members were never united in point of view and never worked together as a monolithic interest, even though, after 1920, industry critics often spoke of them so.

By the late 1920s, the U.S. petroleum industry began to suffer from some very real problems, rooted in the production of more crude oil than markets could absorb. When both industry members and federal and state officials tried to approach these problems, however, contradictory ideas in discourse held them back. In terms of conservationist thinking, the industry needed to cooperate to keep more oil in the ground. In the antimonopoly view, this was combination in restraint of trade and price-fixing. Once the New Deal was launched, federal bureaucrats pushed for a federal resolution of problems but could get no industry consensus on what should be done. Federal involvement, however, evoked a strong states' rights discourse about the industry, a discourse opposing federal regulation. In the end, as we have seen, all the New Deal discussion of oil resulted in but two lasting measures, the Connally Hot Oil Act and the Interstate Oil Compact. Industry members had better luck at resolving their problems with state regulators, for state officials were willing to share understandings rooted in industry operations and economics.

In the later Roosevelt years, antimonopoly discourse once again dominated public discussion of the petroleum industry, continuing into World War II. During the war, such thinking guided the responses of the Office of Price Administration to the industry's request for price hikes to offset increased costs; the OPA set prices low, with consumers in mind, but with a disregard for industry economics and the need for more oil. Conservationist thinking also resurfaced in an especially unconstructive way; at a time when all-out production and exploration were needed to support the war effort, federal regulators initially fell back on the thinking geared to keeping more oil in the ground. Federal regulators and industry members worked together, but in an uneasy partnership that industry members were anxious to be rid of at the end of the war. As for state regulatory officials, they were even more dedicated to ending federal regulation. No wonder that, once the war ended, there was little to show for wartime experience of government-industry cooperation.

We have concluded our case study of public discourse and the U.S. petroleum industry in 1945, but as anyone familiar with the subsequent history of the industry and policy making will recognize, the same channels of discourse continued on, and those who spoke to petroleum-related issues repeated familiar ideas. Anyone who lived through the energy crisis of the 1970s, for example, will recall the revival of "running out of oil" as a theme in public discussion. The most significant change in discourse after 1945 was the emergence of an environmental channel, in part an outgrowth of old conservationist ideas but also shaped by new ecological concerns. Exploration of this area of discourse would make a case study on its own, but, beyond a doubt, environmental discourse made it even more difficult to map out consensus on energy-related issues because of the inherent conflicts of elements of it with elements in older channels. Take the example of substituting coal for oil and natural gas, as a boiler fuel; this ideal, embodied in President Jimmy Carter's National Energy Plan, which Carter called "the moral equivalent of war," had the old conservationist virtue of prioritizing petroleum use and conserving petroleum but aggravated the problem of acid rain, about which environmentalists sounded the alarm. The National Energy Plan also swore at the objectives of the Clean Air Act of 1970, which sought to move away from coal as a boiler fuel in order to cut air pollution.

If one looks for examples of elements of discourse being inappropriately applied to yield unworkable policy, the problem is selecting only a few out of a staggering range of postwar examples. Certainly federal price controls on natural gas produced for interstate markets and on crude oil in the 1970s had grandly dysfunctional results and aggravated problems of supply; indeed, as Richard H. K. Vietor has pointed out, federal natural gas pricing policy amounted to a "formula for shortage." As Vietor also points out, once natural gas shortages appeared in interstate markets, industry critics were quick to look for oil company conspiracy to raise rates by creating artificial shortage in order to gouge consumers; they assumed that the natural gas industry was inherently monopolistic. In vain did gas companies point to the operational realities of lease obligations, state conservation regulation, and long-term contracts to argue such "conspiracy" could have no rational basis. Similar conspiracy theories also tied to old antimonopoly discourse surfaced in oil supply dislocations of the 1970s, creating an atmosphere of "misinformation and mistrust," as Vietor puts it, that was hardly conducive to constructive policy making. They helped fuel attacks on vertical integration in the industry as promoting monopoly.[5]

Overall, what can the history of public discourse about the petroleum industry tell us about the industry and policy making? And what are the implications with respect to other industries? Looking at how discourse worked explains a great deal about why policies misfired or never passed the discussion stage. Policy makers built or tried to build policies and regulations out of the channels and elements of discourse with which they were familiar, ideas that were available to them sustained by moralistic references. They, like everybody else, understood the industry in terms of what was said about it, in terms of its culturally constructed identity. Few of them were in a position to reflect on whether that identity corresponded with how the oil industry operated. The embedding of political discourse in moral discourse made weak positions seem plausible because of the "fit" with dominant cultural values, leading the public and policy makers to overlook distortions produced by interest and inadequate data. When that culturally constructed industry identity mandated an adversarial or even punitive role for government, it was hard for lawmakers or regulators to take a different course, even when events such as wartime crisis warranted doing so. But, as we have seen, where discourse did not translate into effective policy, one could seek remedy in more discourse. One could find a scapegoat for failure — or simply argue that, appearances notwithstanding, policy had been a splendid success. Or as Harold Ickes and his subordinates did at the end of World War II, one could do both.

Beyond explaining policy failures, looking at discourse relating to competition and regulation in the U.S. petroleum industry raises another issue, the cost to the American public of decades of adventures in trying to reify parts of moral discourse into industry operation. It would be easy to smile at some of the counterproductive absurdities of these endeavors had they they not cost the American taxpayers millions of dollars and saddled the industry with unwonted costs. Because we have not outgrown contradictions in discourse, we can expect to see more counterproductive responses to problems in the future. For example, activists such as Ralph Nader and Edwin Rothschild continue to keep alive the old antimonopolist idea that the oil industry is "ripping off" consumers. That implies that every American's birthright is cheap fuel — an idea incompatible with conservationist or environmental thinking. In sum, it is surely time that the old ideas in public discourse about the petroleum industry, not to mention the cultural construction of the industry itself, were reexamined and their appropriateness reassessed.

With respect to other industries, obviously the petroleum industry has not been alone in being saddled by dysfunctional policies. For example, transportation and communication sectors of the economy long bore the burden of dysfunctional regulation. Railroads were long refused the right to compete, and, within states — Texas, notably — trucking firms were regulated as public utilities. Well after safety and technical considerations were no longer mandated, airlines and radio and television stations were also regulated. New industries and reconfigurations of old business have often been targeted in public attention in much the same way and for many of the same reasons as railroads and petroleum. In computer software, for example, Microsoft has been subject to charges by competitors of raising barriers to entry; like the Pennsylvania oilmen of a century ago, they have also received strong political support in Congress. Most notable is Netware, headquartered in Utah, whose senior United States senator, Orrin Hatch, has used his chairmanship of the Senate Judiciary Committee to advance the attack on Microsoft. The Clinton administration, liberally supported by Microsoft competitors in California, has pursued antitrust proceedings with vigor as well. Public discourse worked for Roger Sherman, and it still works as a competitive strategy.

In a different sector, that of health and medical care, public discourse has raised a host of questions with implications for industries involved, health care providers, and policy makers. For example, breakthroughs in medical research have opened up a tremendous range of treatment options over the past few decades. As new medications are developed, this raises the question of how much of the cost of research and development should be passed along to consumers by the pharmaceuticals industry, as well as whether insurers should pay for new and sometimes experimental treatments. Recent controversy over Viagra is a case in point. Then there is the controversy over policies of health maintenance organizations: whether they are seen as consumer-friendly providers or, as is more current on the nightly news, as heartless abusers of the sick, there will be pressure on policy makers to respond to complaints about them. Finally, medical research has tied tobacco to the incidence of cancer and coronary disease; this has already resulted in regulation and litigation, and a study of discourse about this industry is overdue.

In short, what one can learn from study of the petroleum industry and public discourse raises many questions, most particularly of interest to business historians, scholars of policy and regulation, policy makers and regulators, business and industry. Though we only generalize here on the basis of the oil industry, if the past remains a guide to the future, unless the role of public discourse is addressed and assessed and old ideas are reexamined, it will be hard to avoid misperceptions and misfires in future public policy relating to petroleum. It is time to reconstruct the cultural construction of oil.

# Notes

PREFACE

1. Daniel Yergin, *The Prize: The Epic Quest for Oil, Money, and Power* (New York: Simon & Schuster, 1991), 39, 41, 44, 54; Ron Chernow, *Titan: The Life of John D. Rockefeller, Sr.* (New York: Random House, 1998), 254, 259, 257.

2. Gerald D. Nash, *United States Oil Policy, 1890–1964* (1968; reprint, Westport, Conn.: Greenwood Press, 1976), 249, 111, 155, 201, 207.

3. John G. Clark, *Energy and the Federal Government: Fossil Fuel Policies, 1900–1946* (Urbana: University of Illinois Press, 1987), 381–84.

4. Richard H. K. Vietor, *Energy Policy in America since 1945* (Cambridge: Cambridge University Press, 1984), 7.

5. Thomas K. McCraw, *Prophets of Regulation* (Cambridge: Harvard University Press, Belknap Press, 1984), viii, 138–42.

6. We have found useful suggestions and some direction in the works of scholars in a number of fields. See, for example, Christopher Norris, *Deconstruction: Theory and Practice*, rev. ed. (London: Routledge, 1991); Barbara Herrnstein Smith, *Contingencies of Value: Alternative Perspectives for Critical Theory* (Cambridge: Harvard University Press, 1988); Clifford Geertz, *Local Knowledge: Further Essays in Interpretative Anthropology* (New York: Basic Books, 1983); Mary Douglas, *Purity and Danger: An Analysis of the Concepts of Pollution and Taboo* (London: Ark Publications, 1966); Carroll Smith-Rosenberg, "Domesticating Virtue: Coquettes and Revolutionaries in Young America," in *Literature and the Body: Essays on Population and Persons*, ed. Elaine Scarry (Baltimore: Johns Hopkins University Press, 1988).

7. These kinds of conflicts and contradictions in discourse have been observed in a different context, that of early American national consciousness, in Carroll Smith-Rosenberg, "Dis-Covering the Subject of the 'Great Constitutional Discussion,' 1786–1789," *Journal of American History* (December 1992): 842.

1. Stephen Foster, *Their Solitary Way: The Puritan Social Ethic in the First Century of Settlement in New England* (New Haven: Yale University Press, 1971), 99–129; J. E. Crowley, *This Sheba, Self: The Conceptualization of Economic Life in Eighteenth-Century America* (Baltimore: Johns Hopkins University Press, 1974), 56–61, 76–97; David E. Shi, *The Simple Life: Plain Living and High Thinking in American Culture* (New York: Oxford University Press, 1985), 29–32; Benjamin Franklin, "Positions to Be Examined, Concerned National Wealth," in *The Writings of Benjamin Franklin*, ed. Albert Henry Smyth (New York: Macmillan, 1905–7), 5:202.

2. For the classic biographical study of the production of the legend of Jay Gould, see Maury Klein, *The Life and Legend of Jay Gould* (Baltimore: Johns Hopkins University Press, 1986).

3. J. G. A. Pocock, *The Machiavellian Moment: Florentine Political Thought and the Atlantic Republican Tradition* (Princeton: Princeton University Press, 1975), 465; J. G. A. Pocock, *Virtue, Commerce, and History: Essays on Political Thought and History, Chiefly in the Eighteenth Century* (Cambridge: Cambridge University Press, 1988), 114; Carroll Smith-Rosenberg, "Domesticating Virtue: Coquettes and Revolutionaries in Young America," in *Literature and the Body: Essays on Population and Persons*, ed. Elaine Scarry (Baltimore: Johns Hopkins University Press, 1988); Toby L. Ditz, "Shipwrecked: Imperiled Masculinity and the Representation of Business Failures among Philadelphia's Eighteenth-Century Merchants," *Journal of American History* (June 1994): 51–80; Crowley, *This Sheba, Self*, 60–61, 76–97; Shi, *Simple Life*, 17–20, 25, 29–32, 51–53.

4. Pocock, *Machiavellian Moment*, 446–47, 464; Pocock, *Virtue, Commerce, and History*, 107–11, 120–30; Lance Banning, *The Jeffersonian Persuasion: Evolution of a Party Ideology* (Ithaca, N.Y.: Cornell University Press, 1978), 68–69, 56–57; Bernard Bailyn, *The Ideological Origins of the American Revolution* (Cambridge: Harvard University Press, Belknap Press, 1967), 48.

5. Bernard Bailyn, "The Cultural Themes of the American Revolution: An Interpretation," in *Essays on the American Revolution*, ed. Stephen G. Kurtz and James H. Hutson (Chapel Hill: University of North Carolina Press, 1973), 26–27; Bailyn, *Ideological Origins*, 56; Banning, *Jeffersonian Persuasion*, 58–61, 67–68, 74–75.

6. Pocock, *Machiavellian Moment*, 468, 546; Gordon S. Wood, *The Creation of the American Republic* (Chapel Hill: University of North Carolina Press, 1969), 14; Gordon S. Wood, "Conspiracy and the Paranoid Style: Causality and Deceit in the Eighteenth Century," *William and Mary Quarterly* (July 1982): 401–41; Bailyn, *Ideological Origins*, 19, 43, 291; Banning, *Jeffersonian Persuasion*, 75, 92; Drew R. McCoy, *The Elusive Republic: Political Economy in Jeffersonian America* (New York: W. W. Norton, 1980), 42, 49–60, 64–65, 97–99, 102; Leo Marx, *The Machine in the Garden: Technology and the Pastoral Ideal in America* (New York: Oxford University Press, 1964), 118–35; Smith-Rosenberg, "Domesticating Virtue," 163–64; Smith-Rosenberg, "Dis-Covering the Subject," 850.

7. Lawrence Frederick Kohl, *The Politics of Individualism: Parties and the American Character in the Jacksonian Era* (New York: Oxford University Press, 1989), 16, 22, 31.

8. John L. Thomas, "Antislavery and Hope," in *The Antislavery Vanguard: New Essays on the Abolitionists*, ed. Martin Duberman (Princeton: Princeton University Press, 1969), 240–47.

9. T. S. Arthur, *Ten Nights in a Bar-Room* (1853; Philadelphia: John E. Potter and Co., 1860), 14.

10. Quoted in W. P. and F. J. Garrison, *William Lloyd Garrison, 1805–1879* (1885–89; reprint, New York: Arno Press, 1969), 4:21.

11. Rondo Cameron, *A Concise Economic History of the World: From Paleolithic Times to the Present* (New York: Oxford University Press, 1989), 226.

12. Charles Spahr, *An Essay on the Present Distribution of Wealth in the United States* (New York: T. Y. Crowell & Co., 1896), 69.

13. Alfred D. Chandler Jr., *The Visible Hand: The Managerial Revolution in American Business* (Cambridge: Harvard University Press, 1977), 87–88.

14. George Rogers Taylor, *The Transportation Revolution, 1815–1860* (New York: Harper Torchbooks, 1968); *The Statistical History of the United States*, (Stamford, Conn.: Fairfield Publishers, 1965), 427.

15. Chandler, *Visible Hand*, 249–51.

16. Ibid., 254–56.

17. Richard S. Tedlow, *New and Improved: The Story of Mass Marketing in America* (New York: Basic Books, 1990), 5.

18. Ibid., 373; Alfred D. Chandler Jr., *Scale and Scope: The Dynamics of Industrial Capitalism* (Cambridge: Harvard University Press, 1990), 35–36.

19. U.S. Bureau of the Census, "Agriculture, General Report and Analytical Tables," in *Fourteenth Census of the United States* (1920) (Washington, D.C.: U.S. Government Printing Office, 1922), 5:38–43.

20. U.S. Commissioner of Labor, *Thirteenth Annual Report: Hand and Machine Labor* (Washington, D.C.: U.S. Government Printing Office, 1899), 1:6.

21. Robert A. McGuire, "Economic Causes of Late Nineteenth Century Agrarian Unrest," *Journal of Economic History* (December 1981): 17.

22. George H. Miller, *Railroads and the Granger Laws* (Madison: University of Wisconsin Press, 1971).

23. D. C. Cloud, *Monopolies and the People* (Davenport, Iowa: Day, Egbert, and Fidlar, 1873), 13, 259, 275.

24. Lee Benson, *Merchants, Farmers, and Railroads: Railroad Regulation and New York Politics, 1850–1877* (Cambridge: Harvard University Press, 1957), 29, 68; "Pennsylvania Railroad Scheme," *New York Times*, September 8, 1875; "Freight Problems," *New York Times*, December 9, 1875.

25. Ari and Olive Hoogenboom, *A History of the ICC — From Panacea to Palliative* (New York: W. W. Norton, 1976), 3.

26. "The Railway Monopoly," *New York Times*, February 25, 1878.

27. Ibid.

28. Charles W. McCurdy, "American Law and the Marketing Structure of the Large Corporation, 1875–1890," *Journal of Economic History* (September 1978): 641.

29. Mary Yeager, *Competition and Regulation: The Development of Oligopoly in the Meat Packing Industry* (Greenwich, Conn.: JAI Press, 1981), 31, 90, 97.

30. J. D. Forrest, "Anti-Monopoly Legislation in the United States," *American Journal of Sociology* (January 1896): 413, 418, 419.

31. John R. Allison, "Survey of the Texas Anti-Trust Laws," *Antitrust Bulletin* (Summer 1975): 217–19, 222, 237, 263, 308.

32. Ibid., 317–18.

33. Chandler, *Visible Hand*, 168, and passim, 145–87.

34. Cameron, *Concise Economic History of the World*, 14.

35. George Ellington [pseud.], *The Women of New York, or The Underworld of the Great City* (New York: New York Book Co., 1869), 572, 59, 210.

36. Quoted by Jack Blicksilver, "Defenders and Defense of Big Business in the United States, 1880–1900" (Ph.D. diss., Northwestern University, 1955), 114.

37. Washington Gladden, "The Problem of Poverty," *Century* (1892–93): 247; Carroll Smith-Rosenberg, *Religion and the Rise of the American City: The New York City Mission Movement, 1820–1870* (Ithaca, N.Y.: Cornell University Press, 1971), 156.

38. *Independent* (May 1, 1902), 1080.

39. For an overview of urban growth and reactions to urban machines, see Charles N. Glaab and A. Theodore Brown, *A History of Urban America*, 2d ed. (New York: Macmillan, 1976), esp. chap. 10: "Bosses and Reformers."

40. See Mark Wahlgren Summers, *The Era of Good Stealings* (New York: Oxford University Press, 1993), chaps. 1 and 6.

41. Summers, *Era of Good Stealings*, 305, 306.

42. Quoted by Paul Lancaster, *Gentleman of the Press: The Life and Times of an Early Reporter, Julian Ralph of the Sun* (Syracuse, N.Y.: Syracuse University Press, 1992), 241, 99.

43. W. A. Swanberg, *Citizen Hearst: A Biography of William Randolph Hearst* (New York: Charles Scribner's Sons, 1961), 81.

44. See Menahem Blondheim, *News over the Wires: The Telegraph and the Flow of Public Information in America, 1844–1897* (Cambridge: Harvard University Press, 1994).

45. Thomas Bender, *New York Intellect: A History of Intellectual Life in New York City, from 1750 to the Beginnings of Our Own Time* (Baltimore: Johns Hopkins University Press, 1987), xv.

46. David M. Chalmers, *The Social and Political Ideas of the Muckrakers* (New York: Citadel Press, 1964), 11, 14, 105.

47. Harry H. Stein and John M. Harrison, "Muckraking Journalism in Twentieth-Century America," in *Muckraking: Past, Present, and Future*, ed. Harrison and Stein (University Park: Pennsylvania State University Press, 1973), 19; Louis Filler, "Muckrakers and Middle America," in Harrison and Stein, *Muckraking*, 25–40.

48. Stanley K. Schultz, "The Morality of Politics: The Muckrakers' Vision of Democracy," *Journal of American History* (December 1969): 530.

49. Thomas L. Haskell, *The Emergence of Professional Social Science: The American Social Science Association and the Nineteenth-Century Crisis of Authority* (Urbana: University of Illinois Press, 1977), vi, vii.

50. Ibid., 12–14.

51. Benjamin G. Rader, *The Academic Mind and Reform* (Lexington: University of Kentucky Press, 1966), 21, 65, 108.

52. E. Benjamin Andrews, "The Combination of Capital," *International Journal of Ethics* (April 1893): 334.

53. Charles F. Beach, "Facts about Trusts," *Forum* (May 1889): 23; Ernest von Halle, *Trusts, or Industrial Combinations and Coalitions in the United States* (New York: Macmillan, 1899), 64.

54. Henry C. Adams, *Relation of the State to Industrial Action* (New York: American Economic Association, 1887), 38, 43.

55. David B. Danbom, *"The World of Hope": Progressives and the Struggle for an Ethical Public Life* (Philadelphia: Temple University Press, 1987), 81.

56. Haskell, *Emergence of Professional Social Science*, vi.

57. Adams, *Relation of the State to Industrial Action*, 43, 44, 47, 65.

58. Quoted by Sidney Fine, "Richard T. Ely, Forerunner of Progressivism, 1880–1901," *Mississippi Valley Historical Review* (March 1951): 620.

59. Richard T. Ely, "The Nature and Significance of Monopolies and Trusts," *International Journal of Ethics* (1900): 273–74.

60. *New York Times*, February 2, 1887.

61. *New York World*, September 2, 1896.

62. Thomas Elmer Will, "A Menace to Freedom: The College Trust," *Arena* (September 1901): 246–48. In this regard, Will was, perhaps, representative of the mass of college and university professors who continued to defend small-producer perspectives, reflecting their fear of disorder and revolution and their unhappiness with tensions within their institutions; the shift in emphasis from humanities and social sciences to applied studies and of institutional power from professors to trustees and, hence, to administrators made the professors natural allies of the rural and small-town resisters of some aspects of the new industrial age. Walter P. Metzger, "College Professors and Big Businessmen: A Study of American Ideologies, 1880–1915" (Ph.D. diss., University of Iowa, 1950).

63. P. K. Edwards, *Strikes in the United States, 1881–1974* (New York: St. Martin's Press, 1981), 264; Dorothea Sneider, "The New York Cigarmakers' Strike of 1877," *Labor History* 26, no. 3 (Summer 1985): 13–27.

64. David Montgomery, *The Fall of the House of Labor: The Workplace, the State, and American Labor Activism, 1865–1925* (Cambridge: Cambridge University Press, 1987), 47.

65. Dee Brown, *The Year of the Century: 1876* (New York: Charles Scribner's Sons, 1966), 129.

66. See, for example, John S. Gilkenson Jr., *Middle-Class Providence, 1820–1940* (Princeton: Princeton University Press), 52–55.

67. Quoted in Stuart M. Blumin, *The Emergence of the Middle Class: Social Experience in the American City, 1760–1900* (Cambridge: Cambridge University Press, 1989), 242.

68. On antebellum and postbellum worker ideology, see Edward Magdol, *The Antislavery Rank and File: A Social Profile of the Abolitionists' Constituency* (Westport, Conn.: Greenwood Press, 1986), and Sean Wilentz, "The Rise of the American Working Class, 1776–1877: A Survey," and Leon Fink, "Looking Backward: Reflections on Workers' Culture and Certain Conceptual Dilemmas within Labor History," in *Perspectives on American Labor History: The Problems of Synthesis*, ed. J. Carroll Moody and Alice Kessler-Harris (DeKalb: Northern Illinois University Press, 1990). It was noteworthy that the first nationwide strike was that of the railroad workers in 1877. With this event, the railroads were inextricably associated with blue-collar discontent and violence for large segments of the American population.

69. John G. Sproat, *"The Best Men": Liberal Reformers in the Gilded Age* (New York: Oxford University Press, 1968), 3, 148.

70. George Davis Herron, *The Christian Society* (1894; reprint, New York: Johnson Reprint Corp., 1969), 16, 83, 102.

71. Gladden, "Problem of Poverty," 247, 253, 256.

72. C. T. Russell, *Studies in the Scriptures: Series IV, The Battle of Armageddon* (Brooklyn, N.Y.: International Bible Students Association, 1897), 274–78, 359, 413.

73. Michael S. Kimmel, "The Contemporary 'Crisis' of Masculinity in Historical Perspective," in *The Making of Masculinities: The New Men's Studies*, ed. Harry Brod (Boston: Allen & Unwin, 1987), 138.

74. Tom Lutz, *American Nervousness, 1903: An Anecdotal History* (Ithaca, N.Y.: Cornell University Press, 1991), 4, 11.

75. E. Anthony Rotundo, *American Manhood: Transformation in Masculinity from the*

*Revolution to the Modern Era* (New York: Basic Books, 1993), 177. See also Kevin White, *The First Sexual Revolution: The Emergence of Male Heterosexuality in Modern America* (New York: New York University Press, 1993).

76. Gladden, "Problem of Poverty," 256.

77. Quoted by Blicksilver, "Defenders and Defense of Big Business," 122.

78. Speech on August 12, 1908, quoted in Nick Salvatore, *Eugene V. Debs: Citizen and Socialist* (Urbana: University of Illinois Press, 1982), 228.


CHAPTER TWO

1. Harold F. Williamson and Arnold R. Daum, with Ralph L. Andreano, Gilbert C. Klose, and Paul A. Weinstein, *The American Petroleum Industry: The Age of Illumination, 1859–1900* (Evanston: Northwestern University Press, 1959), 57, 88–89. The authors make the point that farmers seldom joined the rush to become oil producers.

2. Charles Burr Todd, "In the Oil Region," *Lippincott's Magazine* 34 (December 1884): 558; Professor Owen, "On Petroleum and Oil Wells," *Fraser's Magazine* 92 (October 1875): 444.

3. Thomas A. Gale, *The Wonder of the Nineteenth Century! Rock Oil in Pennsylvania and Elsewhere* (Erie, Pa.: Sloan and Griffeth, 1860); B. Franklin [pseud.], "The Petroleum Regions of America," *Harper's New Monthly Magazine* 30, no. 179 (April 1865): 562; "Prospectuses and Pamphlets of Various Petroleum Companies in New York and Philadelphia, 1865," *National Quarterly Review* (March 1865): 342; "Petroleum and the Oil Fields," *Every Saturday* (April 7, 1866); see Gerald T. White, "The Case of the Salted Sample: A California Oil Industry Skeleton," *Pacific Historical Review* (May 1966): 153–84.

4. Franklin, "Petroleum Regions of America," 569.

5. J. S. Schooley, "After Petroleum," *Harper's New Monthly Magazine* 30, no. 175 (December 1864): 58–59.

6. Ibid., 54.

7. Quoted by Barbara Welter, *Dimity Convictions: The American Woman in the Nineteenth Century* (Athens: Ohio University Press, 1976), 31.

8. Franklin, "Petroleum Regions of America," 565; Schooley, "After Petroleum," 63.

9. "Pa Has Struck Ile," quoted in *The Great Oildorado: The Gaudy and Turbulent Years of the First Oil Rush, Pennsylvania, 1859–1880*, by Hildegard Dolson (New York: Random House, 1959), 48–49.

10. Paul H. Giddens, *The Birth of the Oil Industry* (New York: Macmillan, 1938), 115–18. Quoted in Paul H. Giddens, *Early Days of Oil: A Pictorial History of the Beginnings of the Industry in Pennsylvania* (Princeton: Princeton University Press, 1948), 32.

11. Giddens, *Early Days of Oil*, 33.

12. "A Carpet Bagger in Pennsylvania," *Atlantic Monthly* (June 1869): 734.

13. Quoted in John G. Sproat, *"The Best Men": Liberal Reformers in the Gilded Age* (New York: Oxford University Press, 1968), 149.

14. William Wright, *The Oil Regions of Pennsylvania: Showing Where Petroleum Is Found: How It Is Obtained and at What Cost with Hints for Whom It May Concern* (New York: Harper & Brothers, 1865), 4.

15. Ibid., 38, 35.

16. Ibid., 27, 35, 38, 44, 45, 93–96.

17. Giddens, *Birth of the Oil Industry*, 85–86.

18. Wright, *Oil Regions of Pennsylvania*, 4, 215.

19. Ibid., 212–15, 201–10, 218, 227.

20. Dolson, *Great Oildorado*, 187–88.

21. Samuel Morris, *Derrick and Drill* (New York: J. Miller, 1865), 260.

22. E. I. Sears, "Prospectuses and Pamphlets of Various Petroleum Companies, New York and Philadelphia, 1865," *National Quarterly Review* (1865): 336, 342.

23. Wright, *Oil Regions of Pennsylvania*, 4.

24. Prospectus for the Pennsylvania Imperial Oil Company, Philadelphia, 1864. Hagley Museum and Library, Wilmington, Delaware.

25. Ibid. The Du Pont investment, however, was not finally made until 1871. Handwritten comments are appended to the brochure.

26. Williamson et al., *American Petroleum Industry: The Age of Illumination*, 93–97, 157; Schooley, "After Petroleum," 54; Franklin, "Petroleum Regions of America," 876; Wright, *Oil Regions of Pennsylvania*, 78; *The Derrick's Handbook of the Petroleum Industry*, vol. 2 (Oil City, Pa.: Derrick Publishing Co., 1900), 513.

27. Williamson et al., *American Petroleum Industry: The Age of Illumination*, 100, 146–47.

28. Ibid., 112.

29. Ibid., 40, 220; for descriptions of all-too-frequent spills, see *The Derrick's Handbook of Petroleum: A Complete Chronological and Statistical Review of Petroleum Developments from 1859 to 1898* (Oil City, Pa.: Derrick Publishing Co., 1898), 18–121, passim.

30. Williamson et al., *American Petroleum Industry: The Age of Illumination*, 228–30; Giddens, *Birth of the Petroleum Industry*, 92–94.

31. Williamson et al., *American Petroleum Industry: The Age of Illumination*, 118, 373, 378, 566, 594, 631.

32. J. H. Connelly, "An Oil Speculator's Mishaps," *American Magazine* 8 (1894): 47.

33. Giddens, *Birth of the Oil Industry*, 84, 184–87.

34. Ibid.

35. *Oil City Derrick*, January 5, 10, July 26, 1872.

36. Giddens, *Birth of the Oil Industry*, 122.

37. Williamson et al., *American Petroleum Industry: The Age of Illumination*, 184–87, 288.

38. Ibid., 83–86, 106–7; *Derrick's Handbook*, 1:42–45, 104.

39. *Derrick's Handbook*, 1:42–45.

40. Williamson et al., *American Petroleum Industry: The Age of Illumination*, 289–91.

41. *Oil City Derrick*, January 20, February 28, March 1, 2, 9, 12, May 17, 25, 1872.

42. B. Franklin [pseud.], "Petroleum Regions of America," *Harper's New Monthly Magazine* (April 1865): 571; J. T. Henry, *History and Romance of the Oil Industry* (Philadelphia: James B. Rogers, 1873), 107–10; Giddens, *Birth of the Oil Industry*, 93–95.

43. Giddens, *Birth of the Oil Industry*, 98. *Derrick's Handbook*, 63.

44. Giddens, *Birth of the Oil Industry*, 91–130, passim, 150.

45. See the *Titusville Herald*, January 20, February 24, 28, 1872.

46. Thomas J. Schlereth, *Victorian America: Transformation in Everyday Life, 1876–1915* (New York: Harper Collins, 1991), 143; Luc Sante, *Low Life: Lures and Snares of Old New York* (New York: Vintage, 1992), 64.

47. Charles McArthur Destler, *Roger Sherman and the Independent Oil Men* (Ithaca, N.Y.: Cornell University Press, 1967), 20.

48. Ibid., 35–39.

49. Ibid., 58–66.

50. See, for example, the *Titusville Herald*, February 25, 1868.

51. *Derrick's Handbook*, 113, 127.

52. *Titusville Herald*, February 25, 1868; January 23, 30, 1872.

53. Destler, *Roger Sherman*, 20–21.

54. Ralph W. Hidy and Muriel E. Hidy, *Pioneering in Big Business: History of the Standard Oil Company (New Jersey), 1882–1911* (New York: Harper & Brothers, 1955), 24–40.

55. Ibid.

56. Destler, *Roger Sherman*, 21.

57. Sidney Walter Martin, *Florida's Flagler* (Athens: Ohio University Press, 1949), 51.

58. "Rockefeller, Andrews, and Flagler," Roy G. Dun Manuscripts, New York, vol. 236, 1020A, Baker Library, Harvard Business School.

59. Ibid.

60. Ibid.

61. Allan Nevins, *Study in Power: John D. Rockefeller, Industrialist and Philanthropist* (New York: Charles Scribner's Sons, 1953), 1:86.

62. Ibid.; Hidy and Hidy, *Pioneering in Big Business*, 14; John Davidson Rockefeller, *Random Reminiscences of Men and Events* (New York: Doubleday, Page & Co., 1909), 58.

63. Martin, *Florida's Flagler*, 60.

64. Nevins, *Study in Power*, 1:132–36; Rockefeller, *Random Reminiscences*, 19, 60.

65. Nevins, *Study in Power*, 1:136; Rockefeller, *Random Reminiscences*, 60, 83.

66. Richard S. Tedlow, *New and Improved: The Story of Mass Marketing in America* (New York: Basic Books, 1990), 364.

67. See the *Cleveland Leader*, January 25, 1866, March 5 and December 3, 1867, and January 19, 1870, for examples.

68. *Cleveland Herald*, March 2, 1872; *Cleveland Leader*, April 11, 1872.

69. Hidy and Hidy, *Pioneering in Big Business*, 12.

70. David Freeman Hawke, *John D: The Founding Father of the Rockefellers* (New York: Harper & Row, 1980), 71–72; Nevins, *Study in Power*, 1:103–4.

71. *Titusville Morning Herald*, January 21, 22, 24, 1872; Destler, *Roger Sherman*, 35.

72. *Petroleum Centre Record*, February 20, 1872; *Oil City Register*, February 21, 1872; Nevins, *Study in Power*, 1:113.

73. *Titusville Morning Herald*, March 1, 1872.

74. *Oil City Derrick*, March 2, 1872.

75. *Derrick's Handbook*, 172.

76. *Oil City Derrick*, February 27, March 29, 1872.

77. See, for example, the *New York Herald*, March 19, 1872, and the *New York Tribune*, March 8, 1872.

78. *Titusville Morning Herald*, February 24, March 2, 4, 8, 9, 19, 26, 1872.

79. *Titusville Morning Herald*, March 12, April 4, 1872.

80. *New York Daily Tribune*, March 7, 9, 18, 1872.

81. *New York Daily Tribune*, March 18, 1872. See, for example, the *New York Daily Tribune*, March 12, 13, 18, 1872; *New York Times*, March 1, 12, 26, 1872; *New York World*, March 13, 1872.

82. *New York World*, March 18, 20, April 1, 3, 5, 6, July 26, 1872.

83. *Titusville Morning Herald*, March 5, 13, 19, 20, 1872.

84. See, for example, E. C. Bishop, comp., *The History of the Rise and Fall of the South Improvement Company* (Titusville, Pa., Petroleum Producers' Association, 1872); *Printed Testimony* (1879), and [Roger Sherman], *History of the General Council* (Ti-

tusville, Pa.: Petroleum Producers' Association, 1880); *Titusville Morning Herald*, March 18, 1872.

85. *Oil City Derrick*, May 9, 1872.

86. Harold M. Helfman, "Twenty-nine Hectic Days: Public Opinion and the War of 1872," *Pennsylvania History* (April 1950): 128.

87. *Titusville Morning Herald*, June 20, July 10, 1872.

88. *Oil City Derrick*, April 2, 1872; *New York Bulletin*, April 15, 1872.

89. Maury Klein, "Competition and Regulation: The Railroad Model," *Business History Review* (Summer 1990): 322, 314, 316.

90. Quoted in *Derrick's Handbook*, 172.

91. Ibid., 217–18.

92. Herbert G. Gutman, "La politique ouvrière de la grande enterprise américaine de 'l'age clinquant,'" *Le Mouvement Social* (January–March 1978): 77–81, passim.

93. Ibid., 87–88.

94. Ibid., 96.

95. Nevins, *Study in Power*, 1:171.

96. Ibid., 1:163–65.

97. Ibid.

98. Ibid., 1:169–70.

99. Quoted by Nevins, *Study in Power*, 1:171.

100. Ibid.

101. Ibid., 1:229–46, passim.

102. Ibid., 1:248–58, passim.

103. Ibid., 1:186, 249.

104. Joseph D. Potts, *A Brief History of the Standard Oil Company* (Oil City, Pa.: General Council of the Petroleum Producers' Union, 1878).

105. Destler, *Roger Sherman*, 45–50.

106. *New York Sun*, November 13, 23, 1878.

107. Destler, *Roger Sherman*, 125–26.

108. Ibid., 127.

109. Ibid., 126, 134, 142–45, 182.

110. Ibid., 155, 169, 170, 182.

111. Rockefeller, *Random Reminiscences*, 60.

112. Quoted by Nevins, *Study in Power*, 1:115.

CHAPTER THREE

1. Allan Nevins, *Study in Power: John D. Rockefeller, Industrialist and Philanthropist* (New York: Charles Scribner's Sons, 1953), 1:371.

2. Ibid., 1:350; Ralph W. Hidy and Muriel E. Hidy, *Pioneering in Big Business: History of the Standard Oil Company (New Jersey), 1882–1911* (New York: Harper & Brothers, 1955), 203.

3. Nevins, *Study in Power*, 1:379. The Standard Oil–Tidewater contracts were reproduced in Ida M. Tarbell, *The History of the Standard Oil Company* (New York: Macmillan, 1904), 2:300–308. For a full and reliable account of Standard Oil's pipeline construction, the resulting uproar, and the effects on public policy, see Arthur Menzies Johnson, *The Development of American Petroleum Pipelines: A Study in Private Enterprise and Public Policy, 1862–1906* (Ithaca, N.Y.: Cornell University Press, 1956), chaps. 3, 4, and 5.

4. For the general background of New York shipper and producer activities, Sterne activity, and the Cheap Transportation Association, see Lee Benson, *Merchants, Farmers, and Railroads: Railroad Regulation and New York Politics, 1850–1887* (Cambridge: Harvard University Press, 1955).

5. John Foord, *The Life and Public Services of Simon Sterne* (London: Macmillan, 1903), 18–21; Simon Sterne, "Railway Management and New York Prosperity," *Nation* (May 9, 1878): 302–4. In keeping with the convention of the *Nation*, the article was unsigned; Sterne was identified as the author in the index to the volume.

6. New York (State) Legislature, Assembly, Special Committee on Railroads, *Proceedings of the Special Committee on Railroads Appointed under a Resolution of the Assembly to Investigate Alleged Abuses in the Management of Railroads Chartered by the State of New York* (New York: Evening Post Steam Presses, 1879–80), 8 (hereafter cited as "Hepburn, *Proceedings*"). The volumes include a summary report, printed at the beginning of the initial volume, proceedings, and an appendix. The report is subsequently identified as such and "proceedings" designates the actual printed transcript of the committee's meetings.

7. Hepburn, *Proceedings*, 1.

8. Hepburn, *Report*, 41.

9. Hepburn, *Proceedings*, 4.

10. Ibid., 83.

11. Ibid., 135–39, 156–60, 165, 183.

12. Ibid., 256–57, 289.

13. Ibid., 2656, 3944, 3958.

14. Ibid., 709, 724. At the time of his appearance, Lombard had a civil suit pending against Standard Oil based on the same allegation. He allied his company with Tidewater later.

15. Ibid., 799–801.

16. Ibid., 2526–29, 2361, 2623–33.

17. Ibid., 3424, 3420–21, 2601–2, 3553, 3555.

18. Ibid., 1595.

19. Ibid., 1669.

20. Ibid.

21. Foord, *Simon Sterne*, 21–21; Hepburn, *Report*, 5, 48.

22. The remedies contained in the appendix to the *Proceedings*, pp. 7–8, are the most direct statement of this position.

23. Hepburn, *Report*, 51–52.

24. Ibid.

25. Hepburn, *Proceedings*, 3965, 3968.

26. Benson, *Merchants, Farmers, and Railroads*, 124, 135.

27. Ibid., 134.

28. *New York Herald*, June 20, October 14, 18, 1879.

29. Quoted in Benson, *Merchants, Farmers, and Railroads*, 140.

30. Editor's preface, "Story of a Great Monopoly," by Henry Demarest Lloyd, *Atlantic Monthly* (March 1881): 316.

31. Lloyd, "Story of a Great Monopoly," 317, 318, 319, 320.

32. Ibid., 321, 324–25.

33. Ibid., 326–27, 330.

34. Ibid., 330.

35. Ibid., 322, 329, 333, 334.

36. Chester McArthur Destler, *Henry Demarest Lloyd and the Empire of Reform* (Philadelphia: University of Pennsylvania Press, 1963), 132.

37. John L. Thomas, *Alternative America: Henry George, Edward Bellamy, Henry Demarest Lloyd, and the Adversary Tradition* (Cambridge: Harvard University Press, Belknap Press, 1983), 132.

38. Ibid., 121.

39. Steven L. Piott, *The Anti-Monopoly Persuasion: Popular Resistance to the Rise of Big Business in the Midwest* (Westport, Conn.: Greenwood Press, 1985), 106.

40. Lloyd, "Story of a Great Monopoly," 322, 333, 329.

41. Hidy and Hidy, *Pioneering in Big Business*, 204, 655.

42. Quoted in Festus P. Summers, *John Newlon Camden: A Study in Individualism* (New York: G. P. Putnam's Sons, 1937), 225.

43. Quoted in ibid., 183.

44. Quoted in ibid., 200.

45. Ibid., 199; J. T. Henry, *The Early and Later History of Petroleum* (Philadelphia: Jas. B. Rogers Co., 1873), 319. Henry was the editor of the *Titusville Courier*, a Democratic newspaper owned, in part, by Roger Sherman. See Chester McArthur Destler, *Roger Sherman and the Independent Oil Men* (Ithaca, N.Y.: Cornell University Press, 1967), 20.

46. See Tarbell, *History of the Standard Oil Company*, 1:198–99. Joseph D. Potts, *A Brief History of the Standard Oil Company* (Oil City, Pa.: General Council of the Petroleum Producers' Union, 1878).

47. "Marietta Oil and Refining," Roy G. Dun, Reports, Ohio, vol. 193, 210.

48. Nevins, *Study in Power*, 1:220.

49. Ibid., 220–27; O. D. Donnell, "The Petroleum Industry in Ohio," *Northwest Ohio Quarterly* (October 1947): 189; Summers, *Johnson Newlon Camden*, 181–83.

50. *New York Times*, March 1, 1905; Roy G. Dun, Reports, Ohio, vol. 193, p. 538 7/B, Baker Library, Harvard Business School; Donnell, "Petroleum Industry in Ohio," 189.

51. "George Rice," Roy G. Dun, Reports, Ohio, vol. 193, 716.

52. Hidy and Hidy, *Pioneering in Big Business*, 203; Nevins, *Study in Power*, 2:77; "George Rice," 716.

53. Nevins, *Study in Power*, 2:31.

54. On the general question of price-cutting by Standard Oil, see John S. McGee, "Predatory Price Cutting: The Standard Oil (N.J.) Case," *Journal of Law and Economics* (October 1958): 137–69.

55. Hidy and Hidy, *Pioneering in Big Business*, 203; "George Rice," 716; Nevins, *Study in Power*, 2:635; C. Joseph Pusateri, "The 'Turn Another Screw' Affair: Oil and Railroads in the 1880's," *Register of the Kentucky Historical Society* (1975): 346–55.

56. For the 1870s background, see Gerald D. Nash, "Origins of the Interstate Commerce Act of 1887," *Pennsylvania History* (1957): 181–90. The senators were quoted by Ari and Olive Hoogenboom, in *A History of the ICC: From Panacea to Palliative* (New York: W. W. Norton, 1976), 13–15.

57. See, for example, January 18, 25, 1882.

58. George Rice, *Railway Discrimination As Given to the Standard Oil Trust* (Marietta, Ohio: By the author, 1888), 13, 24–25; Tony Freyer, *Regulating Big Business: Antitrust in Great Britain and America, 1880–1990* (Cambridge: Cambridge University Press, 1992), 101.

59. For Lloyd, see Rice's letters of February 18 and March 10, Henry Demarest Lloyd

Papers, Wisconsin State Historical Society, Madison. More than one hundred letters followed through 1904.

60. Interstate Commerce Commission, *Second Annual Report* (December 1, 1888) (Washington, D.C.: U.S. Government Printing Office, 1888), 119.

61. Ibid., 133. See also *Rice, Robinson and Winthrop v. The Western New York and Pennsylvania Railroad Company*, Interstate Commerce Commission, *Fourth Annual Report* (December 1, 1890) (Washington, D.C.: U.S. Government Printing Office, 1890), 99–100, and *The Independent Refiners' Association of Titusville, Pa., et al., v. The Western New York and Pennsylvania Railroad Company et al.*, Sixth Annual Report (December 1, 1892) (Washington, D.C.: U.S. Government Printing Office, 1892), 102.

62. See the *Rice v. Cincinnati, Washington and Baltimore Railroad*, Case #185. Interstate Commerce Commission, *Third Annual Report* (December 1, 1889) (Washington, D.C.: U.S. Government Printing Office, 1889), 173.

63. *New York World*, November 23, 1887.

64. *New York World*, February 24, 1888. The *New York Times* had reported favorably on Rice's challenges to Standard Oil even earlier, on December 11, 1885.

65. *New York World*, November 28, 29, 1887. The fullest accounts of the legal proceedings were published by the *New York Times*, February 2, May 3, 6, 7, 8, 10, 11, 12, 13, 14, 15, 1887.

66. *New York World*, May 8, 9, 1888.

67. Hidy and Hidy, *Pioneering in Big Business*, 204; Nevins, *Study in Power*, 2:127, 141, 224, 237, 238, 334, 341–42.

68. Roy G. Dun, Reports, Ohio, vol. 194, 443; on Rice's later career, see obituaries in the *New York Times*, March 1, 1905, and the *New York World*, March 2, 1905.

69. J. N. Camden, "The Standard Oil Company," *North American Review* (February 1883): 183–85.

70. Ibid., 189–90.

71. John C. Welch, "The Standard Oil Company," *North American Review* (February 1883): 191, 192, 193–94, 194–99, 200.

72. *New York Times*, April 5, 1882, March 17, 1884.

73. J. F. Hudson, *The Railways and the Republic* (New York: Harper & Brothers, 1886), 84, 101.

74. Ibid., 80–82, 84, 90, 91–92, 94–94.

75. Ibid., 477.

76. Richard T. Ely, "Social Studies: The Nature of the Railway Problem," *Harper's New Monthly Magazine* (July 1886): 254, 257.

77. Jack Blicksliver, "Defenders and Defense of Big Business in the United States, 1880–1900" (Ph.D. diss., Northwestern University, 1955), 118; Marc Allen Eisner, *Antitrust and the Triumph of Economics: Institutions, Expertise, and Policy Change* (Chapel Hill: University of North Carolina Press, 1991), 48; Thomas R. McCraw, "Rethinking the Trust Question," in *Regulation in Perspective: Historical Essays*, ed. McCraw (Boston: Division of Research, Graduate School of Business Administration, Harvard University, 1981), 49; Tony Freyer, "The Sherman Antitrust Act: Comparative Business Structure and the Rule of Reason: America and Great Britain, 1880–1920," *Iowa Law Review* (July 1989): 203–5.

78. Herbert Hovenkamp, "Antitrust's Protected Classes," *Michigan Law Review* (1989): 123; May 1987, 564; Naomi R. Lamoreaux, *The Great Merger Movement in American Business, 1895–1904* (Cambridge: Cambridge University Press, 1985), 175–77, 180, 191.

79. Lamoreaux, *Great Merger Movement*, 312–19; Eisner, *Antitrust*, 53, 55.

80. Bruce Bringhurst, *Antitrust and the Oil Monopoly: The Standard Oil Cases, 1890–1911* (Westport, Conn.: Greenwood Press, 1979), 16–19; J. D. Forrest, "Anti-Monopoly Legislation in the United States," *American Journal of Sociology* (January 1896): 413, 418, 419; John R. Allison, "Survey of the Texas Anti-Trust Laws," *Antitrust Bulletin* (Summer 1975): 217–19, 222, 237, 263, 308.

81. *New York World*, March 11, 1892.

82. U.S. House of Representatives, *Investigation of Trusts*, 50th Cong., 1st sess., 1888, 110, 116, 129–207, 228, 257, 573, 644, 729.

83. Blicksilver, "Defenders and Defense of Big Business," 122; Hovenkamp, "Antitrust's Protected Classes," 26.

84. Freyer, "Sherman Antitrust Act," 207, 203.

85. See George Stigler, "Perfect Competition, Historically Contemplated," *Journal of Political Economy* (February 1957): 1, 15.

86. Hovenkamp, "Antitrust's Protected Classes," 123; James May, "Antitrust Practice and Procedure in the Formative Era: The Constitutional and Conceptual Reach of State Antitrust Law, 1880–1918," *University of Pennsylvania Law Review* (March 1987): 564.

87. On the course and consequences of the Sherman Antitrust Act, see Freyer, "Sherman Antitrust Act," 85–87, 97; Blicksilver, "Defenders and Defense of Big Business," 122–28; Martin J. Sklar, *The Corporate Reconstruction of American Capitalism, 1890–1916: The Market, the Law, and Politics* (Cambridge: Cambridge University Press, 1988), 113–15, 116; Allison, "Survey of Texas Anti-Trust Laws"; and Eisner, *Antitrust*, 22.

88. Herbert Hovenkamp, *Enterprise and American Law, 1836–1937* (Cambridge: Harvard University Press, 1991), 242, 244, 245, 251, 258, 264.

89. U.S. Bureau of Corporations, *Report of the Commissioner, 1908* (Washington, D.C.: U.S. Government Printing Office, 1908), 5; J. A. McLaughlin, "Legal Control of Competitive Methods," *Iowa Law Review* (1936): 280.

90. Hovenkamp, *Enterprise and American Law*, 299, 301, 304.

91. Sklar, *Corporate Reconstruction*, 101.

92. Oliver E. Williamson, "Predatory Pricing: A Strategic and Welfare Analysis," *Yale Law Journal* (1977): 337. As Martin J. Sklar expressed it, "Out of court, the debates proceeded in a mixture of legal and lay language, where terms such as monopoly, competition, restraint of trade, trusts and combinations, assumed meanings different from or less precise than, those prevailing in the courts or in professional economic thought." *Corporate Reconstruction*, 184.

93. Thomas K. McCraw, "Regulatory Agencies," in *The Coming of Managerial Capitalism: A Casebook on the History of American Economic Institutions* (Homewood, Ill.: Richard A. Irwin, 1985), 632.

94. Roger Sherman, "The Standard Oil Trust: The Gospel of Greed," *Forum* (July 1892): 605, 606, 607, 614–15.

CHAPTER FOUR

1. Henry Demarest Lloyd, *Wealth against Commonwealth* (New York: Harper & Brothers, 1894), 81–82.

2. Ibid., 244–45, 297, 221, 427.

3. Ibid., 434, 537–44.

4. *New York World*, March 11, 1892; W. A. Swanberg, *Pulitzer* (New York: Charles Scribner's Sons, 1967), 160; Don C. Seitz, *Joseph Pulitzer: His Life and Letters* (New York: Simon & Schuster, 1924), 305, 321–22.

5. Lloyd, *Wealth against Commonwealth*, 196–98, 197.

6. *Arena* (July 1901): 100.

7. Coverage is surveyed by Chester McArthur Destler, *American Radicalism, 1865–1901: Essays and Documents* (New London: Connecticut College, [c. 1945]), 161, and John L. Thomas, *Alternative America: Henry George, Edward Bellamy, Henry Demarest Lloyd, and the Adversary Tradition* (Cambridge: Harvard University Press, Belknap Press, 1983), 306–7.

8. William D. P. Bliss, ed., *The Encyclopedia of Social Reform* (New York: Funk & Wagnalls, 1897), 1285.

9. Ibid., 823.

10. George Davis Herron, *The Christian Society* (1894; reprint, New York: Johnson Reprint Corp. 1969), 16, 83.

11. Chester McArthur Destler, *Roger Sherman and the Independent Oil Man* (Ithaca, N.Y.: Cornell University Press, 1967), 275.

12. Ibid., 223, 227.

13. *Nation* quoted in Allan Nevins, *Study in Power: John D. Rockefeller, Industrialist and Philanthropist* (New York: Charles Scribner's Sons, 1953), 2:332. Ernest von Halle, *Trusts, or Industrial Combinations and Coalitions in the United States* (New York: Macmillan, 1899), xiv.

14. Thomas, *Alternative America*, 306–7.

15. M. W. Howard, *The American Plutocracy* (New York: Holland Publishing Co., 1895), 225–44.

16. Ibid., 8, 37–38, 26.

17. F. F. Murray, *The Middle Ten* (Titusville, Pa.: World Publishing Co., 1897), 84–86, 119.

18. Arthur T. Hadley, "The Good and the Evil of Industrial Combination," *Atlantic Monthly* (March 1897): 391.

19. Edward W. Bemis, "The Trust Problem: Its Real Nature," *Forum* (December 1899): 413–16.

20. Harold Hotelling, "Stability in Competition," *Economic Journal* (March 1929): 44.

21. Bruce Bringhurst, *Antitrust and the Oil Monopoly: The Standard Oil Cases, 1890–1911* (Westport, Conn.: Greenwood Press, 1979), 127–28; *New York Times*, June 28, August 20, 1904.

22. Ohio Trust Investigation Committee, *Proceedings, Trust Investigation of [the] Ohio Senate* (n.p., [1898]), 8–19.

23. Ibid., 214–22, 277–80.

24. Ibid., 327–28, 359–60, 379.

25. See the issues of January 30, February 8, October 12, 13, 14, 16, and December 22–24.

26. *Hearings before the United States Industrial Commission* (Washington, D.C.: U.S. Government Printing Office, 1898) 1:368–83, 384–403, 602–71, 686–755.

27. Transcript, *Hearings of the United States Industrial Commission, 1898* (Oil City, Pa.: Derrick Publishing Co., 1898), 72–73, passim.

28. Ibid., 271, 427, 72–73, 172–74.

29. Bringhurst, *Antitrust and the Oil Monopoly*, 228 n. 48; *New York World*, October 12, 1898; *New York Tribune*, December 22, 1898.

30. See, for example, the *New York Daily Tribune*, January 30, February 8, June 19, October 11, 12, 13, 14, 16, and December 22, 24, 25, and 27, 1898.

31. *New York World*, April 15, 1899; *New York Daily Tribune*, May 9, 10, 1899.

32. Bringhurst, *Antitrust and the Oil Monopoly*, 31; see, for example, Frank S. Monnett, "Transportation Franchises Always the Property of Sovereignty," *Arena* (August 1901): 113–27. The editor of the magazine introduced Monnett as "the incorruptible Attorney-General of Ohio" (130).

33. Howard Horowitz, "The Standard Oil Trust as Emersonian Hero," *Raritan: A Quarterly Review* (1987): 97; Henry James, *A Little Town in France* (1883; New York: Farrar, Straus & Giroux, 1983), 162.

34. Ralph Waldo Emerson, "Self-Reliance," *Essays and English Traits* (1841; New York: P. F. Collier & Son, 1909), 72. *Nation* (December 14, 1916): 556. Ida Tarbell used the same quotation to explain and justify her narrow focus on Rockefeller in her history of Standard Oil.

35. Such references are legion. See, for example, the *New York American*, October 1, 1916, and the *New York Times*, September 29, 1916.

36. See, for example, the *Baltimore Sun*, November 3, 1907; *New York American*, September 20, 1907.

37. Barbara Welter, *Dimity Convictions: The American Woman in the Nineteenth Century* (Athens: Ohio University Press, 1976), 21–41.

38. E. Anthony Rotundo, *American Manhood: Transformations in Masculinity from the Revolution to the Modern Era* (New York: Basic Books, 1993), 172.

39. *New York World*, January 13, 1906.

40. *New York World*, February 15, 1906.

41. *New York Evening Journal*, March 26, 1906; *Brooklyn Eagle*, March 7, 1907.

42. *New York American*, June 1, 1907, August 19, 1904.

43. Kevin White, *The First Sexual Revolution: The Emergence of Male Heterosexuality in Modern America* (New York: New York University Press, 1993), 3, 119; Richard Slotkin, *Gunfighter Nation: The Myth of the Frontier in Twentieth Century America* (New York: HarperCollins, 1992), 292; Toby L. Ditz, "Shipwrecked: Imperiled Masculinity and the Representation of Business Failures among Philadelphia's Eighteenth-Century Merchants" (paper given at the Berkshire Conference on Women's History, Poughkeepsie, N.Y., June 1993), 6, 21.

44. "A Menace to Freedom," *Arena* (September 1901): 250, 255.

45. *New York World*, August 5, December 31, 1900; *New York Tribune*, October 24, 1909.

46. *New York Journal*, January 5, 1901; *New York American*, July 30, 1902.

47. See, for example, Samuel C. T. Dodd, *Combinations: Their Uses and Abuses, with a History of the Standard Oil Trust* (New York: G. F. Nesbitt & Co., 1888); "Trusts, an Address at the Merchants' Association of Boston, January 8, 1889" ([New York]: Standard Oil Company, 1889), 17 ; "A Defense of Trusts," *New York Tribune*, February 2, 1890; "The War against Wealth," *Independent* (March 4, 1897): 4.

48. For a full account of Standard Oil's public relations activities, see Ralph W. Hidy and Muriel E. Hidy, *Pioneering in Big Business: History of the Standard Oil Company (New Jersey), 1882–1911* (New York: Harper & Brothers, 1955), chap. 22.

49. *New York World*, May 6, 1906; *New York Times*, July 8, 1906.

50. See Mary E. Tompkins, *Ida M. Tarbell* (New York: Twayne Publishers, 1974), chaps. 2–3; Harold S. Wilson, *McClure's Magazine and the Muckrakers* (Princeton: Princeton University Press, 1970), 63, 73–74, 137.

51. Kathleen Brady, *Ida Tarbell: Portrait of a Muckraker* (New York: Sea View–Putnam, 1984), 22, 45, 88, 110, 122, 133; Tarbell's version of the writing of the book is

contained in *All in the Day's Work: An Autobiography* (1939; Boston: G. K. Hall & Co., 1985), chaps. 11–12; Peter Lyon, *Success Story: The Life and Times of S. S. McClure* (New York: Charles Scribner's Sons, 1963), 202.

52. David Freeman Hawke, *John D.: The Founding Father of the Rockefellers* (New York: Harper & Row, 1980), 215.

53. Ibid., 214–15.

54. Lyon, *Success Story*, 212–13.

55. Frank Luther Mott, *A History of American Magazines, 1885–1905* (Cambridge: Harvard University Press, 1957), 598.

56. Ida M. Tarbell, *The History of the Standard Oil Company* (New York: Macmillan, 1904), 1:viii, ix.

57. Ibid., 12–13, 216–20; Tarbell, *All in the Day's Work*, 26, 94.

58. Tarbell, *History of the Standard Oil Company*, 1:21, 34.

59. Ibid., 1:37, 38.

60. Ibid., 1:43, 50, 64, 102, 103, 105.

61. Ibid., 1:42, 141, 43, 64.

62. Ibid., 1:145–46, 241.

63. Jeremiah W. Jenks, *The Trust Problem* (New York: McClure, Phillips, 1900), 73–74.

64. Tarbell, *History of the Standard Oil Company*, 2:52.

65. Isabelle Sheifer, "Ida M. Tarbell and Morality in Big Business" (Ph.D. diss., New York University, 1980), 77; Tompkins, *Ida M. Tarbell*, 90; Brady, *Ida Tarbell*, 155.

66. See Tarbell, *History of the Standard Oil Company*, notes to 1:71, for example.

67. Ibid., 2:131; 1:112, xiii; 2:174–76.

68. Ibid., 1:98, 256, 106.

69. Ibid., 1:147, 155, 241–42.

70. Ibid., 2:225, 384–85.

71. Ibid., 1:44, 46, 49, 93, 102, 108; 2:8–11.

72. Ibid., 1:203–8; 2:41–48, 78–80, 46–56; 1:189.

73. Ibid., note to 2:115.

74. Ibid., 2:146–47, 115–16. See Thomas E. Felt, "What Mark Hanna Said to Attorney General Watson," *Ohio History* (1963): 293–302.

75. Tarbell, *History of the Standard Oil Company*, 2:292.

76. *New York Times*, December 31, 1904; "Ida Tarbell's Tale of the Standard Oil," *Gunton's Magazine* (February 1904): 95, 105, 107, 102. (It was later claimed, with some evidential support, that Standard Oil supported Gunton with a large subscription list.)

77. *Nation* (January 5, 1905): 15–16.

78. Gilbert Holland Montague, "The Legend of the Standard Oil Company," *North American Review* (September 1905): 353, 354–55, 368.

79. *Oil Investors' Journal* (December 15, 1902): 5.

80. Frank L. McVey, "The Story of a Great Monopoly," *Dial* (May 1, 1905): 314, 315; "A History of the Master-Trust," *Arena* (October 1905): 436, 437.

81. *Outlook* (February 11, 1905): 395–96.

82. Maxwell Bloomfield, "Muckraking and the American Stage: The Emergence of Realism, 1905–1917," *South Atlantic Quarterly* (1967): 172.

83. Charles Klein, *The Lion and the Mouse: A Story of American Life Novelized from the Play by Arthur Hornblow* (New York: Grosset & Dunlap, 1906), 14–16, 70.

84. Ibid., 13.

85. See Clifford Geertz, *The Interpretation of Cultures* (New York: Basic Books,

1973), chap. 5: "Ethos, World View, and the Analysis of Sacred Symbols," and Mary Douglas, *Purity and Danger: An Analysis of the Concepts of Pollution and Taboo* (1966; London: Ark Paperbacks, 1984).

86. Dorothy Ross, *The Origins of American Social Sciences* (Cambridge: Cambridge University Press, 1991), 471.

87. Quoted in Arthur Johnson, "Theodore Roosevelt and the Bureau of Corporations," *Mississippi Valley Historical Review* (March 1959): 573.

88. *New York World*, February 3, 1903; *Philadelphia North American*, February 12, 1903.

89. Johnson, "Theodore Roosevelt and the Bureau of Corporations," 577.

90. Lewis L. Gould, *Reform and Regulation: American Politics from Roosevelt to Wilson*, 2d ed. (New York: Alfred A. Knopf, 1986), 98.

91. George E. Mowry, *The Era of Theodore Roosevelt and the Birth of Modern America* (New York: Harper & Row, 1962), 133.

92. *New York Times*, August 13, 1907.

93. Quoted in Gould, *Reform and Regulation*, 23.

94. Richard L. Douglas, "A History of Manufacturers in the Kansas District," *Collections of the Kansas State Historical Society* (1910): 137–39; Francis Schruben, *From Wea Creek to El Dorado: Oil in Kansas, 1860–1920* (Columbia: University of Missouri Press, 1972), 54–55, 102.

95. Schruben, *From Wea Creek to El Dorado*, 48–51, 55.

96. Ibid., 57, 75; William E. Connelley, "The Kansas Oil Producers against the Standard Oil Company," *Collections of the Kansas State Historical Society* (1906): 95.

97. The letter was quoted by Schruben, *From Wea Creek to El Dorado*, 80. On the activities of the KOPA, see Schruben, *From Wea Creek to El Dorado*, 102–22; F. S. Barde, "The Oil Fields and Pipelines of Kansas," *Outlook* (May 6, 1905): 25; Isaac F. Marcosson, "The Kansas Oil Fight," *World's Work* (May 1905): 61–62.

98. Philip Eastman, "Going against the Octopus," *Saturday Evening Post* (April 8, 1905): 1–4.

99. See "The Battle of a Commonwealth," *Arena* (April 1905): 435–40; Edward Wallace Hoch [Kansas's governor], "Kansas and the Standard Oil Company," *Independent* (March 2, 1905): 461–63; "After Standard Oil," *Literary Digest* (March 4, 1905): 303–4; Charles Moreau Harger, "Kansas' Battle for Its Oil Interests," *Review of Reviews* (April 1905): 471–74; Barde, "Oil Fields and Pipelines of Kansas"; Marcosson, "Kansas Oil Fight."

100. Schruben, *From Wea Creek to El Dorado*, 109–10, 122.

101. Theodore Roosevelt, "Annual Message, December 5, 1905," quoted in Bernard Schwartz, ed., *The Economic Regulation of Business and Industry: A Legislative History of U.S. Regulatory Agencies*, (New York: Chelsea House, 1973), 1:616.

102. Willard B. Gatewood Jr., *Theodore Roosevelt and the Art of Controversy: Episodes of the White House Years* (Baton Rouge: Louisiana State University Press, 1970), 29.

103. David Graham Phillips, "The Treason of the Senate: Gorman's Chief Lieutenant, Bailey," *Cosmopolitan* (July 1906): 267, 268, 274, 276.

104. See, for example, xvii and xix of the U.S. Bureau of Corporations, "Report on the Place of Standard Oil in the Oil Industry," August 5, 1907.

105. *New York American*, April 10, 1909.

106. *New York American*, January 16, 1906; see, for example, "On the Missouri Case," *Arena* (March 1906): 307.

107. *New York World*, February 10, March 1, 25, 27, 1906.

108. *New York American*, February 13, 14, 1908; the other press reports are in the

"Morgue" files of the *New York American*, in the Hearst Papers, Ransom Humanities Research Center, University of Texas at Austin. The Standard Oil files are all labeled "King Oil."

109. W. A. Swanberg, *Citizen Hearst: A Biography of William Randolph Hearst* (New York: Charles Scribner's Sons, 1961), 258–62.

110. *New York World*, September 26, 1908.

111. *New York World*, August 4, 1907.

112. *New York Journal*, July 24, 1908.

113. Quoted in Charles Musser, *The Emergence of Cinema: The American Screen to 1907* (Berkeley: University of California Press, 1990), 479.

114. *New York Times*, May 13, June 3, 13, 24, 26, November 19, December 24, 1908; January 21, February 12, March 15, July 1, November 2, 1909; *Philadelphia North American*, July 1, 1907, November 23, 1909; *New York Tribune*, November 8, 1910; *New York Herald*, March 30, 1905; *New York American*, November 8, 1910.

115. John Davidson Rockefeller, *Random Reminiscences of Men and Events* (New York: Doubleday, Page & Co., 1909), 10, 21, 25, 55–56, 58, 96–107, 107–12, 134, 147, 185–88.

116. *New York American*, September 26, 1908.

117. "Additional Memorandum in Support of Application to Advance Course and Set Same for Argument about March 1," *U.S. v. Standard Oil of New Jersey et al.*, 221 U.S. 1, p. 3. The circuit court case is styled 173 Fed 177 (November 20, 1909).

118. Ibid., "Brief for the United States," 139, 141, 148, 149, 151–60.

119. Joseph D. Potts, *A Brief History of the Standard Oil Company* (Oil City, Pa.: General Council of the Petroleum Producers' Union, 1878).

120. *U.S. v. Standard Oil*, Petitioner's Testimony, 247–48, 463–65, 2671–85, 2773–74.

121. Ibid., 2610.

122. "Opinion of the Supreme Court of the United States," 47, 70, 74, 76 (221 U.S. 1).

123. Tony Freyer, *Regulating Big Business: Antitrust in Great Britain and America, 1880–1990* (Cambridge: Cambridge University Press, 1992), 25, 41.

124. *New York Times*, May 16, 17, 18, June 3, 4, November 15, December 30, 1911.

125. *New York Evening Journal*, May 16, 1911; *New York American*, May 17, 1911.

126. George Sweet Gibb and Evelyn H. Knowlton, *The Resurgent Years: History of the Standard Oil Company (New Jersey), 1911–1927* (New York: Harper & Brothers, 1956), 169–71, 175; Harold F. Williamson, Ralph L. Andreano, Arnold R. Daum, and Gilbert C. Klose, *The American Petroleum Industry: The Age of Energy, 1899–1959* (Evanston, Ill.: Northwestern University Press, 1963), 109, 597. The issue revived with the Great Depression; a pipeline divorcement bill was introduced in every session of Congress, from 1931 to 1941, when the ICC ordered an 8 percent rate of return on pipeline investments.

127. *New York American*, November 24, 1913.

128. Eileen Bowser, *The Transformation of Cinema, 1907–1915* (Berkeley: University of California Press, 1990), 189.

129. *New York American*, February 15, 1915; Graham Adams Jr., "Francis Patrick Walsh," *Dictionary of American Biography*, Supplement 2 (New York: Charles Scribner's Sons, 1958), 690.

130. *New York Journal*, April 16, 1916.

131. *New York World*, December 30, 1915; *New York Journal*, December 30, 1915.

132. Federal Trade Commission, *Report on the Price of Gasoline in 1915* (Washington, D.C.: U.S. Government Printing Office, 1917), 3–5.

133. Ibid., 5–6, 10.

134. Ibid.

135. Ibid., 16–18.

136. Federal Trade Commission, *Report on Pipe-Line Transportation of Petroleum* (Washington, D.C.: U.S. Government Printing Office, 1916), xxvi, xxxi, 15–16.

137. Federal Trade Commission, *Report of the Federal Trade Commission on the Pacific Coast Petroleum Industry, Part II: Prices and Competitive Conditions* (Washington, D.C.: U.S. Government Printing Office, 1922), ix, xi; Federal Trade Commission, *Report of the Federal Trade Commission on the Petroleum Industry of Wyoming* (Washington, D.C.: U.S. Government Printing Office, 1921), 7.

CHAPTER FIVE

1. William Rintoul, *Spudding In: Recollections of Pioneer Days in the California Oil Fields* (Fresno, Calif.: Valley Publishers, 1978), 109–11.

2. In much the same vein, some years ago John A. DeNovo noted the irony of anxiety over United States naval petroleum supplies at the beginning of the twentieth century, a time when the United States alone of the great naval powers had ample domestic reserves; "Petroleum and the United States Navy before World War I," *Mississippi Valley Historical Review* (March 1955): 646.

3. Harold F. Williamson, Ralph L. Andreano, Arnold R. Daum, and Gilbert C. Klose, *The American Petroleum Industry: The Age of Energy, 1899–1959* (Evanston, Ill.: Northwestern University Press, 1963), 6–7, 17–28.

4. W. J. McGee, "The Conservation of Natural Resources," Mississippi Valley Historical Association, *Proceedings for the Year 1909–1910* (Cedar Rapids, Iowa: Mississippi Valley Historical Association, 1910), 367. Roosevelt spoke of the Forest Service taking up the fight to handle the national forests so as "to prevent speculation and monopoly"; Wayne Andrews, ed., *The Autobiography of Theodore Roosevelt* (New York: Octagon Books, 1975), 217. La Follette began a campaign in 1906 to keep public lands bearing coal, oil, and other minerals "from being exploited by monopoly control" in passing to private ownership; Robert M. La Follette, *La Follette's Autobiography: A Personal Narrative of Political Experiences* (Madison: University of Wisconsin Press, 1968), 163–65. But, as Samuel P. Hays has argued, conservationists such as George Otis Smith were less preoccupied with monopoly than with efficiency in the use of resources; *Conservation and the Gospel of Efficiency: The Progressive Conservation Movement, 1890–1920* (Cambridge: Harvard University Press, 1959), 261–62. Hays puts Gifford Pinchot in the same camp as Smith, despite Pinchot's tendency to explain conservation's value as a safeguard against "oppressive private monopoly," to use Pinchot's phrase from "The A B C of Conservation," in *Outlook* (December 4, 1909): 770.

5. On the emergence of the general idea of the United States running out of resources, see our description of the role of George Perkins Marsh, in Diana Davids Olien and Roger M. Olien, "Running Out of Oil: Discourse and Public Policy, 1909–1929," *Business and Economic History* (Winter 1993): 39.

6. J. Peter Lesley, "Letter of Transmittal," *Second Geological Survey of Pennsylvania, 1880 to 1883* (Harrisburg, Pa.: Board of Commissioners for the Second Geological Survey, 1883), x; E. W. Claypole, "The Future of Natural Gas," *American Geologist* (January 1888): 34–35. This opinion left little hope for secondary oil recovery.

7. Geological Survey of Pennsylvania, *Seventh Report on the Oil and Gas Fields of Western Pennsylvania for 1887* (Harrisburg, Pa.: Board of Commissioners for the Geological Survey, 1890), 9, 1, 12–13.

8. Claypole, "Future of Natural Gas," 32–34; editorial, *American Geologist* (December 1889): 374.

9. Geological Survey of Pennsylvania, *Seventh Report*, 23; *Annual Report of the Geological Survey of Pennsylvania for 1885* (Harrisburg, Pa.: Board of Commissioners for the Geological Survey, 1886), 31; *Second Geological Survey*, xiii–xiv.

10. J. Leonard Bates, "Fulfilling American Democracy: The Conservation Movement, 1907 to 1921," *Mississippi Valley Historical Review* (June 1957): 30–31, 38–39, 41.

11. For the close connection of USGS and Forestry Service experts, see Hays, *Conservation and the Gospel of Efficiency*, 15, 24–26, 69–70.

12. Ibid., 85–87.

13. Ibid., 88–90; J. Leonard Bates, *The Origin of Teapot Dome: Progressives, Parties, and Petroleum, 1909–1921* (Urbana: University of Illinois Press, 1963), 18–21; Gerald T. White, *Formative Years in the Far West: A History of Standard Oil Company of California and Predecessors through 1919* (New York: Appleton-Century-Crofts, 1962), 434–35.

14. United States Geological Survey, Bulletin 623, Max W. Ball, *Petroleum Withdrawals and Restorations Affecting the Public Domain* (Washington, D.C.: U.S. Government Printing Office, 1917).

15. Ibid., 104, 113–14.

16. Hays, *Conservation*, 128–33.

17. *New York Times*, May 14, 15, 1908.

18. David T. Day, "The Petroleum Resources of the United States," in 60th Cong., 2d sess., S. Doc. 676: *Report of the National Conservation Commission*, 3: Accompanying Papers (Washington, D.C.: U.S. Government Printing Office, 1909), 446–64. David T. Day, "The Petroleum Resources of the United States," *American Review of Reviews* (January 1909): 49–56; United States Geological Survey, Bulletin 394: *Papers on the Conservation of Mineral Resources* (Washington, D.C.: U.S. Government Printing Office, 1909), 30–61. Jameson W. Doig and Erwin C. Hargrove call innovative government officials such as Pinchot entrepreneurs, noting that such persons indentify new missions for their organizations and try to cultivate through "rhetorical leadership" extended constituencies for their programs; *Leadership and Innovation: A Biographical Perspective on Entrepreneurs in Government* (Baltimore: Johns Hopkins University Press, 1987), 7–16. By these criteria, George Otis Smith and Bureau of Mines head Joseph Austin Holmes certainly qualify as "entrepreneurs."

19. Day, "Petroleum Resources," in *American Review of Reviews*, 50.

20. Day, "Petroleum Resources," in *Report*, 34–35. He gave the National Conservation Commission a higher minimum of fifteen billion barrels; *Report*, 460.

21. Ibid., 45; Day, "Petroleum Resources," in *American Review of Reviews*, 50, 52.

22. Day, "Petroleum Resources," in *American Review of Reviews*, 50.

23. Ibid., 56.

24. Ibid., 54–56.

25. Charles Richard Van Hise, "Patriotism and Waste," *Collier's*, September 18, 1909, 23, 36.

26. Charles Richard Van Hise, *The Conservation of Natural Resources in the United States*, 2d ed. (New York: Macmillan, 1921), 47–60, 376, 52.

27. United States Geological Survey, *Mineral Resources of the United States*, 1913, pt. 2: Nonmetals, David T. Day, "Petroleum" (Washington, D.C.: U.S. Government Printing Office, 1914), 929, 931–32, 946; *Mineral Resources of the United States*, 1915, pt.

2: Nonmetals, John D. Northrop, "Petroleum" (Washington, D.C.: U.S. Government Printing Office, 1917), 559–60.

28. Ball, *Petroleum Withdrawals*, 19–21.

29. *Dictionary of American Biography*, 9:167–68; Irving C. Allen, *Problems of the Petroleum Industry: Results of Conferences at Pittsburgh, Pa., August 1 and September 10, 1913*, Department of the Interior, Bureau of Mines, Technical Paper 72 (Washington, D.C.: U.S. Government Printing Office, 1914), 3, 5; J. A. Holmes, "Production and Waste of Mineral Resources and Their Bearing on Conservation," *Annals of the American Academy of Political and Social Science* (May 1909): 206–13.

30. Raymond S. Blatchley, *Waste of Oil and Gas in the Mid-Continent Fields*, Department of the Interior, Bureau of Mines, Technical Paper 45 (Washington, D.C.: U.S. Government Printing Office, 1913), 46.

31. Ibid., 37–38, 20. Given the size of the field at issue, Blatchley's figures are absurdly high, as petroleum engineer Robert M. Leibrock has noted to the authors.

32. Ibid., 47.

33. Ibid., 48–52.

34. Ibid., 19, 53.

35. Allen, *Problems of the Petroleum Industry*, 9–10, 15–16; Ralph Arnold and V. R. Garfias, *The Cementing Process of Excluding Water from Oil Wells As Practiced in California*, Technical Paper 32 (Washington, D.C.: U.S. Government Printing Office, 1913); Ralph Arnold and F. G. Clapp, *Wastes in the Production and Utilization of Natural Gas and Means for Their Prevention*, Technical Paper 38 (Washington, D.C.: U.S. Government Printing Office, 1913); Ralph Arnold and V. R. Garfias, *The Prevention of Waste of Oil and Gas from Flowing Wells in California, with a Discussion of Special Methods Used by J. A. Pollard*, Technical Paper 42 (Washington, D.C.: U.S. Government Printing Office, 1913).

36. Ball, *Petroleum Withdrawals*, 133–49, 283–84, 290; Bates, *Teapot Dome*, 24, 28–29.

37. Bates, *Teapot Dome*, 28–31. As Mansel G. Blackford has pointed out, California oil producers during this period faced problems of overproduction, falling prices, and purchaser cutbacks; oilmen as a group were divided on questions of production limitations and designation of pipelines as common carriers. Federal land withdrawals added an additional complication to industry operations, to be confronted by an already divided set of participants. See Mansel G. Blackford, *The Politics of Business in California, 1890–1920* (Columbus: Ohio State University Press, 1977), 40–59.

38. Bates, *Teapot Dome*, 26–27, 166–75.

39. *Annual Report of the Secretary of the Navy*, 1913 (Washington, D.C.: U.S. Government Printing Office, 1914), 15; DeNovo, "Petroleum and the United States Navy," 648–50; Bates, *Teapot Dome*, 37–38, 82. On Daniels, see also Joseph L. Morrison, *Josephus Daniels: The Small-d Democrat* (Chapel Hill: University of North Carolina Press, 1966).

40. U.S. Senate, *Oil Land Leasing Bill: Hearings before the Committee on Naval Affairs on the So-Called Relief Provisions of the Leasing bill relative to the California Naval Petroleum Reserve*, 64th Cong., 2d sess. (Washington, D.C.: U.S. Government Printing Office, 1917), 7, 12, 13–15.

41. See Bates, *Teapot Dome*, and Burl Noggle, *Teapot Dome: Oil and Politics in the 1920s* (Baton Rouge: Louisiana State University Press, 1962).

42. U.S. House of Representatives, *Hearings before a Subcommittee of the Committee on Interstate and Foreign Commerce on H. Res. 441*, 73d Cong., recess (Washington, D.C.: U.S. Government Printing Office, 1934), 2667–68. It was not uncommon for promoters of mineral ventures to switch back and forth from hard-rock mineral promotion to

oil or vice versa. Once one mastered the art of promotion, it could be used in a wide variety of endeavors, legal or otherwise. For examples, see Roger M. Olien and Diana Davids Olien, *Easy Money: Oil Promoters and Investors in the Jazz Age* (Chapel Hill: University of North Carolina Press, 1990), 19, 27, 31, 169, and passim. Blackford notes that Requa took an active part in the controversy over legislative action to limit water incursion in California oil fields in 1910–12; *Politics of Business*, 57.

43. U.S. Senate, *Senate Document No. 363*: Mark L. Requa, "Petroleum Resources of the United States," 64th Cong., 1st sess. (Washington, D.C.: U.S. Government Printing Office, 1916), 3, 5, 17.

44. Ibid., 17–18, 3.

45. Ibid., 18. Bates, *Teapot Dome*, 104.

46. Requa, "Petroleum Resources," 15–16.

47. *Dictionary of American Biography*, 11: Supplement 2 (New York: Charles Scribner's Sons, 1958), 552–53.

48. Mark L. Requa, "Report of the Oil Division," in H. A. Garfield, *Final Report of the United States Fuel Administrator, 1917–1919* (Washington, D.C.: U.S. Government Printing Office, 1921), 264, 267.

49. Bates, *Teapot Dome*, 104, 107, 109–11.

50. Williamson et al., *American Petroleum Industry: The Age of Energy*, 287–91; Requa, "Report," 261, 267–71.

51. Requa, "Report," 261, 271.

52. Ibid., 272.

53. George Otis Smith, "Industry's Need of an Oil Supply," *Oil and Gas Journal* (hereafter cited as "*OGJ*") (May 28, 1920): 56, 60; George Otis Smith, "A Foreign Oil Supply for the United States," in *Transactions of the American Institute of Mining and Metallurgical Engineers* (New York: AIME, 1921), 93; "Exhaustion of Our Oil Supply in Twenty Years Predicted," *New York Times*, May 17, 1920.

54. "Are We Wasting Our Petroleum?," *Scientific American* (April 20, 1918): 372; "How Long the Oil Will Last," ibid., May 3, 1919, 459.

55. *Congressional Record*, 69th Cong., 1st sess., February 11, 1926, 3761–66; Frank M. Burke Jr. and G. Houston Hall, "America's Oil and Gas Tax Incentives," *Petroleum Independent* (February 1981): 38–39.

56. *OGJ* (October 10, 1921): 1; (March 2, 1916): 2; (February 24, 1916): 2.

57. *OGJ* (December 3, 1920): 3.

58. *OGJ* (December 16, 1921): 78.

59. *OGJ* (April 23, 1920): 54.

60. *OGJ* (June 14, 1918): 47; (May 17, 1921): 3; (January 25, 1923): 10.

61. John A. DeNovo, "The Movement for an Aggressive American Oil Policy Abroad, 1918–1920," *American Historical Review* (July 1956): 859.

62. *OGJ* (January 24, 1919): 54; (June 11, 1920): 85.

63. Bennett H. Wall and George S. Gibb, *Teagle of Standard* (New Orleans: Tulane University Press, 1974), 120; *OGJ* (November 26, 1920): 71; (November 19, 1920): 62.

64. *OGJ* (April 18, 1919): 2; (January 30, 1920): 2; (February 6, 1920): 2; (February 20, 1920): 2; (August 20, 1920): 2; (October 18, 1920): 2.

65. *OGJ* (June 18, 1920): 68.

66. Olien and Olien, "Running Out of Oil," 53–55; Stephen J. Randall, *United States Foreign Oil Policy, 1919–1948: For Profits and Security* (Kingston, Canada: Queens University Press, 1985), 13–42.

## CHAPTER SIX

1. Harold F. Williamson, Ralph L. Andreano, Arnold R. Daum, and Gilbert C. Klose, *The American Petroleum Industry: The Age of Energy, 1899–1959* (Evanston, Ill.: Northwestern University Press, 1963), 468–69; Roger M. Olien and Diana Davids Olien, *Easy Money: Oil Promoters and Investors in the Jazz Age* (Chapel Hill: University of North Carolina Press, 1990), 26.

2. William Rintoul, *Spudding In: Recollections of Pioneer Days in the California Oil Fields* (Fresno, Calif.: Valley Publishers, 1978), 147–63; Olien and Olien, *Easy Money*, 19–21; *New York Times*, August 19, 1923.

3. Olien and Olien, *Easy Money*, 27.

4. Albert Atwood, "Mad from Oil," *Saturday Evening Post* (July 14, 1923): 11, 92. On the general subject of promoters and such investors, see Olien and Olien, *Easy Money*, 18–24, 55–59, 75–89, and passim.

5. *Oil and Gas Journal* (hereafter cited as "*OGJ*") (January 11, 1923): 10.

6. See, for example, Paul H. Giddens, "The Naval Oil Reserve, Teapot Dome, and the Continental Trading Company," *Annals of Wyoming* (1981): 14–27, and J. Leonard Bates, "Watergate and Teapot Dome," *South Atlantic Quarterly* (Spring 1974): 145–59.

7. Bruce Bliven, "Oil-Driven Politics," *New Republic* (February 13, 1924): 302–3; Gifford Pinchot, "Ships, Oil, and the Ten Commandments," *Saturday Evening Post* (May, 17, 1924): 6–7, 105. It should be noted that the main focus of Bliven's article is corruption, not conservation. The editors of the *New Republic* mixed concern for conservation with outrage over corruption; February 4, 1924, 266–67.

8. *New York American*, February 13, 1, 2, 19, 1924; "Sinister Shadows behind the Oil Scandal," *Literary Digest* (March 1, 1924): 1–4; "Distributing the Blame for Teapot Dome," *New Republic* (February 13, 1924): 297–98.

9. "What Is to Become of Our Reservoir of Oil?," *Outlook* (March 13, 1918): 403.

10. *New York Times*, May 3, 17, 19, June 20, 1920; "To Find Out How Much Oil We Have Left," *Literary Digest* (March 6, 1920): 111; Edward G. Acheson, "Our Vanishing Coal and Oil," *Forum* (November 1920): 299; John K. Barnes, "Crisis in Our Oil Supply," *World's Work* (May 1920): 29, 32; J. W. Gregory, "The Future of Oil Supply," *Contemporary Review* (June 1921): 778–80; P. W. Wilson, "Crisis in Oil," *American Review of Reviews* (July 1921): 37–38.

11. M. L. Requa, "Some Fundamentals of the Petroleum Problem," *Saturday Evening Post* (August 28, 1920): 29, 57–58; "The Petroleum Problem of the United States," ibid. (September 4, 1920): 30, 170–73; "The Petroleum Problem of the World," ibid. (October 30, 1920): 18, 46.

12. Gregory Mason, "America's Empty Oil Barrel," *Outlook* (March 31, 1920): 549; Acheson, "Our Vanishing Coal and Oil," 299–300; Edward Mead Earle, "Oil and American Foreign Policy," *New Republic* (August 20, 1924): 356.

13. "Huge Wastes of Oil and Gas," *Literary Digest* (December 8, 1923): 70; R. G. Skerrett, "America's Fuel Resources: What We Have and How We Are Using Them," *Scientific American* (February 1922): 87; J. G. Gregory, "The Future of Oil Supply," *Contemporary Review* (June 1921): 778.

14. Walter N. Polakov, "Oil," *New Republic* (June 14, 1922): 68–70.

15. Walter N. Polakov, "The Newer Nationalism," *New Republic* (January 29, 1916): 321. Thorstein Veblen took a similar point of view of the antisocial role of oil in "The Timber Lands and Oil Fields," *Freeman* (May 30, 1923) 272–73.

16. Richard T. Ely, Ralph H. Hess, Charles K. Leith, and Thomas Nixon Carver, *The

*Foundations of National Prosperity* (New York: Macmillan, 1918), v. Ely also used the work to praise Van Hise, whose place in conservation he put on a par with Roosevelt and Pinchot and whose work Ely termed "epoch-making."

17. Ibid., v, 25–26, 3–4, 6–7, 27, 33–44.

18. Ibid., 39–40.

19. Ibid., 37, 42, 60, 120, 106–7, 155–56, 157–58.

20. *Congressional Record*, 66th Cong., 1st sess., September 3, 1919, 4741; for La Follette's repetition of this description of Standard Oil, see also ibid., 67th Cong., 2d sess., April 28, 1922, 6047.

21. *Congressional Record*, 66th Cong., 1st sess., September 3, 1919, 4741.

22. Richard H. K. Vietor, *Energy Policy in America since 1945: A Study of Business-Government Relations* (Cambridge: Cambridge University Press, 1984), 9. On La Follette's campaign, see Fred Greenbaum, *Robert Marion LaFollette* (Boston: Twayne Publishers, 1975), 204–19.

23. U.S. Senate, Committee on Manufactures, *High Cost of Gasoline and Other Petroleum Products, Hearings on S. Res. 295 amending S. Res. 292*, 67th Cong., 2d sess. (Washington, D.C.: U.S. Government Printing Office, 1922), 2, 8–10, 21.

24. Ibid., 34, 55, 103, 574; *OGJ* (July 19, 1923): 60. Nicholas was a perennial witness in oil-related congressional hearings of the 1920s and 1930s.

25. U.S. Senate, Committee on Manufactures, *High Cost of Gasoline and Other Petroleum Products*, Report No. 1263, 67th Cong., 4th sess. (Washington, D.C.: U.S. Government Printing Office, 1923), 3, 4, 27–28.

26. Ibid., 67–69.

27. *Nation* (April 4, 1923): 378; Ida M. Tarbell, "Is the Standard Oil Crumbling?," *New Republic* (November 14, 1923): 300–301.

28. *OGJ* (April 5, 1923): 121; *New York Times*, October 7, 19, 1923; *OGJ* (October 25, 1923): 82; (November 22, 1923): 20, 22; *New York Times*, November 13, 1923.

29. *OGJ* (December 14, 1922): 78; (October 23, 1924): 34; (March 8, 1923): 54; (March 29, 1923): 10; (October 18, 1923): 54; (October 4, 1923): 20; (October 11, 1923): 20.

30. *OGJ* (September 6, 1923): 15.

31. *OGJ* (September 20, 1923): 18; (February 14, 1924): 27, 29; (February 21, 1924): 24, 28; (March 6, 1924): 24; (December 14, 1922): 86; (June 14, 1923): 82.

32. *OGJ* (April 24, 1924): 24; (May 8, 1924): 28; (December 11, 1924): 116.

33. *OGJ* (February 22, 1923): 11; (December 14, 1922): 86; (July 26, 1923): 118; (October 11, 1923): 122.

34. *OGJ* (June 14, 1923): 82.

35. *OGJ* (August 16, 1923): 18.

36. Ibid.

37. *OGJ* (October 4, 1923): 66; (October 11, 1923): 122; (May 15, 1924): 28.

38. For a more extensive legal history, see Nicholas George Malavis, *Bless the Pure and Humble: Texas Lawyers and Oil Regulation, 1919–1936* (College Station: Texas A&M Press, 1996), 12–16.

39. Robert Livingston Schuyler, ed., *Dictionary of American Biography* 9: pt. 2, Supplement 2 (New York: Charles Scribner's Sons, 1958), 154.

40. Ibid. On promotional techniques, see Olien and Olien, *Easy Money*, 73–122. Robert E. Hardwicke, *Antitrust Laws, et al., v. Unit Operation of Oil or Gas Pools* (New York: American Institute of Mining and Metallurgical Engineers, 1948), 1; *Dictionary of American Biography*, 9:155.

41. Hardwicke, *Antitrust Laws*, 13, 185.

42. Ibid., 1–2, 179.

43. Ibid., 180–81, 182–83, 179, 186; Henry L. Doherty, "Suggestions for Conservation of Petroleum by Control of Production," in *Production of Petroleum in 1924: Papers Presented at the Symposium on Petroleum and Gas, at the New York Meeting, February, 1924* (New York: American Institute of Mining and Metallurgical Engineers, 1925), 9.

44. Hardwicke, *Antitrust Laws*, 185–86, 181, 184.

45. Ibid., 181, 183–84, 185, 189, 188.

46. Ibid., 186.

47. Federal Oil Conservation Board, *Complete Record of Public Hearings, February 10 and 11, 1926* (Washington, D.C.: U.S. Government Printing Office, 1926), 2–3. Coolidge had set up the naval reserves commission earlier in the year; Gerald D. Nash has noted the political expedience of both bodies; *United States Oil Policy, 1890–1964* (Westport, Conn.: Greenwood Press, 1976), 81.

48. *New York Times*, November 6, 1924; February 19, May 9, May 21, July 19, October 14, November 18, 1925; January 5, 19, 1926.

49. *Austin American-Statesman*, April 12, 1942; *Daily Texan*, December 4, 1935; *Texas Digest*, April 19, 1941, 16; *Town and Country Review* (April 1935): 36; in George Ward Stocking Biographical File, Eugene C. Barker Texas History Center, University of Texas at Austin; George Ward Stocking, *The Oil Industry and the Competitive System: A Study in Waste* (Boston: Houghton Mifflin, 1925), 4. His dissertation won the Hart, Schaffner, and Marx prize for the best study in economics, giving Stocking one thousand dollars and subsidized publication by Houghton and Mifflin. That meant national notice—the *New York Times* quoted him approvingly on its editorial page—and an associate professorship at his Austin alma mater, a rank from which he advanced to full professor in 1926. To say that *The Oil Industry and the Competitive System* made Stocking's reputation would be a considerable understatement. *New York Times*, May 29, 1926.

50. Stocking, *Oil Industry*, 116.

51. Ibid., 13–15, 19–23, 24–25, 32–33, 48, 116, 113, 100–102, 61.

52. Ibid., 125–26, 128, 246.

53. Ibid., 156–57, 165, 179.

54. Ibid., 138, 140–41, 144–45.

55. Ibid., 139, 116, 168, 140.

56. Ibid., 311–12, 264.

CHAPTER SEVEN

1. Joseph A. Pratt, "Oil and Public Opinion: The American Petroleum Institute in the 1920s," in *Energy and Transport: Historical Perspectives on Policy Issues*, ed. George H. Daniels and Mark H. Rose (Beverly Hills, Calif.: Sage Publications, 1982), 127.

2. *Oil and Gas Journal* (hereafter cited as "*OGJ*") (April 2, 1925): 26–28.

3. American Petroleum Institute, *American Petroleum Supply and Demand, A Report to the Board of Directors of the American Petroleum Institute by a Committee of Eleven Members of the Board* (New York: McGraw-Hill, 1925), 43–45.

4. Ibid., 31, 7–11.

5. Ibid., 12–13, 4, 11.

6. Ibid., 2–3.

7. Ibid., 23–26.

8. Ibid., 24.

9. Federal Oil Conservation Board, *Public Hearing, May 27, 1926* (Washington, D.C.:

U.S. Government Printing Office, 1926), 42–43, 48. Lampoons by Snider and De-Golyer appeared in *National Petroleum News* in February and April 1926 and are reprinted in Robert E. Hardwicke, *Antitrust Laws, et al. v. Unit Operation of Oil or Gas Pools* (New York: American Institute of Mining and Metallurgical Engineers, 1948), 193–201.

10. "Nation in No Peril from Oil Shortage," *New York Times*, August 7, 1925; J. Bernard Walker, "Uncle Sam, Spendthrift," *Scientific American* (August 1926): 101.

11. Ibid., 70.

12. *OGJ* (November 26, 1925): 26; Federal Oil Conservation Board, *Complete Record of Public Hearings*, February 10 and 11, 1926 (Washington, D.C.: U.S. Government Printing Office, 1926), 4–5, 10.

13. Ibid., 149–50, 29, 14, 16, 147.

14. Ibid., 8, 19–21, 12–16, 79–80, 82.

15. Ibid., 129–30.

16. Federal Oil Conservation Board, *Public Hearing, May 27, 1926*, 6–9, 11–12, 17–21.

17. Ibid., 44–47, 51, 49.

18. Federal Oil Conservation Board, *Report of the Federal Oil Conservation Board to the President of the United States* (Washington, D.C.: U.S. Government Printing Office, 1926), 4, 6–10.

19. Ibid., 12, 8–9.

20. Ibid., 16, 14.

21. Ibid., 15, 20–22, 13, 24–25.

22. *OGJ* (September 16, 1926): 40.

23. *Nation* (October 20, 1926): 390.

24. Stuart Chase, "Gasless America," *Nation* (December 8, 1926): 587. For more on Chase, see David E. Shi, *The Simple Life: Plain Living and High Thinking in American Culture* (New York: Oxford University Press, 1985), 231.

25. John Ise, *The United States Oil Policy* (1926; reprint, New York: Arno Press, 1972), vii, 169, 174. Ise greatly admired Gifford Pinchot and thought Interior Secretary F. K. Lane a "most dangerous man" (336). For him, the best place for oil was underground, where it was "stored without risk or expense" (112).

26. Ibid., 184, 205, 35, 40, 49, 239–40, 48.

27. Ibid., 209–20, 168–70, 218, 252.

28. Ibid., 176, 179, 164, 494–99, 605–7, 509, 174.

29. *Petroleum Development and Technology in 1927*, in *Transactions of the American Institute of Mining and Metallurgical Engineers: Petroleum Division* (New York: American Institute of Mining and Metallurgical Engineers, 1928), 732.

30. Arthur Knapp, "The Place of Petroleum Industry," *Petroleum Development and Technology in 1926*, in *Transactions of the American Institute of Mining and Metallurgical Engineers: Petroleum Division* (New York: American Institute of Mining and Metallurgical Engineers, 1927), 771–72.

31. J. B. Umpleby, "Production Engineering in 1927," *Petroleum Development and Technology in 1926*, in *Transactions of the American Institute of Mining and Metallurgical Engineers: Petroleum Divison*, 12–13.

32. Walter van der Gracht's comment on Umpleby, ibid., 193–94.

33. J. B. Umpleby, "Changing Concepts in the Petroleum Industry," in *Petroleum Development and Technology 1932* (New York: Petroleum Division, American Institute of Mining and Metallurgical Engineers, 1932), 38–45.

34. Ibid., 40.

35. E. L. Estabrook, "Unit Operation in Foreign Fields," in *Petroleum Development*

*and Technology 1931* (New York: Petroleum Division, American Institute of Mining and Metallurgical Engineers, 1931), 46; Umpleby, "Changing Concepts," 41; L. C. Snider, "Propositions and Corollaries in Petroleum Production," in *Petroleum Development and Technology 1932*, 54–55.

36. Snider, "Propositions," 53–54, 63–65.

37. *OGJ* (May 26, 1932): 14.

38. Earl Oliver, "Stabilizing Influences for the Petroleum Industry," in *Petroleum Developments and Technology 1932*, 22, 26; Earl Oliver, "Oil Industry's Problems and Remedies," *OGJ* (April 21, 1932): 14.

39. Earl Oliver, "Present Methods Encourage Big Waste," *OGJ* (December 17, 1931): 21.

40. Earl Oliver, "Why Adequate Oil Legislation Failed," *OGJ* (September 17, 1931): 15.

41. Ibid.

42. Oliver, "Why Adequate Oil Legislation Failed," 15; Oliver, "Oil Industry's Problems and Remedies," 15; Earl Oliver, "Plans for the Petroleum Division in 1932," *Petroleum Developments and Technology 1932*, 6–7; Hardwicke, *Antitrust Laws*, 49–55.

43. J. Howard Marshall and Norman L. Meyers, "Legal Planning of Petroleum Production," *Yale Law Journal* (November 1931): 33, 35–36, 37; Donald H. Ford, "Controlling the Production of Oil," *Michigan Law Review* (June 1932): 1203. For similar perspectives from other legal scholars, see Andrew A. Bruce, "The Oil Cases and the Public Interest," *American Bar Association Journal* (1933): 82–86, 168–72, and Charles G. Haglund, "The New Conservation Movement with Respect to Petroleum and Natural Gas," *Kentucky Law Journal* (1933–34): 543–81. As a group, lawyers subscribed to the view that there had been reckless waste of petroleum in the past. Indeed, Dean Henry M. Bates of Michigan Law School sounded like Edward MacKay Edgar when he said that what had happened to petroleum represented "the most reckless, extravagant waste of natural resources that even the American people have been guilty of." "Some Constitutional Aspects of the Oil Problem," *Petroleum Development and Technology 1935* (New York: American Institute of Mining and Metallurgical Engineers, 1935), 199.

44. W. P. Z. German, "Compulsory Unit Operation of Oil Pools," in *Petroleum Development and Technology 1931*, 13.

45. Ibid., 31.

46. Roger M. Olien and Diana Davids Olien, *Wildcatters: Texas Independent Oilmen* (Austin: Texas Monthly Press, 1984), 45–54; John G. Clark, *Energy and the Federal Government: Fossil Fuel Policies, 1900–1946* (Urbana: University of Illinois Press, 1987), 209.

47. Olien and Olien, *Wildcatters*, 54; Clarence L. Linz, "Washington Withholds Approval of Institute's Conservation Plan," *Oil Weekly* (April 12, 1929): 37; "Limitation of Crude Output Up to Regional Committees," *Oil Weekly* (March 22, 1929): 19; "Wilbur Outlines Government Policy," *Oil Weekly* (December 6, 1929): 41. The best recent survey of the Hoover administration's policy is Clark's *Energy and the Federal Government*, 208–13.

48. "IPAA's Tenth Anniversary," *IPAA Monthly* (June 1939): 9; George Elliott Sweet, *Gentleman in Oil* (Los Angeles: Science Press, 1966), 49–63.

49. Wirt Franklin, "Economic Situation Demands Immediate Action," *Oil Weekly* (January 23, 1931): 31. Other scholars have also noted the traditionalist framework of these arguments; see, for example, Clark, *Energy and the Federal Government*, 194.

50. Wirt Franklin, "Statement Presented to Congress to Support Tariff Plea," *Oil Weekly* (February 14, 1930): 31–33.

51. Ibid.

52. "A Carpet Bagger in Pennsylvania: Part V, The Oil Region," *Atlantic Monthly* (June 1869): 736; W. C. Franklin, "The Petroleum Industry under the X-ray," *Oil Weekly* (April 10, 1925): 44.

53. L. G. Bignell, "Effect of Shutting in Producing Wells," *OGJ* (June 25, 1932): 17; H. H. Power and C. H. Pishny, "Effect of Proration on Decline, Potential, and Ultimate Production of Oil Wells," *Petroleum Development and Technology 1931*, 115. From the early 1930s, however, petroleum engineers argued that shut-ins might not be calamitous after all. If a well was in good operating condition when shut in, engineers thought it could probably resume production of prior or even increased levels; wells that could not be put back into production because of problems such as water incursion probably would have had such problems regardless of the shut down. In short, by the early thirties, technological discourse no longer maintained that shutting in producing wells meant irretrievable physical loss. Of course, it must be noted that engineers advancing this argument were scarcely disinterested. Many worked for large companies favoring production cutbacks, and restoring a shut-in well to production involved engineering expertise, so engineers were in effect advancing the interests of their employers and themselves.

54. Roger M. Olien, *From Token to Triumph: The Texas Republicans since 1920* (Dallas: Southern Methodist University Press, 1982), 62–63; *Regulating Importation of Petroleum and Related Products, Hearings before the Committee on Commerce*, U.S. Senate, 71st Cong., 3d sess. (Washington, D.C.: U.S. Government Printing Office, 1931), 5–6, 18, 21.

55. Russell B. Brown, "Is There an Independent Oil Industry?," *IPAA Monthly* (September–October 1932): 10. In 1931 Congressman Homer Hoch introduced an IPAA-endorsed bill for pipeline divestiture; *Pipelines, Hearings before the Committee on Interstate and Foreign Commerce*, U.S. House of Representatives, 71st Cong., 3d sess. (Washington, D.C.: U.S. Government Printing Office, 1931).

56. "Regulating Importation of Petroleum," 16; "Texas Independents Voice Disapproval of New Law," *Oil Weekly* (December 12, 1932): 43; "I.P.A. of Texas Opposes Present Proration Rules," *OGJ* (December 15, 1932): 32; Olien and Olien, *Wildcatters*, 59–61; H. J. Struth, "Oklahoma Proration Opponents Making Appeal to the Public," *Oil Weekly* (January 16, 1931): 52–55.

CHAPTER EIGHT

1. Leonard M. Logan Jr., *Stabilization of the Petroleum Industry* (Norman: University of Oklahoma Press, 1930), 2–4.

2. Ibid., 13–15, 21, 43.

3. Ibid., 128, 212, 125, 193–94, 71, 217, 191.

4. Ibid., 4, 128, 193, 5, 99.

5. Ibid., 191–94, 217.

6. *Oil Weekly* (February 22, 1929): 109; (January 31, 1930): 48, 50, 68–69; (August 1, 1930): 132, 134; (April 11, 1930): 52.

7. *Oil Weekly* (July 4, 1930): 31; (August 1, 1930): 132, 134.

8. Roger M. Olien and Diana Davids Olien, *Wildcatters: Texas Independent Oilmen* (Austin: Texas Monthly Press, 1984), 56–58; Roger M. Olien and Diana Davids Olien, *Easy Money: Oil Promoters and Investors in the Jazz Age* (Chapel Hill: University of North Carolina Press, 1990), 32–33, 57–58.

9. James McIntyre, "East Texas Blocking Proration Plans," *Oil and Gas Journal* (here-

after cited as "*OGJ*") (May 21, 1931): 28; "East Texas Price Slash Brings Daily Revenue Down to $75,000," *Oil Weekly* (May 1, 1933): 42; Henrietta M. Larson and Kenneth Wiggins Porter, *History of Humble Oil and Refining Company: A Study in Industrial Growth* (New York: Harper & Brothers, 1959), 453–54.

10. Olien and Olien, *Wildcatters*, 58–59; "Demand for Special Session in Texas," *OGJ* (June 25, 1931): 13. Ironically, one of the most flagrant violators of the order was the Arkansas Fuel Oil Company, a part of conservationist Henry L. Doherty's Cities Service group.

11. "Governor Sterling Urges Action in Message to Texas Legislature," *OGJ* (July 16, 1931): 39. Humble Oil, Sterling's former affiliation, was not tottering on its foundations.

12. Andrew A. Bruce, "The Oil Cases and the Public Interest," *Journal of the American Bar Association* (1933): 85; Charles E. G. Haglund, "The New Conservation Movement with Respect to Petroleum and Natural Gas," *Kentucky Law Journal* (1933–34): 558; "Court Decides Proration Orders of Commission Are 'Usurpations,'" *OGJ* (July 30, 1931): 155; Olien and Olien, *Wildcatters*, 58–60; Larson and Porter, *History of Humble*, 458–59; "Troops Shut Down East Texas Fields," *OGJ* (August 20, 1931): 15.

13. *Oil Weekly* (July 31, 1931): 100; (August 7, 1931): 40; (August 14, 1931): 16.

14. "Governor to Enforce East Texas Regulation Despite Federal Order," *Oil Weekly* (October 16, 1931): 41; Neil Williams, "Martial Law Case before Federal Court," *OGJ* (January 7, 1932): 13; "East Texas Proration Almanac," *Oil Weekly* (December 12, 1932): 18.

15. "Commission Holds East Texas to 325,000 Bbls. on Bottom Hole Pressure-Acreage Basis," *OGJ* (December 1, 1932): 9. With the benefit of hindsight, it is possible to see the conservationist clamor about oil left unrecovered as reflective of the limitation of the technology of oil recovery and the desire of operators for minimal recovery costs. With today's enhanced recovery technology oil can be recovered economically on a scale unimaginable for a producer of 1930. Operators no longer see such problems as water incursion as devastating but simply face the cost of separation and water disposal. As Robert M. Leibrock has pointed out to the authors, it no longer makes sense to talk of damage to a reservoir as was common with reference to the East Texas field during the 1930s. For a highly useful account of the legal decisions in the East Texas controversy, see Nicholas George Malavis, *Bless the Pure and Humble: Texas Lawyers and Oil Regulation, 1919–1936* (College Station: Texas A&M Press, 1996).

16. Writing for *Oil Weekly* in 1936, Leonard M. Logan estimated that in 1933 East Texas operators produced some thirty million of barrels of oil illegally and in excess of field proration (January 27, 1936, 13).

17. "East Texas Proration Almanac," 16–18; "East Texas Price Slash Brings Daily Revenue Down to $75,000," *Oil Weekly* (May 1, 1933): 42.

18. Larson and Porter, *History of Humble*, 475; I. L. "Pinkie" Edwards (pipeline walker), interviewed by Roger M. Olien, February 23, 1979, Midland, Tex.

19. Roger M. Olien, *From Token to Triumph: The Texas Republicans since 1920* (Dallas: Southern Methodist University Press, 1982), 19, 54–55.

20. Owen P. White, "Drilling for Trouble," *Collier's*, June 27, 1931, 14, 55; "You Pay for the Oil War," *Collier's*, July 8, 1933, 13; "Piping Hot," *Collier's*, January 12, 1935, 11.

21. "Black Gold and Red Ink," *World's Work* (November 1931): 20.

22. John T. Flynn, "War to the Last Drop," *Collier's*, May 24, 1930, 10, 70.

23. John T. Flynn, *God's Gold: The Story of Rockefeller and His Times* (New York: Harcourt, Brace, 1932), 388, 179–80, 182, 482.

24. Ibid., 125, 204, 227, 221, 202–3, 136–37, 392–93, 262.

25. Ibid., 393, 404.

26. T. H. Watkins, *Righteous Pilgrim: The Life and Times of Harold L. Ickes, 1874–1952* (New York: Henry Holt, 1990), 240–42, 277, 319; Graham White and John Maze, *Harold Ickes of the New Deal: His Private Life and Public Career* (Cambridge: Harvard University Press, 1985), 106.

27. Linda J. Lear, "Harold L. Ickes and the Oil Crisis of the First Hundred Days," *Mid-America* (January 1981): 5. The idea of an oil czar predated Ickes's appointment. In February 1931, Leonard Logan suggested to the AIME that the industry have "a dictatorship" similar to that in organized baseball, thinking in terms of a three-person body consisting of an API representative, an IPAA representative, and a presidential appointee. In August of that year, the *Times* reported that some "independent interests" had been circulating pamphlets suggesting an "oil czar." See the *New York Times*, February 20, 22, 1931.

28. U.S. House of Representatives, Committee on Ways and Means, *Conservation of Petroleum: Hearings on H.R. 5720*, 73d Cong., 1st sess. (Washington, D.C.: U.S. Government Printing Office, 1933), 23, 49, 44, 45, 43. On Margold's connection to Frankfurter, see Ellis W. Hawley, *The New Deal and the Problem of Monopoly: A Study in Economic Ambivalence* (Princeton: Princeton University Press, 1974), 283.

29. Lear, "Harold Ickes," 7–8; "Few Oil Producers Included in List Sent by Governors to Washington," *Oil Weekly* (March 27, 1933): 8; Charles E. Kern, "Text of the Bill Introduced in Congress to Compel Divorce of Pipe Lines from Oil Companies," *OGJ* (April 13, 1933): 10.

30. Lear, "Harold Ickes," 8–9.

31. U.S. Senate, Committee on Finance, *Bills to Encourage National Industrial Recovery, to Foster Fair Competition, and to Provide for the Construction of Certain Useful Public Works, and for other Purposes, Hearings on S. 1712 and H.R. 5755, 1933*, 73d Cong., 1st sess. (Washington, D.C.: U.S. Government Printing Office, 1933), 38, 42; *New York Times*, June 4, 1933. Ickes picked up on the turn-of-the-valve imagery used by the *World's Work* two years earlier.

32. *OGJ* (September 8, 1933): 8; (September 21, 1933): 9.

33. U.S. House of Representatives, Committee on Interstate and Foreign Commerce, *Petroleum Investigation: Hearings on House Resolution 441*, 73d Cong., recess, 1934 (Washington, D.C.: U.S. Government Printing Office, 1934), 1136–38; see also John G. Clark, *Energy and the Federal Government: Fossil Fuel Policies, 1900–1946* (Urbana: University of Illinois Press, 1987), 223–25.

34. U.S. House, *Petroleum Investigation*, 1136–39. For a useful look at beginning of the NRA's oil codes, see Douglas R. Brand, "Corporatism, the N.R.A., and the Oil Industry," *Political Science Quarterly* (Spring 1983): 99–118.

35. U.S. House, *Petroleum Investigation*, 1139.

36. Malavis, *Pure and Humble*, 175.

37. James A. Clark, *Three Stars for the Colonel* (New York: Random House, 1954), 94, 107. Inasmuch as one can speak of a mythology of Ernest Thompson, Clark's book embodies it. For a scholarly appraisal of Thompson, see William R. Childs, "The Transformation of the Railroad Commission of Texas, 1917–1940: Business-Government Relations and the Importance of Personality, Agency Culture, and Regional Difference," *Business History Review* (Summer 1991): 285–344.

38. *Petroleum, Hearing before a Subcommittee of the Committee on Interstate and Foreign Commerce on S. 790 and H.R. 5366*, April 27, 28, 29, and May 3, 1937. U.S. House of Representatives, 75th Cong., 1st sess. (Washington, D.C.: U.S. Government Printing Office, 1937), 57. Note that in the instance of personifying nations, like virtues, femi-

nized gendering of language has a positive rather than negative connotation. Clark, *Three Stars*, 106. Childs points out that the ideology of the "lost cause" had early been part of agency culture at the TRC; "Transformation," 296.

39. Clark, *Three Stars*, 108.

40. Ibid., 114. Childs sees Thompson as using Ickes as a "symbolic magnet," as someone he could rally opposition to from a wide variety of quarters and thus unite behind himself; "Transformation," 324.

41. Ibid., 169.

42. *Federal Petroleum Act: Hearings before the Subcommittee of the Committee on Mines and Mining*, May 21, 22, 1934. U.S. Senate, 73d Cong., 2d sess. (Washington, D.C.: U.S. Government Printing Office, 1934), 21; U.S. House, *Petroleum Investigation*, 1610, 706, 2159, 2033, 377. For a different interpretation of industry positions, see Brand, "Corporatism, the N.R.A., and the Oil Industry."

43. U.S. House, *Petroleum Investigation*, 1415, 2315, 1508, 1730–31, 1442, 1521.

44. *Federal Petroleum Act*, 108, 111, 125, 128; U.S. House, *Petroleum Investigation*, 779.

45. U.S. House, *Petroleum Investigation*, 209, 355, 514, 1426, 1430, 1433, 1396–97.

46. *Oil and Oil Pipe Lines: Hearings before the Committee on Interstate and Foreign Commerce*, May 30, 31, June 1, 5, 6, and 7, 1934. U.S. House of Representatives, 73d Cong., 2d sess. (Washington, D.D.: U.S. Government Printing Office, 1934), 27.

47. U.S. House, *Petroleum Investigation*, 514, 706, 1398, 1508, 705.

48. Ibid., 706, 698, 1402, 2322.

49. Ibid., 190, 1424, 1699.

50. Ibid., 1397.

51. U.S. House, *Petroleum Investigation*, 384.

52. Ibid., 220, 2203, 2197–98.

53. *Oil and Oil Pipe Lines*, 145, 144, 142; *Federal Petroleum Act*, 80–81.

54. U.S. House, *Petroleum Investigation*, 1812; *Oil and Oil Pipe Lines*, 145–46; *Federal Petroleum Act*, 80.

55. Larson and Porter, *History of Humble*, 471, 478; Edward W. Constant II, "Cause or Consequence: Science, Technology, and Regulatory Change in the Oil Business in Texas, 1930–1975," *Technology and Culture* (April 1989): 434–35.

56. David F. Prindle, *Petroleum Politics and the Texas Railroad Commission* (Austin: University of Texas Press, 1981), 47–49.

57. Larson and Porter, *History of Humble*, 532–34.

58. Clark, *Energy and the Federal Government*, 213.

59. William R. Childs, "Texas, the Interstate Oil Compact Commission, and State Control of Oil Production: Regionalism, States' Rights, and Federalism during World War II," *Pacific Historical Review* (November 1995): 592–94.

60. Clark, *Energy and the Federal Government*, 239.

61. Gerald D. Nash, *United States Oil Policy, 1890–1964: Business and Government in Twentieth-Century America* (1968; Westport, Conn.: Greenwood Press, 1976), 131.

62. Richard Lowitt, *The New Deal and the West* (Bloomington: Indiana University Press, 1984), 100, 103–4, 248, 61. Curiously, Lowitt does not describe how Ickes's ideas translated into oil policy on federally owned or Indian lands. Lowitt sees Ickes accomplishing less on oil than in other areas, but in programs he sees as more successful, such as land use and soil conservation, he admits not as much was accomplished as might have been hoped for.

63. Clark, *Energy and the Federal Government*, 243.

64. Ickes liked to claim that under the Petroleum Administration Board, East Texas

had been brought under control; Lowitt, *New Deal*, 105. But this claim completely disregards the growing authority of the Texas Railroad Commission once courts began ruling for it; the growing influence of major property owners in the field; and the fact that problems with hot oil did not go away as soon as the federal government intervened in East Texas. What the NRA did was not unimportant, but it was no more important than other developments.

CHAPTER NINE

1. Matthew Josephson, *The Robber Barons: The Great American Capitalists, 1861–1901* (1934; New York: Harcourt Brace Jovanovich, 1962), 192, 216, 451, 453.

2. Jordan A. Schwarz, *The New Dealers: Power Politics in the Age of Roosevelt* (New York: Vintage, 1994), 101, 103.

3. René de Visme Williamson, *The Politics of Planning in the Oil Industry under the Code* (New York: Harper & Brothers, 1936), v.

4. Ibid., 32, 27, 38–39, 28.

5. Ibid., 30–32, 38, 23, 27. Examples of unfamiliarity with the industry on Williamson's part are strewn throughout the text and are not limited to technology. He thought, for example, that low crude oil prices "placed a terrible hardship" on refiners, an observation that can only be described as mystifying (39).

6. Ibid., 34–35.

7. Stocking named one of his sons for Watkins. In 1946 they worked together with Alfred E. Kahn and Gertrude Oxenfeldt on *Cartels in Action: Case Studies in International Business Diplomacy* (New York: Twentieth Century Fund, 1946), a work warning that, despite being overshadowed by war, the problem of monopoly would "rise up to plague the world again after the fighting stopped" (vii). The case study on oil argued that Standard Oil (New Jersey) had used control of petroleum-refining technology to control the United States industry and eliminate competition; it had then given synthetic rubber technology to the Germans before World War II, retarding the American synthetic rubber industry and jeopardizing national security (116–17, 493–94). This garbled account of the patents controversy reflects Stocking's continued animus against Standard Oil. In 1951 Stocking and Watkins published *Monopoly and Free Enterprise* (New York: Twentieth Century Fund), in which they condemned economic concentration, speculation, high finance, oil conservation programs that only served monopoly, and John D. Rockefeller Sr. — a veritable litany of old economic bogeys.

8. Myron W. Watkins, *Oil: Stabilization or Conservation: A Case Study in the Organization of Industrial Control* (New York: Harper & Brothers, 1937), 24–27, 29, 89, 248–49, 251.

9. Ibid., 36–37, 89, 248, 39. On management of risk by independents, see Roger M. Olien and Diana Davids Olien, *Wildcatters: Texas Independent Oilmen* (Austin: Texas Monthly Press, 1984), chaps. 1–6. Ruth Sheldon Knowles entitled her largely laudatory view of wildcatters *The Greatest Gamblers: The Epic of American Oil Exploration* (New York: McGraw-Hill, 1959).

10. Watkins, *Oil*, 173, 242–43.

11. Ibid., 256.

12. *Petroleum Shipments: Hearing before a Subcommittee of the Committee on Interstate and Foreign Commerce on S. 1302, H.R. 4547, and H.R. 2308*, House of Representatives, 76th Cong., 1st sess. (Washington, D.C.: U.S. Government Printing Office, 1939), 84, 106; *Oil Marketing Divorcement: Hearings before Subcommittee Number 3 of the Committee*

on the Judiciary on H.R. 2318, House of Representatives, 76th Cong., 1st sess. (Washington, D.C.: U.S. Government Printing Office, 1939), 72, 96; U.S. House, *Petroleum Investigation*, 2765, 2793; "Improvement Seen in Outlook for Oil," *New York Times*, March 13, 1932; "Oil Industry's Ills Laid to Proration," *New York Times*, April 16, 1933. John T. Flynn gave Kemnitzer's book a glowing review, calling it "of rare authority and interest," despite Kemnitzer's markedly different perspective on the oil industry that Flynn's in 1932; *New Republic* (August 17, 1938): 47.

13. U.S. House, *Petroleum Investigation*, 2766, 2776; *Oil Marketing Divorcement*, 96, 73; William J. Kemnitzer, *Rebirth of Monopoly: A Critical Analysis of Economic Conduct in the Petroleum Industry of the United States* (New York: Harper & Brothers, 1938), 9, 34.

14. Ibid., 9, 45, 40.

15. Ibid., 55–56, 27, 30–31, 70, 52, 206, 210–11; *Petroleum Shipments*, 108.

16. Kemnitzer, *Rebirth of Monopoly*, 60–62, 66, 73, 211; *Petroleum Shipments*, 100, 85, 96, 90–91.

17. Kemnitzer, *Rebirth of Monopoly*, 1, 226–27, 140, 229, 91, 224.

18. *Petroleum Shipments*, 92–111.

19. *New York Times*, February 8, 1934, December 8, 1935, June 12, 1936.

20. Ellis W. Hawley, *The New Deal and the Problem of Monopoly: A Study in Economic Ambivalence* (Princeton: Princeton University Press, 1974), 302.

21. Harold F. Williamson, Ralph L. Andreano, Arnold R. Daum, and Gilbert C. Klose, *The American Petroleum Industry: The Age of Energy, 1899–1959* (Evanston, Ill.: Northwestern University Press, 1963), 693–94; John G. Clark, *Energy and the Federal Government: Fossil Fuel Policies, 1900–1946* (Urbana: University of Illinois Press, 1987), 233–34; Simon N. Whitney, *Antitrust Policies: American Experience in Twenty Industries* (New York: Twentieth Century Fund, 1958), 1, 113–15.

22. *New York Times*, February 3, 1934; Clark, *Energy and the Federal Government*, 234.

23. *New York Times*, August 4, 5, 1936; Williamson et al., *American Petroleum Industry: The Age of Energy*, 597.

24. Schwarz, *New Dealers*, 189.

25. Ferdinand Lundberg, *America's Sixty Families* (New York: Vanguard Press, 1937), 3–4, 8–9, 449, 490. Lundberg explained the penchant of the rich for having many luxurious bathrooms in their homes as the result of subconscious need to wash guilt away (417).

26. "Attack on Oligopoly," *Time*, January 10, 1938, 12–13; *New York Mirror*, May 6, 1939; *New York Journal-American*, March 30, 1939, January 2, 1938; *New York Times*, November 30, 1937; *New York Herald Tribune*, March 30, 1939.

27. Joseph Alsop and Robert Kintner, "Trust Buster: The Folklore of Thurman Arnold," *Saturday Evening Post* (August 12, 1939): 5–7, 30, 33; *P. M.* (February 10, 1943); *New York Herald Tribune*, July 31, 1949; *New York Times*, March 11, 1938.

28. Thurman W. Arnold, *The Folklore of Capitalism* (New Haven: Yale University Press, 1937), 392, 211–20.

29. Ibid., 345–46, 393.

30. Ibid., 345.

31. Alsop and Kintner, "Trust Buster," 7; *New York Journal*, October 5, 1948.

32. Williamson et al., *American Petroleum Industry: The Age of Energy*, 597–600. For a meticulously thorough discussion of the Elkins cases, see Arthur M. Johnson, *Petroleum Pipelines and Public Policy, 1906–1959* (Cambridge: Harvard University Press, 1967), 286–301. See also Hawley, *New Deal*, 374.

33. Hawley, *New Deal*, 412; *Petroleum-Industry Hearings before the Temporary National*

Economic Committee (New York: American Petroleum Institute, 1942), 6–7; *Investigation of Concentration of Economic Power: Hearings before the Temporary National Economic Committee Pursuant to Public Resolution No. 113 of the Seventy-fifth Congress*, pts. 14, 15, 16, 17. 76th Cong., 2d sess. (Washington, D.C.: U.S. Government Printing Office, 1940), 7232; Johnson, *Petroleum Pipelines*, 271.

34. Hawley, *New Deal*, 415–16; *Petroleum-Industry Hearings*, 7; *Investigation of Concentration of Economic Power*, 7103.

35. *Investigation of Concentration of Economic Power*, 7101–3, 7108–9, 7111.

36. Ibid., 7102.

37. Ibid., 7370, 7345, 910–80, 9142.

38. *Petroleum-Industry Hearings*, 282.

39. John T. Flynn, "Two Incidents and Some Advice," *New Republic* (July 6, 1938): 250; "Wall Street Myths," *New Republic* (August 17, 1938): 46; "Monopoly and Oligopoly," *New Republic* (May 3, 1939): 377–78; "The Attack on Monopoly," *Nation* (January 8, 1938): 31–33; "Light on the Oil Trust," *Nation* (June 3, 1939): 633; I. F. Stone, "The TNEC Recommends — What?," *Nation* (April 19, 1941): 462–64; Dwight MacDonald, "The Monopoly Committee: A Study in Frustration," *American Scholar* (July 1939): 308.

40. "Twilight of TNEC," *Time*, April 14, 1941, 37; Roy C. Cook, *Control of the Petroleum Industry by Major Oil Companies*, Monograph No. 39, Temporary National Economic Committee (Washington, D.C.: U.S. Government Printing Office, 1941), iii, ix; William S. Farish and J. Howard Pew, *Review and Criticism on Behalf of Standard Oil Co. (New Jersey) and Sun Oil Co. of Monograph No. 39 with Rejoinder by Monograph Author*, Monograph 39-A (Washington, D.C.: U.S. Government Printing Office, 1941).

41. Cook, *Control of the Petroleum Industry*, 53–54, xi. Cook's thinking was very much in the same channel as Kemnitzer's. John Ise reviewed Cook for the *American Economic Review* (September 1941–March 1942) and called it "a genuine service" to economists (600–601).

42. Cook, *Control of the Petroleum Industry*, xi, 2, 4–7, 11–12, 32–39, 13–14, 28, 36, 50–52.

43. Farish and Pew, *Review and Criticism*, 3, 10–12, 5, 13, 16, 27–28.

44. Ibid., 31–33.

45. Ibid., 18, 20.

46. Cook, *Rejoinder*.

47. *Petroleum-Industry Hearings*, iii.

48. "Compounds and Concoctions," *Time*, September 2, 1935, 34–35; "Oil-Supply: Dwindling or Gushing?," *Literary Digest* (November 23, 1935): 38; Department of the Interior, *Report on the Cost of Producing Crude Petroleum* (Washington, D.C.: U.S. Government Printing Office, 1936), pt. 4: 131–33.

49. *Energy Resources and National Policy: Report of the Energy Resources Committee to the National Resources Committee* (Washington, D.C.: U.S. Government Printing Office, 1939), 2, 11, 5, 8, 21, 12, 3–4.


CHAPTER TEN

1. *New York Journal American*, April 11, 1941; *New York Times*, April 13, 1941; *New York World Telegram*, April 12, 1941. For data on which organizations were created when, one can use John W. Frey and H. Chandler Ide's *A History of the Petroleum Administration for War, 1941–1945* (Washington, D.C.: U.S. Government Printing Of-

fice, 1946), though this source must be considered an important part of wartime discourse in its own right.

2. Harold L. Ickes, *Fightin' Oil* (New York: Alfred A. Knopf, 1943), 71, 76; Raymond Moley, "Oil in Troubled Waters," *Newsweek*, June 23, 1941, 68; Ray L. Dudley, "A Pledge and a Code," *Oil Weekly* (June 9, 1941): 9; B. F. Linz, "Washington Roundup," *Oil Weekly* (August 18, 1941): 15. As John G. Clark has pointed out, oilmen distrusted both Ickes and the idea of federal involvement in oil, and their cooperation was "under duress"; *Energy and the Federal Government: Fossil Fuel Policies, 1900–1946* (Urbana: University of Illinois Press, 1987), 317, 323–24.

3. William R. Childs, "Texas, the Interstate Oil Compact Commission, and State Control of Oil Production: Regionalism, States' Rights, and Federalism during World War II," *Pacific Historical Review* (November 1995): 583–95.

4. *Oil Weekly* (June 9, 1941): 8; Moley, "Oil in Troubled Waters," 68; "Harold L. Ickes on THAT Oil Shortage," *Collier's*, October 18, 1941, 80.

5. Moley, "Oil in Troubled Waters," 68; *New York Journal American*, June 15, 1941; *Oil Weekly* (June 9, 1941): 8; Leonard J. Logan, "One Hundred More Tankers for Britain Seen in Shortage Forecast," *Oil Weekly* (August 18, 1941): 9–10; "Famine Closer," *Time*, July 28, 1941, 66.

6. "Harold L. Ickes on THAT Oil Shortage," 13, 80; *Oil Weekly* (August 18, 1941): 8–9, (August 25, 1941): 10; Ray L. Dudley, "Whose Emergency?," ibid., 11; *New York Journal American*, August 1, 1941; *New York Daily News*, August 20, 1941.

7. B. F. Linz, "Washington Oil Round Up," *Oil Weekly* (August 18, 1941): 13–14; *Oil Weekly* (September 8, 1941): 15; U.S. Senate, *Gasoline and Fuel-Oil Shortages: Hearings before the Special Committee to Investigate Gasoline and Fuel-Oil Shortages*, 77th Cong., 1st sess. (Washington, D.C.: U.S. Government Printing Office, 1941), pt. 1, 10.

8. Walter Davenport, "Holy Harold," *Collier's*, October 11, 1941, 49; "Harold L. Ickes on THAT Oil Shortage," *Collier's*, October 18, 1941, 13, 80, 81.

9. Harold L. Ickes, "Plenty of Oil — But How to Get It?," *Collier's*, August 8, 1942, 14; Harold L. Ickes, "The Battle of Oil," *Nation* (August 1, 1942): 86–87; Harold L. Ickes, *Fightin' Oil*, 22–23, 49, 52.

10. *New York Times*, April 24, 1942; *New York Mirror*, December 9, 1942; *New York Journal American*, April 24, May 18, 1942.

11. For a thorough examination of the Standard Oil–IG connection and the development of synthetic rubber, see Henrietta M. Larson, Evelyn N. Knowlton, and Charles S. Popple, *History of Standard Oil Company (New Jersey): New Horizons, 1927–1950* (New York: Harper & Row, 1971), 153–60, 170–74, 406–8, 412–18.

12. Ibid., 428–33.

13. *New York Herald Tribune*, March 28, 1942; *New York Times*, March 27, 1942; *New York Sun*, May 27, 1942; *New York Times*, June 6, 1942; Thurman Arnold, "How Monopolies Have Hobbled Defense," *Reader's Digest* (July 1941): 51–55.

14. *New York Sun*, March 31, 1942; *New York Journal American*, March 28, 31, 1942; *New York Daily News*, April 3, 1942.

15. *New York Journal American*, June 2, 1942; *New York Daily News*, June 3, 1942. As a result of these events, SONJ saw the need for investment in public relations; see Larson, Knowlton and Popple, *New Horizons*, 447–49.

16. *New York Times*, April 1, July 9, 1942; *New York Journal American*, July 6, 1942; "Dinner-Table Treason," *Time*, April 6, 1942, 15–16; *New York Mirror*, April 18, 1942; Michael Straight, "Standard Oil: Axis Ally," *New Republic* (April 6, 1942): 450–51; Guenter Reimann, "Standard Oil and I. G. Farben," *New Republic* (August 4, 1941): 147–49; I. F. Stone, "Not to Be Attributed," *Nation* (April 11, 1942): 415–16.

17. *Oil Weekly* (September 1, 1941); "Oil Placed under Virtual Wartime Federal Control," *Oil Weekly* (December 29, 1941): 10–13; "New Orders Give Oil Priority Ratings, Force Unitization," *Oil Weekly* (January 19, 1942): 13–15. Fear of federal controls led Illinois to pass its first conservation law; Blakely M. Murphy, ed., *Conservation of Oil and Gas: A Legal History, 1948* (Chicago: American Bar Association Section on Mineral Law, 1949), 111–13.

18. Edgar Kraus, " 'MER' — A History," in *Drilling and Production Practice* (New York: American Petroleum Institute, 1947), 108–9.

19. D. Thomas Curtin, *Men, Oil, and War* (Chicago: Petroleum Industry Committee for District No. 1, 1946), 65.

20. Kraus, " 'MER,' " 109.

21. R. G. Hiltz, J. V. Huzarevich, and R. M. Leibrock, "Performance Characteristics of the Slaughter Field Reservoir," in Texas Petroleum Research Committee, *Proceedings for Secondary Oil Recovery Conference: A Symposium on Carbonate Reservoirs* (Austin: Texas Petroleum Research Committee, 1951), 146–53. In 1951, having demonstrated that the rate of recovery did not impact the percentage of recovery in the Slaughter Field, one scientist was told by Texas Railroad commissioner Bill Murray, "Well, it sounds convincing, but I still believe in MER." Robert M. Leibrock, interviewed by Roger M. Olien and Diana Davids Olien, March 2, 1991, Midland, Tex.

22. Curtin, *Men, Oil, and War,* 69–72.

23. "Oil Placed under Virtual Wartime Federal Control," 12; "Paper Work Wasting Many, Many Hours," *Oil Weekly* (April 13, 1942): 10; Rivers Reaves, "Many Factors behind Current Flurry of Producing Lease Sales," *Oil Weekly* (July 12, 1943): 42–44; *New York Herald Tribune,* March 1, 1943.

24. Warren L. Baker, "The Trend of Oil in War," *Oil Weekly* (June 8, 1942): 34–36; *Oil Weekly* (June 1, 1942): 45; M. G Cheney, "Advances in Oil Prices Needed to Avert Threatened Shortage," *Oil Weekly* (November 2, 1942): 15.

25. On well spacing, see Lester Charles Uren, *Petroleum Production Engineering: Oil Field Development* (New York: McGraw-Hill, 1946), 70–73. See also Frederic H. Lahee, "Wildcat Drilling in 1940 More Successful," *Oil Weekly* (March 31, 1941): 37; *Oil Weekly* (June 8, 1942): 8.

26. Reaves, "Many Factors," 43; Arch H. Rowan, "Drilling Efficiency under Wartime Conditions," *Petroleum Engineer* (December 1943): 92, 94, 96. On independents' problems in wartime, see Roger M. Olien and Diana Davids Olien, *Wildcatters: Texas Independent Oil Men* (Austin: Texas Monthly Press, 1984), 84–86. See also the testimony of Lloyd Noble in *Development of Mineral Resources of the Public Lands of the United States: Hearings before a Subcommittee of the Committee on Public Lands and Surveys pursuant to S. Res. 53,* 77th Cong., 1st sess., pt. 1: 1198–1201. On casing, see Curtin, *Men, Oil, and War,* 67.

27. Reaves, "Many Factors," 42; Cheney, "Advances in Oil Prices," 15, 19; *Development of Mineral Resources,* 1419, 1421–22.

28. Reaves, "Many Factors," 43; Rowan, "Drilling Efficiency," 94; D. J. Jones, "Wildcatters' Spirit Must Not Be Crushed," *Oil Weekly* (April 26, 1943): 14, 28; Warren L. Baker, "Independents Declare Only Higher Prices Will Prevent Oil Shortage," *Oil Weekly* (October 26, 1942): 42–44; Frank B. Taylor, "Independents Are Facing Situation Realistically and with Energy," *Oil Weekly* (May 10, 1943): 59–60; U.S. Senate, *Development of Mineral Resources on the Public Lands of the United States: Hearings by a Subcomittee of the Committee on Public Lands,* 77th Cong., 2d sess. (Washington, D.C.: U.S. Government Printing Office, 1942). Only on stripper well production was the OPA willing to make some concessions and then in the form of subsidies to stripper

operators rather than a general price rise; Clark, *Energy and the Federal Government*, 328–29.

29. Ickes, *Fightin' Oil*, ix.

30. Ibid., 2–41, 49–50, 63.

31. Ibid., 76, 69–70, 78–79.

32. Ibid., 41, 48, 81–82, 89, 93.

33. Harold L. Ickes, "Hitler Reaches for the World's Oil," *Collier's*, August 15, 1942, 18–20; Harold L. Ickes, "Oil and Peace," *Collier's*, December 2, 1944, 21, 59.

34. Herbert Feis, *Petroleum and American Foreign Policy* (Palo Alto, Calif.: Food Research Institute, Stanford University, 1944), 12, 14–16, 36.

35. For a scholarly account of these developments, see Michael B. Stoff, *Oil, War, and American Security: The Search for a National Policy on Foreign Oil, 1941–1947* (New Haven: Yale University Press, 1980), esp. 80, 115–17, 140–42.

36. William R. Childs, "Texas, the Interstate Oil Compact Commission, and State Control of Oil Production: Regionalism, States' Rights, and Federalism during World War II," *Pacific Historical Review* (November 1995): 579–95; James A. Clark, *Three Stars for the Colonel* (New York: Random House, 1954), 180–89.

37. Gerald D. Nash, *United States Oil Policy, 1890–1964* (Westport, Conn.: Greenwood Press, 1968), 158; Clark, *Energy and the Federal Government*, 346–47.

38. Curtin, *Men, Oil, and War*, 116.

39. Clark, *Energy and the Federal Government*, 302.

40. For example, by 1944 Humble Oil faced delays of up to a year on pipe and drilling equipment. Henrietta M. Larson and Kenneth Wiggins Porter, *History of Humble Oil and Refining Company: A Study in Industrial Growth* (New York: Harper & Brothers, 1959), 579.

41. Curtin, *Men, Oil, and War*, 74–76.

42. Williamson et al., *American Petroleum Industry: The Age of Energy*, 778–80.

43. Frey and Ide, *Petroleum Administration*, xvii, 170–79, 189.

44. Ibid., 107, 226, 116–17.

45. Ibid., 276–79, 394–401.

46. Ibid., 293, 296–97.

47. Curtin, *Men, Oil, and War*, 16–18, 326, 329–30.

48. Ibid., 28–29, 16; on the emergence of self-made man ideology and its tenet that getting ahead was part of "manly" independence, see Mary P. Ryan, *Cradle of the Middle Class: The Family in Oneida County, New York: 1790–1865* (Cambridge: Cambridge University Press, 1981).

49. The Editors of Look, *Oil for Victory: The Story of Petroleum in War and Peace* (New York: Whittlesey House, 1946), 68–69, 71.

50. Ibid., 48–49, 71–72, 68.

51. Ibid., 252–53, 257.

CONCLUSION

1. As Thomas McCraw has noted of regulation, it is a versatile tool manipulated for particular interests as well as the general public interest. Thomas K. McCraw, *Prophets of Regulation* (Cambridge: Harvard University Press, Belknap Press, 1984), 1.

2. As McCraw has noted with respect to regulation, ideas about what was regulated did not have to be "demonstrably true" to exert strong influence on policy makers. Ibid., 304.

3. *Literary Digest* (August 12, 1911): 227.

4. See, for example, Joseph A. Pratt, "The Petroleum Industry in Transition and the Decline of Monopoly Control in Oil," *Business History Review* (December 1980): 815–37.

5. Richard H. K. Vietor, *Energy Policy in America since 1945: A Study of Business-Government Relations* (Cambridge: Cambridge University Press, 1984), 146–62, 202–4, and passim.

# Index

Buffalo Lubricating Oil Company, 74–75
Bullington, Orville, 180–81
Bureau of Corporations, 81, 108, 159
Bush, Rufus T., 60
Butts, Mrs. G. C., 89

Camden, Johnson Newlon, 69–70, 76
Cameron, Rondo, 5
Carll, John F., 121
Carver, Thomas Nixon, 145
Cassatt, Alexander, 51, 59, 66, 90
Chase, Stuart, 170
Childs, William, 229, 243
Clark, J. I. C., 93–94
Clark, John G., x–xi, 208, 243
Clark, Maurice B., 39
Clark, Payne, and Company, 40
Clarke, John Bates, 87
Claypole, E. W., 121
*Cleveland Herald*, 41
*Cleveland Leader*, 41
Cloud, D. C., 7
Cole, William P., 200
*Collier's*, 231, 242, 193
*Commercial and Shipping List*, 49
Commons, John R., 93–94
Connally Hot Oil Act, 207
Conservation: estimates of reserves,
    121, 125, 164–65, 169; and depletion,
    121, 135, 143; Progressives' opinions,
    122; morality of, 122, 136; and monop-
    oly, 122, 132, 173, 178–79, 197–98; and
    waste, 124, 127–28, 133, 144; prioritiza-
    tion of use, 126, 169; naval reserves,
    129–30; and imported oil, 135, 138,
    169; and supply, 137, 144, 169, 229–30;
    and over-production, 150–51, 187–93;
    unit operation, 155, 169, 174–77, 186–
    93; and East Texas oil field, 188–93;
    and maximum efficient rate of pro-
    duction, 235–37
Cook, Roy C., 223–24
Coolidge, Calvin, 156–57
Corcoran, Thomas G., 217–18
Crowley, Karl A., 221
Cullen, H. R., 201
Curtin, D. Thomas, 247

*Daily Socialist*, 254
Danbom, David, 15

Danciger, Joseph, 204
Daniels, Josephus, 131
Davies, Ralph K., 230–31
Dawes, Henry M., 237
Day, David Talbot, 124–26, 127, 164
Debs, Eugene V., 19
DeGolyer, Everette, 158, 166, 222
Destler, Chester M., 52
Ditz, Toby, 92
Dodd, Samuel C. T., 93
Doheny, E. L., 134, 142
Doherty, Henry L., 154–55, 158, 165–66,
    167–69
Douglas, Mary, 261 (n. 6)
Douglas, William O., 218
Downer, Samuel, 32–33
Drake, Edwin L., 21
Du Pont family: 26, 267 (n. 25)

East Texas oil field, 188–93
Edgar, Edward MacKay, 138
Elk Hills Naval Reserve, 129–30
Ely, Richard T.: professional connec-
    tions, 14; and Henry Demarest Lloyd,
    14; and Ida Tarbell, 14; and American
    Economic Association, 16; on monop-
    oly, 16; *Harper's Magazine* articles,
    77–78; and *The Railways and the
    Republic*, 77–78; and Joseph D. Potts,
    78; and Simon Sterne, 78; and *Wealth
    against Commonwealth*, 87; *Founda-
    tions of National Prosperity*, 145–46; on
    private development of public lands,
    161
Ely, Smith, 8
Emery, Lewis, Jr., 79, 86, 89, 108, 113

Fall, Albert B., 142
Fanning, L. M., 163
Fanning, H. O., 150
Farish, W. S., 167, 200, 202, 203, 220–25,
    233
Federal Oil Conservation Board, 157,
    167–70
Federal Power Commission, 229
Federal Tender Board, 205
Federal Trade Commission, 115–17, 222
Feis, Herbert, 242
Fell, H. B., 201
Flagler, Henry, 39, 47, 97